CANADIAN
URBAN REGIONS

CANADIAN URBAN REGIONS
Trajectories of Growth and Change

EDITED BY LARRY S. BOURNE, TOM HUTTON, RICHARD G. SHEARMUR, AND JIM SIMMONS

OXFORD
UNIVERSITY PRESS

OXFORD
UNIVERSITY PRESS

8 Sampson Mews, Suite 204, Don Mills, Ontario M3C 0H5
www.oupcanada.com

Oxford University Press is a department of the University of Oxford.
It furthers the University's objective of excellence in research, scholarship,
and education by publishing worldwide in

Oxford New York
Auckland Cape Town Dar es Salaam Hong Kong Karachi
Kuala Lumpur Madrid Melbourne Mexico City Nairobi
New Delhi Shanghai Taipei Toronto

With offices in
Argentina Austria Brazil Chile Czech Republic France Greece
Guatemala Hungary Italy Japan Poland Portugal Singapore
South Korea Switzerland Thailand Turkey Ukraine Vietnam

Oxford is a trade mark of Oxford University Press
in the UK and in certain other countries

Published in Canada
by Oxford University Press

Copyright © Oxford University Press Canada 2011

The moral rights of the author have been asserted

Database right Oxford University Press (maker)

First Published 2011

Library and Archives Canada Cataloguing in Publication

Canadian urban regions : trajectories of growth and change / Larry S. Bourne ... [et al.].
Includes bibliographical references and index.

ISBN 978-0-19-543382-1

1. Cities and towns—Canada—Growth. 2. Sociology, Urban—Canada.
I. Bourne, L. S. (Larry Stuart), 1939–

HT384.C3C35 2011   307.760971   C2011-900503-4

Oxford University Press is committed to our environment. This book is printed on
Forest Stewardship Council certified paper, harvested from a responsibly managed forest.

Mixed Sources
Product group from well-managed
forests and other controlled sources
www.fsc.org Cert no. SW-COC-000952
© 1996 Forest Stewardship Council

FSC

Printed and bound in Canada.

1 2 3 4 — 14 13 12 11

Contents

Part III: Perspectives on Theory, Policy, and Practice   329

List of Figures

List of Tables

Preface

This book had its origin in two research grants obtained in 2004 and 2005 under the National Research Cluster (NRC) program administered by the Social Sciences and Humanities Research Council (SSHRC). The mission of the NRC program was to mobilize researchers and intellectual capital in Canadian universities in the interest of more effectively addressing exigent national issues, including public policy challenges as well as fundamental research questions.

Accordingly, we invited colleagues to join us in a collaborative venture with the theme 'urban transformation in Canada: beyond the post-industrial city'—an enterprise designed to address theoretical, empirical, and normative dimensions of the complex (and in many respects problematic) processes shaping urban growth and change in Canada and in recognition of the increasingly central role of metropolitan cities in the economic, social, cultural, and political life of the nation. The economic welfare of Canada will be increasingly determined by the capacity of its cities to function as sites of innovation, creativity, skilled labour formation, specialized production, and global–local interaction. A larger proportion of the national population will live in metropolitan areas, with signal implications for social identity, citizenship, and political affiliation. International immigrants are transforming both the material realities (notably housing and labour markets) and imageries of Canadian cities while shaping the distinctive business practices of an entrepreneurial economy. Canada's gateway cities encompass the transnational social milieus conducive to the 'productive diversity' essential to the operation of a new cultural economy of production, consumption, and spectacle.

The largest Canadian cities also host global hallmark events (Olympics in Montreal [1976], Calgary [1988], and Vancouver [2010]), as well as international expositions (notably Montreal in 1967 and Vancouver in 1986), and meetings of the G20 (Toronto). These events require major injections of capital and may generate transformative benefits as well as commensurately massive costs (including financial, social, environmental, and opportunity costs). Hallmark events are therefore a subject of controversy, as well as an object of vigorous public protests and global media attention—another signifier of difference between the largest cities in Canada and smaller urban centres.

While much of the burden of national development will therefore fall upon Canadian cities, there are daunting constraints on the capacity of urban regions to deliver on national agendas while maintaining an adequate level of service for their citizens. The constitutional division of powers and responsibilities that recognizes only the federal and provincial levels as true orders of government—an outdated legacy of a nineteenth-century post-colonial nation still largely agrarian in nature—represents a serious limiting factor on the scope of local policy-making. The dramatic growth of demand in such areas as transit, housing, education, welfare, and economic development infrastructure has not been matched by a corresponding enhancement of revenue sources, leading to severe stresses on local service provision. Globalization has placed a premium on innovation and the search for international competitive advantage for Canada's larger cities, driving insistent processes of restructuring for industries, firms, and labour markets, as the recent experiences of urban economies among advanced societies have vividly demonstrated. Cities in Canada are also subject to the destabilizing effects of periodic shocks, including, notably, the deep economic downturn beginning in 2008 and a host of natural hazards, as well as the constant stresses of competition, migration, demographic change, and the policy shifts of senior governments. The fragmentation (and/or under-bounded administrative territories) of local government and governance that characterize many Canadian cities represents a further limiting factor on the articulation of effective responses to the problems of twenty-first-century urban development.

To address the challenges of urban development in Canada, an initial meeting of our project was convened at the University of Toronto in 2006 with a view to scoping the dimensions of the project, identifying research problems, and investigating alternative methodologies. We participated in a SSHRC-sponsored meeting at the Fairmont Chateau Laurier in Ottawa in 2006, designed in part to showcase NRC projects to the larger Canadian research and policy communities.

Workshops in the spring and fall of 2007 at the University of Toronto—the intellectual incubator of the project over the past five years—engaged a team of Canadian scholars we enlisted to join us in developing (first) a program of research and (second) an outline of a prospective edited volume on trajectories of urban change in Canada, with a thematic emphasis on processes of economic growth and employment restructuring. These lively sessions drew an enthusiastic group of Canadian urban scholars from a complementary range of disciplines (including political science, anthropology, regional science, and planning, as well as urban, social, and economic geography) and institutions. Our sessions, often intense but always highly stimulating, included debates concerning the selection of cities for detailed study, the need to incorporate aspects of urban 'place' as well as 'space' as foci of analysis, and the choice of research methodologies and techniques. The workshops generated a lively discourse of urban scholarship as well as contributing to the prospectus we subsequently submitted to Oxford University Press—the template for the present volume.

We elected to focus on employment, labour market, and occupational trends as intrinsically critical metrics of urban growth and change and also as entrées to important and related processes of urbanization and urbanism. These corollary themes included land use and urban structure, transportation choice, social class and identity, and an array of complex equity issues and problems. We therefore chose to deploy a multi-scalar research lens, incorporating analysis at the urban system, metropolitan, and district or community levels, to enable us to bring these issues into focus at their appropriate level of resolution.

As for our target audience, we oriented the book toward the interests and needs of the graduate student constituency, as well as to scholars and policy-makers studying Canadian cities both in this country and internationally.

At a meta-research level, we aspired from the outset to combine the strengths of two broad traditions of social sciences and humanities scholarship in the interests of producing a compelling treatment of urban change in Canada, as well as doing justice to important elements of nuance and texture that animate the human experience in the city. Broadly speaking, the research framework incorporated (first) concepts derived from regional science and economic geography to generate a powerful analytical program for assessing change at the urban system level, including analyses of trends and tendencies in comparison with US cities and reference to the positionality of Canadian cities within the global arena. The first section of the book is also intended as a presentation of processes of divergence and convergence in the changes taking place among city-regions at the national level and including reference points for a profile of representative cities derived from an appreciation of structural trends and processes. This initial section provides both critical context for the case studies and a significant statement of urban–regional scholarship in its own right while offering our readership an incisive profile of growth and change within an instructive national urban system.

Second, we deployed a blend of urban studies traditions and research methods to examine detailed aspects of change within individual cities. Here, our choice was to select the five metropolitan city-regions occupying the upper echelons of the Canadian urban hierarchy—increasingly the principal drivers of growth and change in the Canadian economy—to test for commonality and contingency in developmental trajectories and to probe for intersections between global processes and local factors as played out in the nation's major cities. These five city-regions—Toronto (including Hamilton and Oshawa), Montreal, Ottawa, Calgary, and Vancouver—have achieved growth rates and urban–regional scale thresholds that set them apart from others in the Canadian urban system, the result of a mélange of financial, industrial, political, and socio-cultural advantages, signalling the advent of the 'power metropolis' in the twenty-first century. That said, the specific mix (and international projection) of these 'power attributes' differs widely among the cities within our sample, underscoring the salience of exceptionalism in the national urban system, which in turn produces defining contrasts in industrial and employment structure, urban form and social morphology, and policy responses. Accordingly, the five case study teams were asked to report on basic metrics of economic structure, employment, and labour force change, but each was otherwise free to select the approach and methodology that would best disclose the rich complexity of urban change in twenty-first-century Canada. The case studies incorporate heuristic methods in the development of compelling narratives of growth and change in metropolitan areas while including substantive empirical components.

Finally, we wanted to draw upon both the detailed empirical analyses presented in the first part of the volume and the five city-region case studies to identify the principal theoretical implications of our research and to offer some observations on the role of governance, planning, and public policy. This synthesis is presented as an extended essay in the final chapter (Chapter 13). Contemporary discourses concerning the implications of innovation and restructuring are acknowledged as features informing the design of the book. These discourses include debates on the influence of globalization, the rise of a new cultural economy of creative industries and institutions, and the very different forms of local–regional governance and planning systems that shape economic growth and change in Canadian cities. We very much hope that this volume may be seen as a distinctive contribution to the literature on the instructive Canadian experiences of urbanization and urbanism.

Larry S. Bourne, Tom Hutton, Richard Shearmur, and Jim Simmons
Toronto, Vancouver, Montreal, and Victoria: August 2010

Acknowledgements

This volume is a collective effort, not only among the four editors and the other chapter contributors but with the myriad of people who supported the research and helped put the volume together. We first acknowledge the contributions of those colleagues who participated in the original project design and the two subsequent workshops held at the University of Toronto but are not included in the volume: Mark Brown, William Coffey, Gunter Gad, Meric Gertler, Shauna Brail, Damian Dupuy, and Rick DiFrancesco. Their substantial advice and insights are reflected in the quality of the chapters in this book. We are also pleased to acknowledge the enriching value of relationships with other Canadian urban research projects with whom we shared ideas and significant cross-membership, notably the Innovation Systems Research Network (ISRN) directed by David Wolfe and Meric Gertler at the University of Toronto and the Multilevel Governance in Canada project directed by Robert Young and Andrew Sancton at the University of Western Ontario.

We also depended on the efforts of our research assistants, both past and present, as well as those of technical and administrative staff. This is particularly true of programming and cartographic personnel at the five host institutions: the University of Toronto, the University of British Columbia, the University of Ottawa, the University of Calgary, and the Institut national de la recherche scientifique (INRS). At the University of Toronto, Pat Doherty of the Cities Centre helped to organize the two workshops, and Richard Maaranen finalized the maps in Chapter 10. Shizue Kamikihara prepared the maps and graphs in the chapters written by Jim Simmons in Victoria. Byron Miller and Alan Smart want to acknowledge the cartographic work of Scott Bennet in the preparation of maps for the Calgary chapter. At UBC, Eric Leinberger of the Department of Geography prepared the maps for the Vancouver case study chapter.

We also owe a considerable debt to Carolyn Bell in the Centre for Human Settlements (CHS) at UBC, who prepared a composite manuscript by integrating and standardizing chapters and incorporating illustrations and tables submitted in a wide range of formats. Without her efforts and talents, the manuscript would have lacked the polish it needed. Karen Zeller at CHS expertly administered the grants on our behalf over the term of the project.

We are grateful to the Social Sciences and Humanities Research Council (SSHRC) of Canada, who provided crucial support through two grants awarded under the National Research Cluster program in 2004 and 2005 (SSHRC grant numbers 857-2004-0069 and 857-2005-1008). The editors and contributors hope that this volume represents a satisfactory output of this important national research enterprise.

The editors would also like to thank all the contributing authors, each of whom devoted considerable intellectual resources and effort, first in preparing initial drafts and then in committing to the program of revisions required to realize the potential of the project.

Finally, we want to offer due acknowledgement to our editors at Oxford University Press, Kate Skene, Phyllis Wilson, Rebecca Ryoji, and Dorothy Turnbull, who supported the project from the beginning and who elicited critical assessments of the draft manuscript from peer reviewers, which enabled us to improve the book significantly.

Larry S. Bourne, Tom Hutton, Richard Shearmur, and Jim Simmons

List of Contributors

Caroline Andrew, University of Ottawa

Trevor Barnes, University of British Columbia

Larry S. Bourne, University of Toronto

John N.H. Britton, University of Toronto

Cedric Brunelle, INRS, Université du Quebéc

Guy Chiasson, Université du Québec en Outaouais

Tom Hutton, University of British Columbia

Deborah Leslie, University of Toronto

David Ley, University of British Columbia

Byron Miller, University of Calgary

Markus Moos, University of Waterloo

Mario Polèse, INRS, Université du Québec

Norma Rantisi, Concordia University

Brian Ray, University of Ottawa

Richard Shearmur, INRS, Université du Québec

Jim Simmons, Ryerson University

Alan Smart, University of Calgary

R. Alan Walks, University of Toronto

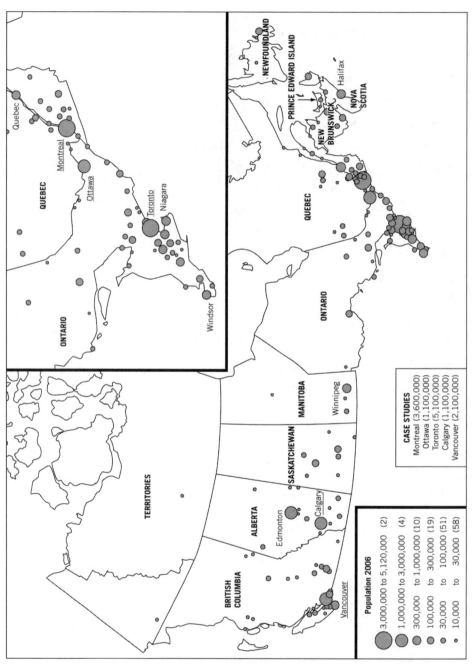

The Canadian Urban System, Population 2006

Population 2006

- ⬤ 3,000,000 to 5,120,000 (2)
- ⬤ 1,000,000 to 3,000,000 (4)
- ● 300,000 to 1,000,000 (10)
- ● 100,000 to 300,000 (19)
- ● 30,000 to 100,000 (51)
- · 10,000 to 30,000 (58)

CASE STUDIES

Montreal (3,600,000)
Ottawa (1,100,000)
Toronto (5,100,000)
Calgary (1,100,000)
Vancouver (2,100,000)

BRITISH COLUMBIA
ALBERTA
SASKATCHEWAN
MANITOBA
TERRITORIES
Edmonton
Calgary
Vancouver
Winnipeg

NEWFOUNDLAND
PRINCE EDWARD ISLAND
NEW BRUNSWICK
NOVA SCOTIA
Halifax
QUEBEC
ONTARIO

QUEBEC
ONTARIO
Quebec
Montreal
Ottawa
Toronto
Niagara
Windsor

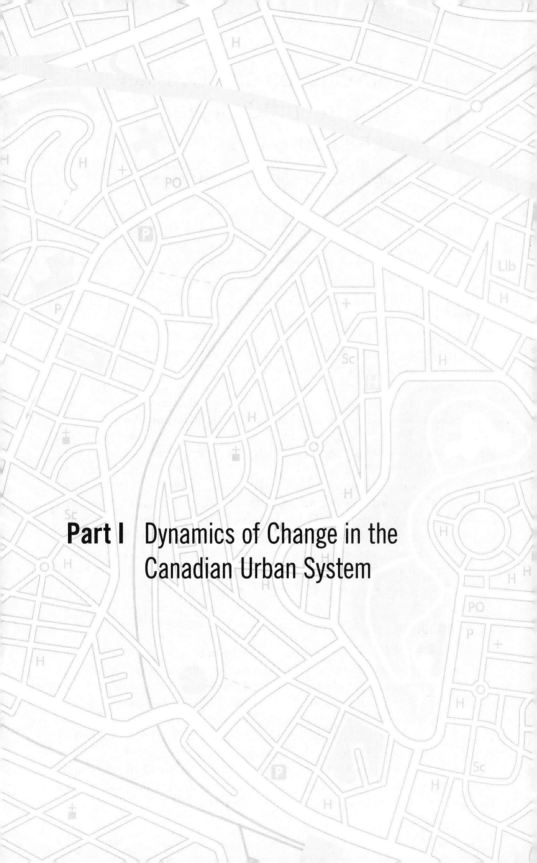

Part I Dynamics of Change in the
Canadian Urban System

1

Introduction and Overview: Growth and Change in Canadian Cities

Larry S. Bourne, Tom Hutton, Richard Shearmur, and Jim Simmons

Introduction

This book explores the patterns, trends, causes, and consequences of urban growth and change in Canada. Our aim is to offer a critical and timely analysis of the country's continuing economic transformation, viewed principally through the lens of changes in employment, labour markets, and occupations in the largest metropolitan areas. These metrics of change serve as entry points for discussions on related domains of social, cultural, and physical development in the contemporary metropolis.

The approach here is *multi-scalar* in that it embraces several distinct but interrelated spatial scales. First, it offers an empirical analysis of national trends, using the urban system as a way of generating insights on structural change in all Canadian cities, set within an international context. Second, it provides detailed case studies of the five dominant city-regions that increasingly shape the evolution of Canada's economy, polity, global connectivity, media, and urban imageries—Toronto, Montreal, Vancouver, Ottawa-Gatineau, and Calgary. Because these city-regions attract significant interest from urban scholars and political decision-makers nationally and within the global community, this volume is intended to appeal to both domestic and international constituencies as well as to scholarly research and policy-making communities.

Why Canada?

Canada is a country of more than 34 million (2010) inhabitants—a small to medium-sized state in terms of population but huge in geographic extent—situated in the northern half of the North American continent. The study of its cities has much to offer researchers and decision-makers not only in Canada but further afield. The trajectory of its cities, the specific contingencies that shape them, and the geography of the country—so unlike Europe, for instance—make Canadian cities distinctive and useful prisms through which to read and understand changes in urban development occurring elsewhere.

There are a variety of ways in which the transformation of Canadian cities can tell us about wider issues that cities throughout the developed world, their citizens, and their governments face. Some of these issues are detailed below. If Canadian cities can inform us about certain processes at work in other contexts, then this implies that certain elements highlighted in this book can be, to some extent at least, generalized. If, on the other hand, each country and each city within it develop along lines primarily determined by historical contingency, local culture, and their unique attributes, then generalization is more problematic.

The authors of this book, each from his or her own methodological perspective, draw out overarching trends (relying on statistics, graphs, and maps) as well as the contingent factors that make Canadian cities unique (through case studies and detailed descriptions of particular cities). Canada itself can, of course, be considered a particular case, uniquely influenced by its history and position in the world, but it is also a country caught in the wider ebb and flow of global economic and technological change, social pressures, and cultural and political tensions (OECD 2002; Savitch and Kantor 2004). The way in which Canadian cities have evolved, and are evolving, in this wider context can provide important insights into the evolution of other cities. The dynamics described in this volume will not be exactly replicated elsewhere, but the general lessons learned from this study of Canadian cities can inform students and policy-makers in Europe, the United States, and beyond.

For instance, Canada's economy is highly integrated with the US economy: more than 80 per cent of all its exports (and about 27 per cent of Canada's GDP) is destined for the US market. From the beginnings of European settlement until the end of World War II, Canada had a colonial economy that was closely integrated with that of Britain, its larger trading partner. Tariffs and transportation routes oriented imports and exports toward the United Kingdom and other Commonwealth countries. The industrial core stretching from Windsor to Quebec City benefitted from a stable and protected market, while the commodity-producing economies in the rest of the country suffered the uncertainty of volatile and competitive markets. But as Britain turned toward Europe, Canada responded by increasing its ties with the US. International trade theory suggests that this type of integration leads to increased aggregate wealth—and there is little to suggest that Canada and Canadians have not, by and large, benefitted from their economy's openness to external influences. However, as will be evident in forthcoming chapters, the price that Canadian cities have paid for this reliance on trade and foreign capital includes volatility and high levels of foreign ownership. Periods of boom and bust characterize these cities. In a world where trade and globalization are still widely understood as beneficial despite recent setbacks, the distributional consequences of openness, in particular the uneven distribution of wealth accumulation and income generation over time and space, is a theme that runs throughout the book.

Europe has recently introduced a single currency, at least in the Euro zone, which is thought to ease trade and facilitate integration and economic growth. Canada, on the other hand, despite its high degree of integration with the US economy, has retained its own currency. The consequences of this are made clear throughout the contributions to this volume. First, it can be argued that the very existence of a Canadian financial sector—primarily centred in Toronto—owes much to the fact that Canada has its own currency. Even in the context of a secular trend toward consolidation of financial markets, each currency area still requires its own financial system (Shearmur 2001). Second, during some periods (the late 1990s, for instance) a low Canadian dollar has helped to maintain Canada's manufacturing sector, encouraging exports (and out-sourcing to lower-cost developing countries notwithstanding). During other periods, however (such as the resource boom of the early 2000s), Canada has suffered from a type of Dutch disease: a high currency, based on its substantial resource endowments, has stifled manufactured and service exports. While this has led to increased efficiency among Canadian exporters, it has also contributed to losses of market share. Indeed, it is an example of the type of economic volatility that globalization brings about and that cities, relatively powerless in the short and medium term in the face of such fluctuations, ultimately need to absorb.

The Canadian dollar and its local financial sector also provided a degree of protection during the 2008–9 recession. As financial markets and banks failed in the US and Europe, Canada's financial system remained fairly stable (although not without its own challenges, including portfolios of loss-making loans). This stability is largely attributable to the conservative financial regulatory framework that governs Canada's banking sector. A common currency diminishes the capacity of nation-states to enforce their own regulations, and the recent crisis amply demonstrated that big (or integrated) is not always beautiful. Such considerations may inform European analysts as they consider the long-term consequences of the common currency and of other forms of harmonization.

As suggested above, Canada's currency—and by extension much of its economy—fluctuates with the value of and demand for resources. Notwithstanding efforts to diversify, and notwithstanding the relatively small number of people directly employed in primary sectors, Canada's staples economy (Howlett and Brownsey 2008; Innis 1933) remains to this day an important, perhaps even in some regions the dominant, driver of economic growth. The story of Canada's regions and cities is in many ways the story of attempts to escape from an economic trajectory that relentlessly draws them back to their resource base (Stelter and Artibise 1986; Stelter 1990). This is true of Calgary, an oil city; Vancouver, which derives a significant share of its employment and revenues from seaborne exports, notably to markets of the Asia-Pacific; Toronto, whose financial sector is ultimately dependent on wealth derived from resource extraction; Ottawa, which governs this resource-based state; and, to a lesser extent, Montreal, where the headquarters of aluminum producers and wood products firms still reside. Few other Western nations, except perhaps Australia, are as dependent on their primary sectors. However, it could be argued that many countries are similarly trapped in their own his-toric roles, which are somewhat more difficult to identify than Canada's because they are a combination of functions, products, and culture (examples here might include Britain and the Netherlands' dependence on finance and ancillary mercantile roles). Thus, Canada presents as a simplified example of long-term path-dependency (Arthur 1994) and of efforts to escape this trajectory: as such, it serves as an interesting study and a reminder of the power of these paths (Davis and Weinstein 2002).

Canada, and in particular its largest cities, have welcomed very large numbers of immi-grants: in Toronto and Vancouver, for instance, more than 40 per cent of the population is foreign-born, and in the other large cities at least 20 to 25 per cent of the population was born abroad (see Chapters 2 and 3). This vast experiment in cohabitation—Canada's cities are often referred to as mosaics rather than melting pots—has by and large taken place peacefully. After initial periods of adaptation, immigrants have on the whole, with some important exceptions (including some recent immigrant groups; see Chapter 12), been able to catch up to the income levels of other Canadians and take on professional and other leadership roles in a manner increasingly indistinguishable from theirs.

Two important questions run through the chapters of this book. First, how has this integration been achieved? Although this question is not directly dealt with in all chapters, the description of Canada's five major cities provides some insight into how they function and how immigration, far from being perceived as problematic, is usually understood as highly beneficial to the receiving areas. The second key question is: will this situation last? In this regard, Chapter 6 looks more closely at poverty and inequality in Canadian cities and suggests that the pressures of 'social globalization'—the homogenization of health, poverty, and social assistance policies toward a neo-liberal norm—may be restricting Canada's ability to peacefully and successfully integrate immigrants. Considering the ghettos often observed in US cities, the reasons that Canada's cities have so far avoided such intense racial and ethnic segregation are worth exploring. In contrast to European cities, where tensions between immigrants and the host nationals are more marked than in Canada, Canadian cities may offer examples that demonstrate the possibility of integration— or at least more peaceful coexistence.

Coexistence does not mean assimilation. Quebec's continued cultural and linguistic identity, which has survived for more than 250 years in predominantly English-speaking North America, is an interesting example that may foreshadow the nature of a notionally borderless Europe. As Favell (2007) has pointed out, notwithstanding the great mobility of young people in Europe, many find it difficult to settle in a country other than their own: given a choice, most would prefer to return home when settling down and starting a family. Their story is very similar to that of Quebecers: many leave Quebec for job-related reasons and have little choice about returning. But many more either do not leave or leave for a while and seek to return. Even though Quebec is part of Canada and now part of NAFTA, Quebecers are far less mobile than most other North Americans. This stands in sharp contrast with the US experience (and indeed with that of the rest of Canada), where interstate (or interprovincial) mobility is markedly higher. Thus, in Canada it has not been possible to overcome deep-seated cultural differences between regions that share broadly similar development levels and political stability. Mobility is strongly determined by attachment to the place through culture and language. In view of the Canadian experience, it is therefore unlikely that the European Union will experience the (almost) seamless mobility seen in the Unites States. Rather, the dynamics observed between Quebec and the rest of Canada—recounted in the chapter on Montreal—seem a more likely scenario.

Finally, Canada can also serve as a laboratory for the study of the geography of urbanization: distances are so great and spaces between cities so sparsely populated that the spatial diffusion of agglomeration, the effects of distance and isolation on levels of economic activity, and the possible effects of local factors can be studied free of outside 'noise'. In many other contexts, the suburbs of one city merge with the suburbs of another: as population grows and concentrates spatially, it is difficult to identify the dynamics of outer suburbs or of rural areas and indeed the structuring effects (if any) of central business districts on the form of cities. While the Toronto region, as described in Chapter 10, is rapidly becoming a regional megalopolis (OECD 2009), with numerous previously distinct metropolitan areas merging into a more or less continuous urbanized area, most other Canadian cities are 'free-standing': they are the only sizeable city, often by a size factor of 10 or 20, within a radius of 200 kilometres or more. In this context, urban form is principally generated by internal factors interacting with aspatial global ones rather than through the influence of activity or settlements just beyond the boundary of study.

This study of Canadian cities, proceeding as it does from generalized accounts of transformations in the urban system writ large (Simmons, Bourne, and Cantos 2004) to richly textured case studies of Canada's largest metropolitan areas, provides both a snapshot of recent urban dynamics in Canada and an exploration of five second- or third-tier global cities and the varied ways in which they find a place for themselves in an increasingly globalized and integrated world. The processes described and illustrated are specific to Canada but offer examples that can help to inform and increase understanding of processes and trends occurring in other cities across the globe.

The following sections of Chapter 1 present an overview of the context and conditions of urban growth and change in Canada as an essential introduction to our study of structural processes and drivers of change operating on and within the country's urban system.

Canadian Cities: Contextual Elements

Canadian cities follow, in most important respects, the pervasive trends and experiences of other advanced societies (Heinz 2006). However, these global trends, as noted above, are modified by national circumstances and a number of specific attributes, such as the geography of the country and its small population; by the comparative openness of the national economy (roughly 34 per cent of Canada's GDP is accounted for by international trade, compared to about 10 per cent for the United States); and by the distinctive Canadian experiences with respect to relatively high levels of foreign ownership of firms, immigration, and cultural diversity.

Often, these drivers of change are working in opposite or contradictory directions. Consider, for example, the differential effects of increased trade and higher levels of immigration. Canada's trade is overwhelmingly dominated by exchanges with one country, the US, further extending the country's high level of integration within continental (American) markets. This level of dependence brings greater economic uncertainty, particularly because the US has imposed tighter border restrictions since the events of 11 September 2001. At the same time, trade links with the US differ widely by region (see Chapter 2): the Atlantic region trades primarily with the New England states, Ontario cities are firmly integrated into the midwest market, and Alberta and BC are tied to the western and southern US, as well as to Asia. In contrast, migration and labour flows between the US and Canada are relatively small; there is no continental labour market as such (Simmons and Kamikihara 2006). The country's substantial immigration flows are increasingly drawn from non-traditional sources, and the rate of immigration is among the highest in the world on a per capita basis. These flows in turn have accelerated the country's global integration but in a different direction: linking residents of Canadian cities through 'transnational' communities with populations in countries outside of North America.

Throughout the book, unless otherwise specified, urban growth refers to population growth. It is, for the most part, positive in Canada. Urban change implies all the causal or resultant effects that accompany urban growth, including changes in the urban economy, the demography, and land use. Think of the former as a simple increase in magnitude and the latter as a form of reorganization. In order to illustrate the trajectory of Canada's overall experience in urbanization and the growth of its metropolitan areas, Figure 1.1 provides a longitudinal view of the growth in urban population, and that of metropolitan areas (CMAs) and the five case study cities, as proportions of national population from 1951 to 2006. Figure 1.2 graphs the rate of urban growth for the same cities for each five-year census period. As is clear from these illustrations, Canada's population is now overwhelmingly urban, despite its immense geographic size and popular image of endless wilderness. More than 80 per cent of the population is now urban (compared to 52 per cent in 1951), more than 65 per cent reside in metropolitan areas (compared to just 45 per cent), and more than 40 per cent reside in the five largest city-regions (compared to 25 per cent). Urban growth rates, however, have declined from their peaks in the boom years of the 1950s and 1960s, in line with declining fertility rates, levelling out after 1976 (Figure 1.2). Nevertheless, the case study cities and all CMAs continue to grow more rapidly than the national population. Canada is now not only urban but predominantly metropolitan in terms of its productive capacity and the living environments of its population.

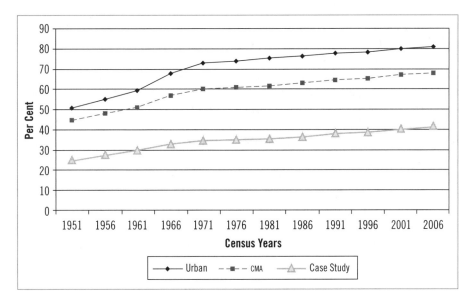

Figure 1.1   Urban shares of Canada's population.

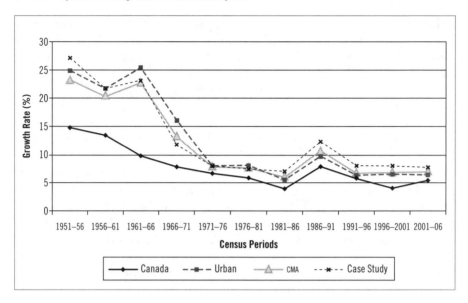

Figure 1.2 Urban growth rates.

Framing the Context: Some Theoretical Perspectives

This, then, is the age of cities—and increasingly, of large city regions, both in Canada and in other developed nations. Cities do matter in terms of economic development, productivity, and innovation (Martin and Florida 2008). They also matter to the nation-state, as we demonstrate later in the chapter, and from a policy perspective (Bradford 2002). From an empirical standpoint, it also seems clear that the larger city-regions exercise increasing dominance in political, economic, and cultural terms at the national level. At the same time, they have become the organizing nodes of a globalizing economy, of its supply chains, and of global trading and financial systems.

At the national scale, the emergence of primate cities is an established condition of development in both transitional societies and in advanced states such as Britain. For the latter, Doreen Massey (1984) was one of the first to write about the consequences of a changing spatial division of labour in the UK that tends to favour London and its larger region. The theme is elaborated by Kevin Morgan (2004) in a paper on the need for 'territorial justice' between the high-growth, London-centred region and the lagging industrial regions of the British periphery. In Canada, a similar debate has focused on the historical relationship between the country's urbanized core and its periphery (Stelter 1990) and most recently on labour market inequalities (see Chapter 6).

While the national system of cities remains a pertinent terrain for the study of urbanization, the emergence of globalization and its consequences has shifted the scale of urban research and analysis to large cities as international hubs. Peter Hall's initial characterization of 'world' cities (1966) acknowledged the importance of urban size (scale), specialization, and industrial concentration as markers of importance within international urban hierarchies. John Friedmann's (1986) and Saskia Sassen's (2006) *global city* hypotheses underscore the centrality of higher-order services—including head offices, intermediate banking and finance, and producer (or business) service industries—to the position of cities within the circuits of economic power and political influence. There is also an element of unevenness and inequality to these processes of globalization and the impact on competitive advantage: certain metropolitan areas attract disproportionate shares of capital, corporate control functions, 'talent', and human capital, exacerbating the differential in growth rates among cities within their respective national urban systems (see Chapters 2 and 3).

Over the past decade or so, other strands of research have contributed to more diverse discourses of urbanization and globalization, within both the advanced societies of the global 'north' and the transitional and developing states of the 'south'. For instance, Taylor, Beaverstock, and colleagues, at the Global and World Cities (GaWC) centre at Loughborough University, UK, have broadened the global cities model to incorporate a more diverse set of measures of interactions and functional specialization, including links to such domains as political and diplomatic power, cultural activity, non-governmental organizations, and education (Taylor et al. 2008). This analysis continues to identify first-tier cities such as London and New York as the 'most-rounded' exemplars of the global city but opens up niche global roles for places such as Geneva, Vancouver, and Nairobi—cities that do not typically figure in global power rankings based on corporate and financial criteria.

Michael Peter Smith (2001) has advanced a model of *transnational urbanism* to complement the overly economic bias of the dominant global cities discourse. In this interpretation, the socio-cultural and ethnic linkages between cities, formed by international immigration and sustained by intense communication within the diaspora, comprise a vital channel of integration within global urban networks, often with substantial economic implications. They suggest the possibility of progressive collaboration among social groups linked either by class or ethnicity, or both. Vancouver's links to Hong Kong, Toronto's to South and Southeast Asia, and Montreal's continued cultural proximity to France provide the most relevant Canadian examples (see Chapters 8, 10, and 12).

These experiences have stimulated renewed conceptual debate in Canada, as in other advanced societies, including a series of universal theories to interpret the restructured city-region and attendant social and cultural features. Notable among these theoretical initiatives is *post-industrialism*, introduced by the American sociologist Daniel Bell (1973) to anticipate the advent of a new knowledge-based social class, with occupations, values, political preferences, and social behaviours differing from those of the old 'industrial class'. Other scholars refer to the 'new middle class' of managers, professionals, and entrepreneurs. *Post-Fordism* emphasizes the decline of traditional, mass-production manufacturing and the concomitant rise of specialized, 'flexible' production systems and labour. The *world* or *global city* concept identifies the growing dominance of major metropolitan regions in corporate control, higher-level banking, and financial power.

In addition to these socio-economic concepts, we also draw on the influential spatial economic processes, notably those of *agglomeration* and *clustering*, derived from urban economic theory and tested through detailed empirical study of industrial location. Cities still thrive largely through their export activities. At the local level, these spatial models take into account the influences of property markets and land values as factors shaping industrial location decisions and the localization of labour skills. Additional emphasis is placed on the advantages of proximity for firms operating within particular production clusters and supply networks and on the economies of transportation logistics and just-in-time delivery. Many of these concepts came together earlier in the theory of the *industrial district*, initially articulated by Alfred Marshall more than a century ago (1890), but have recently generated a lively debate and analysis, including the notion of the 'new industrial district' in the 1980s (Piore and Sabel 1984; Markusen 1996; Martin and Sunley 2003).

These theories and models still have considerable contemporary resonance, as evidenced in the work of Krugman (1995), Porter (2003), and in Canada, Meric Gertler, David Wolfe, and colleagues (Spencer et al. 2010; Wolfe 2009) on clusters of economic activity. This research has extended our understanding of the generation and operation of clusters of enterprises and the social and institutional underpinnings of economic activity and innovation. It also identifies important differences in occupational structure and the quality of work.

A number of other ideas have emerged over the past decade to challenge the conventional wisdom about cities in developed countries,[1] each requiring fresh theoretical investigation. These ideas include a complex, multi-layered set of concepts relating to processes of urban economic and social change. The concept of a technology-driven *new economy* includes not only the episodic rise and fall of the dot.coms but also a more durable 'technological deepening' of production processes, communication systems, and occupations (Beyers 2000; Hutton 2008). The *knowledge-based society* is an extension of Bell's concept of a post-industrial society (1973) based on scientific innovation but with knowledge and learning as the driving forces of change in the economy and society. Others emphasize the social dimensions (e.g., intensely interactive, amenity-sensitive) of economic activity in advanced societies (Thrift and Olds 1996). For example, the *new cultural economy* includes a diverse range of high-value design sectors and industries (media and new media, software design, computer graphics and imaging, film and video production [Scott 1997; Pratt 1997; Hall 2000; Barnes 2001; Florida 2002; Martin and Florida 2008]). An emerging paradigm of the *sustainable city-region* requires a more forward-looking integration of social, environmental, and economic values and interests and incorporates development principles enunciated by advocates of industrial ecology, the 'new urbanism', and the 'liveable city'. Our mandate includes an exploration of these ideas and their applicability to Canada.

City-Regions and the Nation-State

The momentum and widespread consequences of globalization have inexorably shifted the spatial focus of urban studies away from national territories, since the boundaries of the nation-state no longer effectively 'contain' industrial production, supply chains, labour markets, or flows of capital through financial markets. It has also introduced the notion of *accelerated trajectories of innovation and restructuring* within urban–regional space. In an era of intense international competition, immigration, and growing multiculturalism, the nation no longer identifies a mono-cultural foundation, nor does it provide a uniform narrative on either the structure of the economy or the changing social order. The power of telecommunications, and specifically the Internet, while by no means unlimited, has vastly increased the channels of global connectivity among individuals as well as between firms and governments. This trend is likely to subvert efforts to maintain centralized socio-political control over the longer-term within most states, as the conflict (March 2010) between Google and the government of China over privacy and access issues demonstrates. For these (and other) reasons, the dominance of the national state as both a geo-political and an economic unit within global systems and hierarchies has been eroded (Scott 2004).

That said, we can make a strong case that the nation-state still matters (see Chapter 4). It serves as an organizing unit for administration, regulation, taxation, and service delivery; as a repository of socio-cultural values, citizenship, and identity; and as an economic construct—albeit one with many leakages and spill-overs. National governments still set macro-economic and trade policies, though subject to international trade and monetary exchange agreements, and regulate immigration flows. The nation also serves as the territorial setting for the growth of national urban systems and hierarchies, although, as acknowledged, the drivers of growth and change are to a large extent exogenous to the territorial state, especially in an open economy such as Canada's. In developmental terms, too, the nation-state still offers, at the broadest level, scope for applying conveniently simple but useful descriptors of economic trajectories—staples-dependent, industrial/industrializing, post-industrial or post-Fordist, and so on. Each nation provides its own examples (and variants) of these trajectories.

As argued above, even under conditions of accelerated innovation and restructuring, shaped by the forces of globalization, urban and regional economies within the nation tend to reflect certain inherited and path-dependent features. These features are both shaped by natural endowments (e.g., resources, topography, climate) and conditioned by decisions made in earlier eras within which the instruments and agencies of the nation-state imposed a different, perhaps stronger, imprint on local and regional economies. Such state influences on the shape and structures of the economy are transmitted, for example, through transportation and other infrastructural investments, the active regulation of commodity markets, subsidies to particular industries and sectors, commitments to regional development programs, barriers to investment, and a host of indirect instruments such as welfare, employment insurance, and other social policies and programs (Courchene 1997).

To continue along this multi-scalar continuum, while in Canada the federal government sets the national stage—and the boundary conditions—for urban and economic change, its direct role in cities is limited to specific services. As shown in Chapter 4, the provincial governments, in varying degrees, exert the most direct and substantial impact on cities. As readers are aware, cities, at least as political entities (i.e., municipalities), are in constitutional terms the sole responsibility of the provinces. Provincial governments, moreover, have become increasingly powerful. They now collectively spend more than the federal government and deliver most basic social services, including health, education, welfare, and housing. They also regulate urban development and land use and have final approval for municipal expenditure budgets and tax rates, as well as for infrastructure provision and other local services.

Nevertheless, and of critical importance for this volume, city-regions and other empirical urban areas, such as labour markets defined on the basis of commuting data (the census metropolitan area), are not the responsibility of either provincial or federal governments. In most parts of the country, there is little correspondence between municipal boundaries and the geographic living spaces of urban residents, either as labour or housing markets (Shearmur and Motte 2009). This poses an analytical, not to mention political, problem for both researchers and governments. Most of the chapters in this book deal with city regions rather than municipalities and are therefore compelled to confront the mismatch between the political and functional organizations of urban space and to address the conundrum that no one is managing the country's city-regions.

At this regional scale, the past two decades have also witnessed the *internal reconfiguration* of the metropolis—the new city-region. The momentum of growth in population, jobs, and investment has in many cases shifted decisively from the urban core to the suburbs and beyond to include the 'peri-urban' world of the new exurbs (see Chapter 5). The mono-centric or centre-dominated city, reflected in the tenets of both the Chicago School of human ecology and the classic urban economics models, has not disappeared. It has evolved, however, into a 'regional city' or city-region. While traditional city centres often retain key economic functions and remain the focus of transport and communication networks, centrality is no longer uniquely focused there: city-regions are multi-nodal, dispersed, or both, with a preponderant share of population, employment, and business start-ups locating in suburban areas. Within the suburbs, recent growth tends to favour the newer, outer suburbs over the old, established inner suburban districts, many of which are now experiencing a relative decline and/or a 'retrofitting' experience of brownfield redevelopment.

But beyond this bifurcation of growth rates between the metropolitan core and suburbs/exurbs, we can discern increasing complexity and subtle changes that reflect what Scott (1988) has described as the ongoing specialization of the internal spaces of the metropolis. Graham and Marvin (2001) have described the 'splintering (spaces of) urbanism' within the city, driven by new technology in both its material and psychosocial manifestations, while Dear and Flusty (1998) have advanced a more whimsical profile of 'postmodern urbanism', a chaotic, centreless, and almost amorphous urban world. The spatial reality of city-regions, however, is both more complex (Soja 2000) and more orderly (Shearmur et al. 2007).

Although the mix of factors and interdependencies shaping urban space has changed over time, there are still powerful logics at work underlying urban economic development and location decisions, including the persistence of agglomeration economies (and diseconomies) for both industries and labour (Krugman 1995). Within the metropolitan core, the speculative office boom of the late twentieth century has given way in many cities to a more selective experience of purpose-built, often mixed-use, high-value designer office towers. Further, the insidious workings of gentrification within the older residential terrains of the city's inner core, at least in the large and growing cities, have been supplanted by a veritable juggernaut of inner-city redevelopment (also primarily in the large and growing cities). This process has included new, often specialized industrial clusters, notably in the IT and cultural sectors, as well as upscale housing, spaces of leisure and consumption, and investments in the public realm (including sports facilities). As some observers would have it, the inner city—at least in particular city regions—has recently shifted from a 'post-industrial' to a 'neo-industrial' growth trajectory (Hutton 2008) that synthesizes inputs of culture and technology in the fabrication of products and services presenting a richer and more complex construct than that of the office-dominated post-industrial era. In parallel, the economy of the suburbs has experienced rapid change as the outer zones of the metropolis have broadened their role from manufacturing and the provision of retail and personal services to incorporate important industrial/employment clusters such as universities, secondary regional business centres, science parks and R&D centres, and, of course, international airports.

MUN , DT, ST JOHNS.

Toward an Integrated Conceptual Framework

The preceding review raises a number of critical questions for both research and public policy, each question informed (and challenged) by contemporary debates on theory. Yet an additional and complementary framework is still needed to pull some of these ideas together and in so doing to provide specific concepts and measurements that serve as common denominators for our study of growth and change in urban economies. We can begin with a series of basic questions. What do we mean by an urban economy? What does it include? How does that economy actually work in particular places? What principal components and drivers of change are of interest here, and how do they relate to each other? How have these drivers changed over time and why? Is there in fact a 'new' economy emerging in Canada, and if so, how does it relate to elements of the 'old' economy? What are the various processes—or principles—that underlie the sectoral structure, spatial organization, and growth of that economy? What are the most important and relevant measures of economic activity, and how do they relate to other spheres of urban life?

As a starting point, there is little doubt that the economy of a large metropolitan region is immeasurably complex and in constant flux (Simmons, Kamikihara, and Bourne 2009). In Chapter 10, for example, the economy of the Greater Toronto region is shown to have a GDP of almost $300 billion, with 180,000 firms and 3 million employees operating at 80,000 different work locations. In addition, there are some 12 million trips per day moving people and goods and millions of other communications and transactions, notably through the financial system and less formally through the Internet. Investment in the built environment and in infrastructure alone totalled $12 billion annually, at least before the latest recession, and the value of property transactions in the residential and non-residential sectors was even larger.

Despite this complexity, it is possible to identify several of the key indicators of that economy and the factors or variables shaping growth and change. The discussion suggests three basic elements that warrant priority in the case studies to follow:

1. The causes or *drivers of change* may originate from within or more likely from outside the city (and the country). These drivers, of which globalization and technological innovation are two prominent examples, act to create new jobs and stimulate some activities while destroying others (ICP/MPI 2009). They also shift the location of these activities and alter the relationships of each sector or component in the economy with other components (Camagni 1991).

2.   Processes underpin and produce order in the structure and geography of change in each sector of economic activity. These are processes that shape patterns of property values, new investments and technologies, and the location decisions of firms and households. The competitive land market and planning regulations are two examples of these processes (Shearmur and Polèse 2007).

3.   The *linkages* are the complex networks of transactions, exchanges, and flows (of people, goods, capital, ideas, and information) that provide the essential connections and channels among all elements of the modern economy and all parts of the city-region. The journey to work is one example of these linkages; the movement of production inputs and goods is another (Scott 1988).

These three factors can be identified for each of the major components of the economy. The definition of components used here is relatively broad. The economy consists of an integrated set of components: the first is the complex ensemble of production systems, notably of capital goods. This component is typically considered the main determinant of urban form and the principal driver of urban growth in terms of its capacity to generate exports and income. Second is the consumption component, the sum of all income and expenditure flows by individuals, firms, and organizations. Third is the public sector component—that is, the combined activities of governments, public agencies, and the state more generally. Fourth, and increasingly important, is the demographic/population/lifestyle/cultural component that incorporates the economic effects of differences in age structures, migration and immigration, labour force participation, and the changing tastes and preferences of a diverse population. These four components are highlighted in various degrees in each of the following chapters.

While the basic form and structure of both production (Scott 1988) and consumption (Glaeser, Kolko, and Saiz 2001) components are widely recognized and their contributions are relatively clear, the roles of the other two in the urban economy are less obvious and thus warrant some elaboration. The state plays a substantial role, or more accurately a set of roles, in our urban economies. It is both external to the private market as a regulator and policy-maker—in effect setting the rules of the game—and a provider of public goods and services in situations where the market does not work or is inefficient. It is also internal to that market in that it is an integral player in the economy as an employer, an investor, a consumer of private goods and services, and a major owner of land and buildings. In some Canadian cities, notably smaller places and some provincial capitals, the public sector is the largest employer, and in all cities it is the largest landowner.

The varied roles of the demographic component in economic growth are more indirect and less explicit. The rate of population growth in any city, and thus the growth in local demand for goods and services, depends in part on its age structure and on the balance of migration and immigration flows, as Chapters 2 and 3 demonstrate. Moreover, the city's age structure and rate of population growth combined influence the growth and character of the local labour force and determine or influence the labour force participation rate. The latter, in turn, shapes income levels, as Chapter 5 illustrates. At the national scale, Canada's rapidly aging population has elicited widespread public concern over rising health care costs on the one hand and potential labour shortages on the other and has brought to the forefront the increasing importance of immigration in supplying future workers and labour skills. Most of these new workers will end up in cities.

To summarize: urban growth and economic change in Canada and the form of Canadian cities, as elsewhere, are the combined outcomes of a myriad of factors or drivers of change, and these are in turn shaped by underlying structural processes. The city is the cumulative outgrowth of varied production forces, shaped by the logic of agglomeration economies, competition, and clustering. Cities are also the primary settings of public and private consumption, which contributes substantially to the urban economy. The latter sectors are in turn driven by their own consumption functions that reflect the behaviour and preferences of the populations in our cities, populations that are becoming rapidly more diverse. Indeed, the modern city may be seen as a vehicle for individual and collective consumption (think of Niagara Falls or Las Vegas—or even, to a large extent, Vancouver) as much as it is for the production of goods. It is also a *public city* in which the state plays a significant role, directly through the public sphere by providing services and infrastructure and by setting the rules of the development game and indirectly by shaping the market economy through the regulation of markets for jobs, land, and finance. Furthermore, population growth is partly dependent on its demography but increasingly, in the case of Canada's cities, on immigration. All four of these components are tied together by a myriad of transactions, linkages, and flows that serve as the glue of the local economy and the rationale for the concentration of population and productive capacity that we recognize as the city, or more broadly, the city-region.

Organization and Outline of the Book

This volume consists of three principal parts and 13 chapters. Part I—of which this chapter is part—provides the context, an overview of relevant literatures and of alternative theoretical frameworks, and an outline of the drivers of recent trends in the economy and population growth. The earlier sections of this chapter offered an introductory discussion of why the study of urban Canada—of value in its own right—should be of interest to scholars and policy-makers around the world. The chapter also defined the various measures and scales of analysis used, while positioning the book, and Canadian cities, in their national and global contexts. The next five chapters examine changes in all cities in the Canadian urban system, identifying particular attributes of growth and decline. The approach again is multi-scalar: moving from the national/global context to the urban system level, then to the areas of most rapid growth and change within metropolitan areas, and finally to the five city-region case studies.

Chapter 2 positions Canadian cities in a comparative North American and international setting, noting both similarities and differences among cities and in the urban systems to which they belong. Are Canadian cities distinctive? How do they rank (or score) on scales of global cities? Chapter 3 offers an overview of structural changes within the Canadian urban system, first through an examination of employment growth and differences in functional specialization among cities and second through an analysis of the dynamics of population growth and the impacts of the demographic transition and immigration. Chapter 4 then examines the political economy of urbanization in Canada and outlines the ways in which governance and policy are organized and implemented in Canadian cities. Chapter 5 focuses on intra-metropolitan distributions of employment, first at the CMA-wide level, and then examines the most rapidly changing territorial zones of the metropolis, the downtown core and inner city and the outer suburban and peri-urban fringe. Finally, Chapter 6 examines outcomes of changing labour market conditions in terms of emerging social inequalities across all major cities in the country, with particular emphasis on the origins and consequences of wage and income inequalities.

Part II of the book presents the five case studies: the emergent and influential 'power-metropolises' of Canada's urban system. Here we have identified Canada's five major city-regions as a cohort that allows us to examine processes of growth and change at the upper tier of the national urban system, much as Janet Abu-Lughod did for the US (1999) in her treatment of New York, Chicago, and Los Angeles, with (in each national case) their large labour markets, complex industrial structures and innovation systems, and openness to global processes providing rich terrains for empirical analysis and theoretical engagement. Part II opens with Chapter 7, which summarizes the rationale for and design of the case studies and provides a set of comparative indices on the character and growth of the five city-regions, which serve as examples of what the Conference Board of Canada (2006) has called 'hub' cities. This summary is followed by detailed analyses of Canada's 'big five' metropolitan city-regions—Montreal, Ottawa-Gatineau, Toronto, Calgary, and Vancouver—in Chapters 8 through 12. Following the analytical emphases in Part I, each case study chapter examines labour markets, employment structure, and occupational change. But each chapter develops its own narrative on the city in question, reflecting real differences in the places themselves—in their location, history, economy, politics, and culture—and in the richly varied perspectives of the individual authors.

Part III—Chapter 13—brings the varied strands of the discussion back together. It does so first in terms of contemporary issues of public policy and practice, drawing out lessons learned from the case studies. It continues by posing fundamental questions of current political and theoretical debate. The chapter also looks ahead into the economic challenges and opportunities facing Canada's cities and regions under uncertain and rapidly changing global economic, environmental, and political conditions. Given the volatility of economic conditions, this represents not so much a forecast but rather conjectures about evolving scenarios of restructuring in the twenty-first-century metropolis.

References

Abu-Lughod, J. 1999. *New York, Chicago, Los Angeles: America's Global Cities*. Minneapolis: University of Minnesota Press.

Arthur, W. 1994. *Increasing Returns and Path Dependency in the Economy*. Ann Arbor: University of Michigan Press.

Barnes, T.J. 2001. 'Retheorizing economic geography: From the quantitative revolution to the cultural turn'. *Economic Geography* 91: 546–65.

Bell, D. 1973. *The Coming of Postindustrial Society: A Venture in Social Forecasting*. New York: Basic Books.

Beyers, W. 2000. 'Cyberspace or human space: Whither cities in the age of telecommunications'. In J. Wheeler, Y. Aoyama, and B. Warf, *Cities in the Telecommunications Age*. London: Routledge.

Bradford, N. 2002. *Why Cities Matter: Policy Research Perspectives for Canada*. F/23. Ottawa: Canadian Policy Research Network.

Camagni, R., ed. 1991. *Innovation Networks: Spatial Perspectives*. London: Belhaven.

Conference Board of Canada. 2006. *Canada's Hub Cities: A Driving Force of the National Economy*. Ottawa: Conference Board of Canada.

Courchene, T., ed. 1997. *The Nation State in a Global/Information Era: Policy Challenges*. Kingston: Queen's University.

Davis, D., and D. Weinstein. 2002. 'Bones, bombs, and break points: The geography of economic activity'. *The American Economic Review* 92 (5): 1269–89.

Dear, M., and S. Flusty. 1998. 'Postmodern urbanism'. *Annals of the Association of American Geographers* 88: 50–72.

Favell, A. 2007. *Eurostar and Eurocities: Free Moving Urban Professionals in an Integrating Europe*. London: Blackwell.

Florida, R. 2002. *Rise of the Creative Class—and How It's Transforming Work, Leisure, Community and Everyday Life*. New York: Basic Books.

Friedmann, J.F. 1986. 'The world city hypothesis'. *Development and Change* 17: 69–83.

Glaeser, E., J. Kolko, and A. Saiz. 2001. 'Consumer city'. *Journal of Economic Geography* 1: 27–50.

Graham, S., and S. Martin. 2001. *Splintering Urbanism: Networked Infrastructures, Technological Mobilities, and the Urban Condition*. London and New York: Routledge.

Hall, P.G. 1966. *The World Cities*. London: Weidenfeld and Nicolson.

———. 2000. 'Creative cities and economic development'. *Urban Studies* 37: 639–51.

Heinz, A. 2006. *Canada's Global Cities: Socio-economic Conditions in Montreal, Toronto and Vancouver*. Analytical paper 89-613-MIE. Ottawa: Statistics Canada.

Howlett, M., and K. Brownsey. 2008. *Canada's Resource Economy in Transition: The Past, Present and Future of Canada's Staples Industries*. Toronto: Emond Montgomery.

Hutton, T. 2008. *The New Economy of the Inner City: Restructuring, Regeneration and Dislocation in the 21st Century Metropolis*. London and New York: Routledge.

Innis, H. 1933. *Problems of Staple Production in Canada*. Toronto: Ryerson Press.

ICP/MPI (Institute for Competitiveness and Prosperity and Martin Prosperity Institute). 2009. *Opportunity in the Turmoil*. Toronto: ICP and MPI.

Krugman, P. 1995. *Development Geography and Economic Theory*. Cambridge, MA: MIT Press.

Markusen, A. 1996. 'Sticky places in slippery spaces: A taxonomy of industrial districts'. *Economic Geography* 72: 293–313.

Marshall, A. 1890 [1972]. *Principles of Economics*. London: Macmillan.

Martin, R., and R. Florida. 2008. *Ontario in the Creative Age*. Toronto: Martin Prosperity Institute, University of Toronto.

Martin, R., and P. Sunley. 2003. 'Deconstructing clusters: Chaotic concept or policy panacea'. *Journal of Economic Geography* 3: 3–35.

Massey, D. 1984. *Spatial Division of Labour: Social Structure and the Geography of Production*. London: Macmillan.

Morgan, K. 2004. 'Devolution and development: Territorial justice and the north-south divide'. Cardiff: School of City and Regional Planning, University of Cardiff.

OECD. 2002. *Territorial Review: Canada*. Paris: OECD.

———. 2009. *Report on the Metropolitan Toronto Economy*. Paris: OECD.

Piore, M., and M. Sabel. 1984. *The Second Industrial Divide*. New York: Basic Books.

Porter, M. 2003. 'The economic performance of regions'. *Regional Studies* 37 (6, 7): 549–78.

Pratt, A.C. 1997. 'The cultural industries production system: A case study of employment change in Britain'. *Environment and Planning A* 29: 1953–74.

Sassen, S. 2006. *Cities in the World Economy*. 3rd edn. Thousand Oaks, CA: Pine Forge Press.

Savitch, H., and P. Kantor. 2004. *Cities in the International Marketplace: The Political Economy of Urban Development in North America and Western Europe*. Princeton, NJ: Princeton University Press.

Scott, A.J. 1988. *Metropolis: From the Division of Labor to Urban Form*. Berkeley and Los Angeles: University of California Press.

———. 1997. 'The cultural economy of the city'. *International Journal of Urban and Regional Research* 21: 323–39.

———, ed. 2004. *Global City-Regions: Trends, Theory, Policy*. Oxford: Oxford University Press.

Shearmur, R. 2001. 'Financial flows and places: The case of Montreal'. *Canadian Public Policy* 27 (2): 219–33.

Shearmur, R., W. Coffey, C. Dubé, and R. Barbonne. 2007. 'Intrametropolitan employment structure: Polycentricity, scatteration, dispersal and chaos in Toronto, Montreal and Vancouver, 1996–2001'. *Urban Studies* 44 (9): 1713–38.

Shearmur, R., and B. Motte. 2009. 'Weak ties that bind: Do commutes bind Montreal's central and suburban economies?' *Urban Affairs Review* 44: 490–524.

Shearmur, R., and M. Polèse. 2007. 'Do local factors explain local employment growth: Evidence from Canada, 1971–2001'. *Regional Studies* 44 (4): 453–71.

Simmons, J., L.S. Bourne, and J. Cantos. 2004. *How Cities Grow: A Time Series Analysis of Urban Growth in Canada*. Research Paper #204. Toronto: Centre for Urban and Community Studies, University of Toronto.

Simmons, J., and S. Kamikihara. 2006. *The North American Urban System: The Limits to Continental Integration*. Research Paper #205. Toronto: Centre for Urban and Community Studies, University of Toronto.

Simmons, J., S. Kamikihara, and L.S. Bourne. 2009. *The Changing Economy of Urban Neighbourhoods: An Exploration of the Place of Work Data for the Greater Toronto Region*. Research paper #219. Toronto: Cities Centre, University of Toronto.

Smith, M.P. 2001. *Transnational Urbanism: Locating Globalization*. Oxford: Blackwell.

Soja, E. 2000. *Postmetropolis: Critical Studies of Cities and Regions*. Oxford: Blackwell.

Spencer, G., T. Vinodrai, M. Gertler, and D. Wolfe. 2010. 'Do clusters make a difference? Defining and assessing their economic significance'. *Regional Studies* (forthcoming).

Stelter, G., ed. 1990. *Cities and Urbanization: Canadian Historical Perspectives*. Toronto: Copp Clark.

Stelter, G., and A. Artibise, eds. 1986. *Power and Place: Canadian Urban Development in the North American Context*. Vancouver: University of British Columbia Press.

Taylor, P., et al. 2008. *The Way We Were: Command and Control Centres in the Global Space Economy*. GaWC Research Bulletin 289. Loughborough University, UK.

Thrift, N., and K. Olds. 1996. 'Refiguring the economic in economic geography'. *Progress in Human Geography* 20: 311–37.

Wolfe, D. 2009. *The Geography of Innovation*. Toronto: Conference Board of Canada.

Note

1. Since economic functions are increasingly segmented at a global scale (e.g., standardized manufacturing in China and Southeast Asia), it remains an open question whether the concepts described in this section concern only cities in developed countries (which are losing their industrial and mass production functions to developing countries) or all cities (which would imply that economies in all countries are undergoing the same shifts).

2 Canadian Cities in a Global Context

Mario Polèse and Jim Simmons

Introduction

A discussion of the characteristics and growth processes of Canadian cities cannot ignore the continuous interchange of money, goods, ideas, people, and information with the world outside Canada's national borders. In Canada, as in other nations, the distribution and evolution of cities was in large part shaped by a nation's location on the map and by its position within the world economy (Polèse 2009). Canada was first colonized from the east by Europeans in search of land, glory, and riches, a fact that has left a mark on its urban system. From the earliest colonial era, Canada's growth, and especially urban growth, has been driven by the export of resource products and the inflows of immigrants. These processes continue today, although the interchange of goods and people has become rather more complex and the exchange of capital, technologies, and information has increased enormously, with diverse and uneven implications for Canadian cities.

This chapter introduces the reader to the magnitude and variety of these exchanges. We begin with a general discussion of the trade and immigration links, followed by an exploration of the implications of trade for local economies and a discussion of the human capital connections in the form of immigration and education. Finally, we provide a brief historical context for Canadian urbanization.

While Canadian cities do not rank highly on listings of world cities, the influence of other cities and other countries upon Canadian cities is wide-ranging and extremely significant. In particular, these linkages—in terms of global competition and the import of new technologies—affect economic growth through the evolving demand for Canadian products and determine population growth depending on the source of immigration and the preferred destinations within Canada. The graph in Figure 2.1 indicates the increasing importance of these external factors in Canada's growth. Since 1971, the weight of exports has almost doubled from 20 per cent to almost 40 per cent of GDP. During the same period, the contribution of immigration to population growth has increased from 20 per cent to more than 60 per cent of the total. Interestingly, there is more temporal variation in immigration than in exports, and the temporal relationship between the two is quite weak.

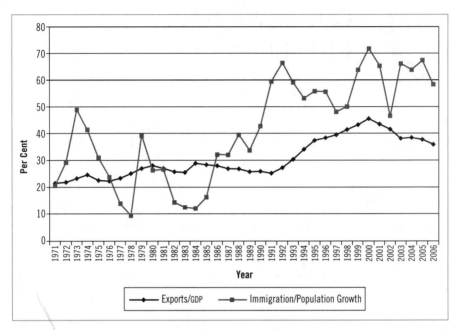

Figure 2.1 External influence: Exports and immigration.

(Source: Statistics Canada. CANSIM database, Tables 510001-510014, Table 3800027. Ottawa: Statistics Canada)

The North American Context

Canadians are intensely aware that they share a continent with a very large, rich, and aggressive neighbour. Canadians—certainly English Canadians—watch American television from their earliest moments ('Sesame Street'), through the teenage years (MTV), adulthood (CNN/ESPN), and through the midlife crisis ('Oprah'). We are infinitely more familiar with their culture than they are with ours. We buy their products (55 per cent of our imports) and sell them ours (81 per cent of our exports). Canadians visit the US regularly, but at present there is surprisingly little migration across the border—a reflection of the restrictions on migration into both countries.

The degree of interdependence reflects ready accessibility, as shown in the map of North America in Figure 2.2. Most Canadian cities are located within 200 kilometres of the US border. Southern Ontario is largely surrounded by US territory. The metropolitan area of Vancouver is in part bounded by the US border, and both Toronto and Montreal fall within an hour's drive. Table 2.1 summarizes the approximate weight of the two nations. The US houses roughly 10 times Canada's total population, urban population, and number of urban areas. The population of the metropolitan area of New York is approximately four times that of Toronto. Many Canadian cities have a large US city nearby, be it New York, Boston, Detroit, Minneapolis, or Seattle.

As in Canada, urbanization in the United States is predominantly concentrated in the eastern portion of the country. Western cities are relatively few and far between, although California and the southwest are growing more rapidly than the east—like Alberta and

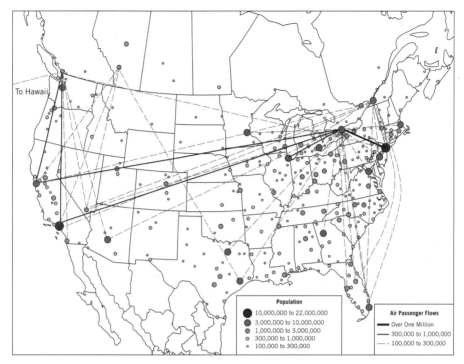

Figure 2.2 Cross-border air passenger destinations for Canadian cities, 2006.

(Source: Statistics Canada. Annual. Air Passenger Origin and Destination: Canada–United States. Catalogue no. 51-205)

BC. Both nations suffer from what might be called the empty middle, as witnessed by the absence of large cities in the Prairie provinces (Saskatchewan, Manitoba) and northwestern Ontario and in the Great Plains states (e.g., the Dakotas, Nebraska). The size and accessibility of American cities means that they provide an important set of destinations for Canadians seeking goods, information, capital, and recreation. Figure 2.2 summarizes the most important cross-border links for Canadian cities, based on air passenger movement (predominantly business). The most important flows link the largest cities, as one would expect. Toronto–New York is the largest link, twice the size of any others. Regional trips for shopping or recreation take place largely by automobile. Canadian winters also generate significant flows to vacation destinations: Florida, Las Vegas, and Phoenix, Arizona. Of course, the size of the airline links pale in comparison with that of travel through land links in the major corridors connecting Canadian and American cities: from Montreal south to New York State, from Toronto through Buffalo or Detroit, from Calgary through Montana, and from Vancouver to Bellingham, with cars, trucks, and rail links all concentrated within a few miles of the crossing points.

However, the international border continues to act as an important constraint on various forms of contact. Table 2.2, based on the most recent data available for Canadian air passengers in 1999, suggests that air travel between Toronto and Montreal is much more frequent than that between Toronto and New York, about the

Table 2.1 Comparing Urban Systems: Canada and the US

	Canada	United States	Ratio (%)
National Population (1000s)			
2006	31,613	298,755	10.6
2001	30,007	282,194	
Growth Rate (%)	5.35	5.87	
Urban Population (1000s)			
2006	21,723	264,711	8.2
2001	20,339	251,101	
Growth Rate (%)	6.80	5.41	
% Urban (over 100,000)	68.7	83.7	
Number of Metropolitan Areas	35	312	11.2
Average Population (1000s)	621	848	
Growth Rate and Log Population (Correlation)	0.275	0.079	

Source: Statistics Canada. *Census of Canada, 2006*. Ottawa: Statistics Canada; US Bureau of the Census. *Census of Population*. Washington: US Bureau of the Census.

same distance away, despite Montreal's much smaller size. Similarly, Vancouver sends many more passengers to Toronto than to Los Angeles, although the latter is much larger and also closer. The same is true for smaller cities. Halifax sends more people to Toronto and Montreal than to Boston and New York.

The Compartmentalized Trade Links

Although the level of both imports (32.7 per cent in 2007[1]) and exports (34.7 per cent) are quite high relative to GDP for Canada, it is the composition of exports that most directly shapes the spatial and temporal variation in growth rates of Canadian cities. Figure 2.3 shows how the content of exports fluctuates over time, providing highly variable growth stimuli to the Canadian economy overall, depending on the region of production. In general, there has been a gradual shift from resource-based exports to manufactured goods in recent years, thus favouring the smaller cities of central Canada over places to the east or west. In this way, cycles of economic growth and decline are imported from elsewhere, following from fluctuations in the demand for the goods and services produced in Canada's cities. Thus, the integration of Canada's automobile industry with that of the US imports the booms and downturns in American demand, directly affecting the Ontario economy. While agricultural exports have undergone

Table 2.2 Cross-Border Air Passenger Flows, 1999

Links		(Population, Distance)	Air Passengers
Toronto (4.7 m)	to	Montreal (3.5 m, 600 km)	1,261,000
		New York (21.4 m, 600 km)	1,048,000
		Vancouver (2.0 m, 4000 km)	947,000
		Los Angeles (16.4 m, 4000 km)	389,000
Montreal (3.5 m)	to	Toronto (4.7 m, 600 km)	1,261,000
		New York (21.4 m, 600 km)	386,000
		Vancouver (2.0 m, 4500 km)	243,000
		Los Angeles (16.4 m, 2600 km)	152,000
Vancouver (2.0 m)	to	Toronto (4.7 m, 4000 km)	947,000
		Montreal (3.5 m, 4500 km)	243,000
		Los Angeles (16.4 m, 2000 km)	453,000
		New York (21.4 m, 4500 km)	157,000
Halifax (0.4 m)	to	Toronto (4.7 m, 1500 km)	333,000
		Montreal (3.5 m, 1000 km)	117,000
		New York (21.4 m, 1200 km)	46,000
		Boston (7.3 m, 800 km)	45,000

Sources: Statistics Canada. Annual. *Air Passenger Origin and Destination: Canada–United States.* Catalogue no. 51-205. Ottawa: Statistics Canada; Statistics Canada. Annual. *Air Passenger Origin and Destinations: Domestic Report.* Catalogue no. 51-204. Ottawa: Statistics Canada.

a long decline, they are now expanding, to the delight of Prairie farmers. As will be underlined in later chapters, the international rise of energy prices benefits Canada's high-cost producers in Alberta—hence the economies of Calgary and Edmonton—while the slump in the value of forest products (newsprint, lumber) has accelerated the decline of small cities across the northern fringe of the urban system.

Over time, the exports of particular economic sectors display greater variation than the aggregate exports so that, for example, while sales of energy products are soaring, forest products are in decline. As we shall see in Chapter 3, many of the smaller Canadian cities are highly specialized in the production of one or two of these products, so export fluctuations directly affect the local economy—not only through direct changes in employment but indirectly by way of the amount of money available for private households and governments to spend in the region.

When the export commodity underpins the economy of a number of urban areas in a region, then regional service centres, and ultimately national centres, are also affected.

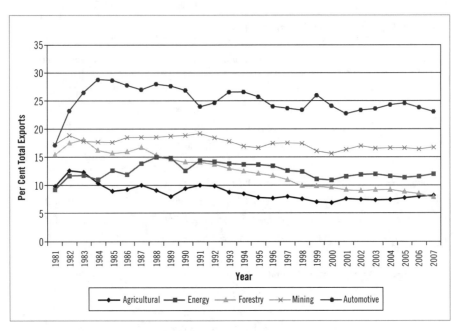

Figure 2.3 Exports by sector, 1981–2007.

(Source: Statistics Canada. CANSIM database, Tables 3860179-3860197. Ottawa: Statistics Canada)

The significant example currently is the fast growth in the energy sector that has driven rapid urban growth in Alberta, especially in Calgary and Edmonton, and has increased the overall political and economic influence of western Canada within the Canadian federation. Exports are inevitably accompanied by flows of capital and technology from the destination countries and elsewhere so that the production-based cities become closely linked with world cities in a financial, corporate, and technological sense. Table 2.3 reconfirms the overwhelming dominance of the US market for Canadian merchandise trade.

Exports to the US dwarf all others for all Canadian regions (Figure 2.5). The US share is lowest—though still dominant—for British Columbia, Quebec, and the two Prairie provinces (Manitoba and Saskatchewan), an indication of BC's Asian and Quebec's European orientation and prairie grain exports to overseas markets. Alberta's close links with the US—some 86 per cent of its exports—are largely due to the massive oil and gas flows south of the border. In sum, exporters, wholesalers, and customs brokers in Vancouver, Montreal, Saskatoon, and Calgary will not necessarily have their eyes on the same markets in terms of either products sold or regions targeted. The most striking aspect of Canada's regional trade structure is Ontario's dependence on the US market—some 84 per cent of its exports—not only because of the size of the Ontario economy but also because of the geographic concentration of its trade due to the weight of a single industry—automobile assembly and auto parts manufacturing—in its trade relationship with its southern partner (Table 2.4).

Table 2.4 shows the percentage of exports going to the five most important American states by Canadian region, as well as the five most important destination states measured

Table 2.3 Canada's Trading Partners

	Per Cent of Canada's Merchandise Trade, 2006 (%)	
	Imports ($396.6 b)	Exports ($440.3 b)
US	54.9	81.6
UK	2.7	2.3
Other European Union	9.4	4.2
China	8.7	1.7
Japan	3.9	2.1
Other Asia	5.1	2.3
Africa	2.0	0.4
Middle East	1.4	0.9
South America	3.2	1.3
Mexico	4.0	1.0

Source: Industry Canada. Trade Data Online@strategis.ic.gc.ca.

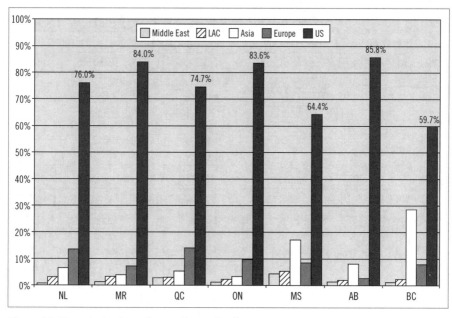

Figure 2.4 Exports by Canadian region to foreign destinations, 2007.

(Source: Industry Canada. Trade Data Online@strategis.ic.gc.ca)

in terms of the ratio between each state's export share and its share of US GDP. Thus, some 30 per cent of Ontario's exports go to a single state—Michigan—whose share

Table 2.4 Exports to the US by Canadian Region

				Newfoundland & Labrador (NL)			
				% per State		To % GDP	
				New Jersey	45.4	New Jersey	13.5
				Pennsylvania	14.6	Delaware	4.3
				Texas	10.1	Pennsylvania	3.8
				Virginia	6.0	Rhode Island	3.3
				Massachusetts	6.0	Connecticut	2.7
				First 5	82.2		
				First 10	93.6		
Maritime Provinces (MR)				**Quebec (QC)**			
% per State		To % GDP		% per State		To % GDP	
New Hampshire	27.6	New Hampshire	65.6	New York	12.8	Vermont	26.8
Massachusetts	20.0	Maine	24.9	Texas	7.6	New Hampshire	5.6
Maine	8.7	Massachusetts	7.8	Pennsylvania	6.3	West Virginia	3.7
Texas	3.9	Rhode Island	2.8	Tennessee	5.4	Maine	3.5
Pennsylvania	3.7	South Carolina	2.0	Illinois	5.4	Tennessee	3.0
First 5	64.0			First 5	37.6		
First 10	78.7			First 10	58.4		
Ontario (ON)				**Manitoba & Saskatchewan (MS)**			
% per State		To % GDP		% per State		To % GDP	
Michigan	30.1	Michigan	10.9	Minnesota	16.7	Montana	42.9
California	11.1	Kentucky	1.7	Illinois	14.2	North Dakota	24.7
New York	8.0	Ohio	1.7	Montana	10.7	Wyoming	10.2
Ohio	5.7	Vermont	1.7	North Dakota	5.1	Minnesota	9.1
Illinois	4.6	Indiana	1.6	Michigan	4.0	South Dakota	4.8
First 5	59.6			First 5	50.8		
First 10	75.2			First 10	67.1		
Alberta (AB)				**British Columbia (BC)**			
% per State		To % GDP		% per State		To % GDP	
Illinois	22.0	Wyoming	13.3	Washington	28.5	Washington	12.6
Washington	12.3	Montana	13.1	California	13.7	Oregon	5.1
New York	10.6	Washington	5.5	Illinois	8.1	Montana	3.2
Minnesota	8.4	Illinois	4.9	Oregon	5.9	Idaho	2.7
Michigan	5.7	Minnesota	4.6	Texas	4.0	Alaska	2.0
First 5	59.0			First 5	60.2		
First 10	79.2			First 10	71.1		

Note: The first column indicates the percentage of each region's exports that are destined to each state. The second column indicates the ratio between the first column and each state's contribution to total US GDP in 2007. If exports to each state were in direct proportion to each state's weight in the US economy, all figures in the second column would be 1. This table therefore illustrates the strong cross-border orientation of Canada–US trade.

Source: Industry Canada. Trade Data Online@strategis.ic.gc.ca.

is some 11 times greater than its GDP would predict. Five US states account for some 60 per cent of Ontario's trade. The comparison with Quebec, Canada's second largest industrial economy, is revealing. Quebec's first market—New York State—accounts for 12.8 per cent of its exports, and the state to which, proportionately, it has the closest links is Vermont, which imports some 27 times more from Quebec than the size of its economy would predict. The Ontario–Quebec comparison reveals not only the difference in the regional (and industrial) orientation of trade of the two provinces but also the importance of cross-border links in shaping that trade. The integration of the North American auto industry goes back to the 1965 Auto Pact, which freed trade within that industry, thus accelerating southern Ontario's integration into the Detroit (Michigan)–centred automobile economy. Need we remind the reader that Windsor, Ontario, is but a stone's throw—across the Detroit River—from its much larger US neighbour?

The 1989 Free Trade Agreement with the US (which became NAFTA in 1994 when Mexico joined) has since accelerated similar relationships along other parts of the border. Southern Quebec is a prime example. The strong ties with Vermont are no accident. Large Montreal-based firms such as Bombardier (aerospace and transportation equipment) have opened plants in Vermont and upstate New York to circumvent 'Buy America' clauses common in US municipal and state tenders[2]—in the Bombardier case, for the delivery of subway cars to New York City. By the same token, the IBM Corporation, based in New York State, has large plants in Burlington, Vermont, and Bromont, Quebec, some 50 miles north, with a constant flow of trucks between the two plants. These are not unique cases. Table 2.4 suggests that each region of Canada—and thus the main cities—has a unique cross-border trade relationship, shaped by proximity and the nature of the goods traded. The Maritime provinces trade first and foremost with New England (the Boston States, in local parlance), while Minnesota and Montana are top of the list for the two Prairie provinces and Washington State for BC. For the two oil-exporting provinces, Alberta and (more recently) Newfoundland and Labrador, major oil refining and consuming states are on top, although proportionally weighted, Alberta's closest links are with Wyoming and Montana.

The picture, at both the global and the continental level, is one of compartmentalized relationships in which each Canadian city is plugged into a different set of trading and business partners. Similar relationships emerge in cities in other nations, but Canada's sheer size and diversity as well as the size of the US market accentuate the situation. The compartmentalized nature of Canada's international and cross-border relationship sheds a different light on Toronto's global standing. Although indisputably number one—certainly in finance—Toronto's primary strength lies in its location at the centre of Canada's largest regional economy and in its role as corporate intermediary with US-based business partners. Toronto, more so than Vancouver or Montreal, looks south before looking west or east. But all three are part of the North American urban system (recall Figure 2.2), to which we now return.

Table 2.5 compares employment shares in manufacturing and in information-rich services for 26 metropolitan areas: nine census metropolitan areas in Canada and 17 metropolitan statistical areas in the US.[3] All have populations of more than 500,000, with the exception of Halifax, which was added to include an urban area from Atlantic

Table 2.5 Employment Shares in Manufacturing and High-Order Services, Selected US and Canadian Metropolitan Areas, 2007

% Manufacturing		% Professional & Scientific Services		% Finance, Insurance, & Real Estate	
Toronto	14.1	Washington, DC	14.9	Boston	11.1
Cleveland	13.3	Boston	12.8	Toronto	10.0
Detroit	13.1	Calgary	11.0	New York	9.3
Boston	13.0	San Francisco	9.6	Quebec City	8.4
Montreal	12.8	Toronto	9.6	Denver	8.0
Winnipeg	12.4	Vancouver	9.5	Dallas	7.9
Portland	12.2	Ottawa	9.2	Minneapolis	7.9
Minneapolis	11.2	Montreal	9.1	Philadelphia	7.8
Los Angeles	11.1	Detroit	8.3	San Francisco	7.5
Seattle	10.8	Denver	7.9	Vancouver	7.4
Chicago	10.6	New York	7.7	Miami	7.4
Dallas	10.1	Philadelphia	7.6	Chicago	7.2
Quebec City	9.8	Edmonton	7.1	Montreal	7.2
Vancouver	9.1	Atlanta	6.9	Halifax	7.1
Pittsburgh	8.8	Los Angeles	6.8	Winnipeg	6.9
Edmonton	8.7	Dallas	6.2	Portland	6.9
Philadelphia	7.8	Quebec City	6.1	Cleveland	6.8
Atlanta	7.2	Pittsburgh	6.0	Los Angeles	6.6
Calgary	6.9	Minneapolis	5.9	Atlanta	6.6
San Francisco	6.8	Winnipeg	5.4	Calgary	6.3
Ottawa	6.7	Cleveland	5.3	Seattle	6.0
Halifax	6.0	Halifax	5.3	Pittsburgh	6.0
Denver	5.7	Portland	5.1	Detroit	5.7
New York	5.3	Chicago	nd	Washington, DC	5.3
Miami	4.0	Miami	nd	Edmonton	4.9
Washington, DC	2.1	Seattle	nd	Ottawa	4.9

Canada. On the whole, very little distinguishes Canadian cities from their sisters to the south, a further indication that these are cities at comparable levels of development, evolving within the same economic universe.

The most significant differences occur between specific cities, whether in the US or in Canada. Within Canada, Toronto's specialization in financial and related services (an employment share close to 50 per cent higher than in Montreal and in Vancouver), when compared to New York's position within the US urban system, owes much to Canada's regulatory framework, which is founded on federally chartered banks. In

Canada, the financial landscape is dominated by a few—federally chartered—institutions, the largest of which are all headquartered in Toronto.[4] In principle, US banks cannot operate outside the state in which they are chartered, producing a more diffuse financial landscape with numerous regional financial centres based on state or locally chartered banks. The downside of this diffuse banking structure is a banking system more volatile and difficult to regulate, to which the 2007–8 sub-prime crisis bears testimony. The primary exceptions to Canada's concentrated banking structure are the francophone-controlled institutions in Quebec. Quebec has spawned a powerful—and provincially chartered—credit union (les Caisses Populaires Desjardins), headquartered in the Quebec City region.

Insurance is traditionally less spatially concentrated, which in large part explains Boston's top spot in the US but also further bolsters Quebec City, home to several francophone-controlled insurance companies. Smaller cities, such as Hartford, Connecticut, and London, Ontario, have specialized in insurance.

Canadian cities have comparatively high shares of employment in professional, scientific, and technical services, with some cities matching the US high-tech stars, Boston and San Francisco. This is not necessarily a sign of a Canadian high-tech advantage but an indication of the high engineering and scientific content of much of Canada's resource economy. Calgary and Montreal are cases in point. The management of oil exploration requires an army of engineers, geologists, and other related professionals and scientists. Calgary is the centre of Canada's oil economy. By the same token, Montreal is home to some of the world's largest engineering consulting firms, nurtured over the years by Quebec's huge hydroelectric projects in the North, an example of global expertise built on an initial resource advantage. The high percentage of professionals and scientists in Washington, DC, and in Ottawa (less so) is explained by their roles as capitals, housing numerous research laboratories. The 'professional and scientific' class includes lawyers, undoubtedly abundant in both capitals.

Some Canadian cities have also developed global niches unrelated to natural resources, notably in the design of software for computer gaming and computer animation,[5] with significant concentrations in Vancouver and Montreal. The former has the advantage of proximity to Seattle and San Francisco and also to Japan (the home of Nintendo), while the second provides a reliable pool of well-educated young programmers who, because of language, are less mobile, implying lower wages than would, for example, be paid in Boston or in San Francisco. The French computer graphics giant Softimage has based its North American operations in Montreal. Within North America—and also globally—labour reliability and labour costs are important considerations in comparative advantage, influenced in turn by the lower continental mobility of Canadian (especially French-speaking) skilled workers, compared to Americans, and by the lower value of the Canadian dollar.

The cost advantage also explains in part why Canadian cities hold their own in manufacturing, despite often fairly peripheral locations from a continental perspective: e.g., Winnipeg and Montreal. At the same time, however, Canada's rich resource endowments can have a detrimental effect on costs, since high resource 'rents' push up wages. This disadvantages Calgary, and perhaps Vancouver, as locations for manufacturing. The majority of manufactured goods are exported to the US, as we have seen, making this sector highly sensitive to fluctuations in the value of the Canadian dollar (compared to the US dollar). In recent years, the Canadian dollar has fluctuated considerably, with an overall trend toward a gradual strengthening, in large part fuelled by the continuing global demand for resources but in turn creating a major challenge for non-resource firms wishing to export goods or services to the US and to global markets.

Here we have the classic Canadian dilemma. While it lasts, resource-led growth brings in money and jobs, especially to peripheral regions, but it also hampers the growth of industries based on know-how rather than resources. In the long run, in Canada as elsewhere, the economic health of cities, especially the largest, depends on their capacity to produce (and export) knowledge-rich and information-rich goods and services. A good indicator of their potential to do so is the educational attainment of the workforce, in turn influenced by immigration, to which we now turn.

The Global Connections

External conditions also shape the flows of immigrants that drive the demographic growth of Canadian cities. Figure 2.5 indicates the changing composition of Canadian immigration. Immigrant flows are linked both to economic conditions in Canada (booms, recessions, policy shifts) and to events in the source countries (e.g., the reunification of Hong Kong with China or refugee crises in the Balkans or the Middle East). The long-term shifts in the place of origin of the immigrants (Table 2.6) also

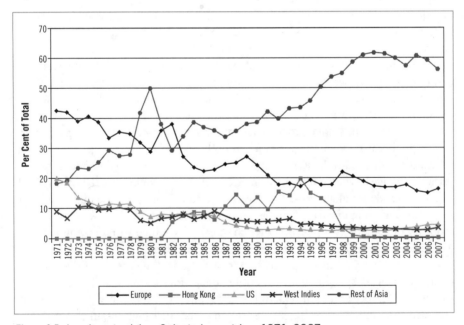

Figure 2.5 Immigrant origins: Selected countries, 1971–2007.

(Source: Statistics Canada. CANSIM database. Table 510006. Ottawa: Statistics Canada)

affect the choice of destination in Canada, hence the city and region that benefit. British and other European immigrants in the postwar period were widely dispersed across the country, but recent immigrants, increasingly from Asia, have preferred to join compatriots in the larger cities such as Toronto and Vancouver, while those originating in francophone countries tend to go to Montreal.

The most notable change is the gradual replacement of European immigrants with Asians, including Chinese, Indians, Koreans, and Filipinos. An upward blip in the graph shows the Hong Kong influx to Vancouver that occurred after China took over the British colony in 1987. The inflow of Americans declined after the Vietnam War. The result, after 60 years of immigration following World War II, is a significant pattern of variation by region and city size, first in the rate of population growth in Canadian cities and second in the degree of complexity of ethnicity, language, and cultures within cities. Overall, some 21 per cent of Canadians were born in other countries, but the variation among urban areas ranges from 0.7 per cent (Dolbeau in Quebec) to 44.7 per cent (Toronto). Most of the larger cities include a substantial foreign-born contingent that maintains a variety of transnational contacts with the countries of origin: visits with relatives or friends and flows of cultural ties, information, and money back and forth; these connections also form the basis for significant community organizations within Canada and may encourage financial investments and trade flows in either direction.

Although Toronto prides itself on having the highest percentage of foreign-born among North American cities, with Vancouver close behind, this characteristic is not unique to Canada. Major US urban areas such as Miami, Los Angeles, and New York

Table 2.6 Sources of Immigration

	% of Immigrants	
	1996 (226,100)	2006 (215,000 through October)
US	2.6	4.1
Americas	8.4	9.2
UK	2.5	2.6
Other Europe	15.2	12.5
China	7.8	13.5
Hong Kong	13.3	0.3
Japan	0.5	0.5
India, Pakistan, Bangladesh	16.6	20.7
Other Asia	16.1	15.1
Africa	6.7	10.5
Middle East	9.6	10.4

Source: Citizenship and Immigration Canada. *Permanent Residents, by Country of Last Permanent Residence, Jan. 1980–Oct. 2006*. Ottawa: Citizenship and Immigration Canada.

also exhibit high shares of foreign-born, above those for Calgary, Montreal, and Ottawa (see Table 2.7). It is difficult to argue that Toronto is more 'international' than New York, but it may well be more diverse in the sense of the numbers and size of immigrant groups. That would certainly be the opinion of the casual visitor to the city.

In addition to the contacts formed by trade and immigration flows, it is possible to identify the continuing contacts of Canadian cities with other major cities that are maintained by various international corporations and organizations. The World Cities group in Britain has identified and mapped the branch offices of 100 of these organizations among the 315 world cities with more than 1 million population (Beaverstock et al. 2009). Among the nine Canadian cities included in this study, Toronto ranks 14 in the number of agencies, Montreal 48, and Vancouver 65.

Table 2.7 Foreign-Born Population, Selected North American Cities

CMA/MSA	% Foreign-Born
Toronto	45.7
Vancouver	39.6
Miami	37.0
Los Angeles	34.2
San Francisco	29.6
New York	28.2
Calgary	23.6
Montreal	21.2
Edmonton	18.5
Ottawa	18.1
Dallas	18.0
Chicago	17.8
Winnipeg	17.7
Seattle	15.8
Atlanta	13.0
Denver	12.9
Washington, DC	12.4
Portland	12.4
Minneapolis	8.9
Philadelphia	8.7
Boston	8.5
Detroit	8.5
Halifax	7.4
Cleveland	5.7
Quebec City	3.7
Pittsburgh	3.0

In global rankings of cities as financial and corporate centres, Toronto again stands out among Canadian cities but with a weighting that generally mirrors the weight of Canada within the world economy (Table 2.8). In terms of GDP, Canada ranked eleventh in 2008 among the world's economies, just before India and following Spain (or Brazil, depending on the source). The Toronto Stock Exchange ranked eighth (after Hong Kong), measured in terms of market capitalization, and fifteenth in terms of the value of shares traded (after Scandinavia's Nordic Exchange) in 2007 (World Federation of Exchanges 2009). The Global Financial Services Index placed Toronto in twelfth place (after Paris) in 2007 (Mainelli 2007). Taylor et al. (2009) were more generous, putting Toronto in fifth place in 2008 (between Paris and Tokyo) on their Financial Command Index but ranked it thirteenth (between Los Angeles and San José, California) on their Business Command Index, which measures the concentration of large headquarters.

The underlying logic of the global positioning of Canadian cities is not fundamentally different from that of most other cities, a reflection of the vigour of domestic economies rather than international services rendered to third parties. London is the exception,

Table 2.8 Top Five Global Accounting Firms and Global Advertising Agencies: Presence in Major Canadian Cities Compared to Other Cities

Accounting Firms						Advertising Agencies		
Rank	City	Practitioners* (2 firms)	Rank	City	Offices** (3 firms)	Rank	City	Offices
1	London	688	1	Paris	26	1	London	33
2	New York	426	2	Brussels	19	2	Paris	29
3	Toronto	343	3	London	15	9	Sydney	12
4	Frankfurt	309	8	Dusseldorf	12	10	Toronto	12
21	Los Angeles	103	9	Toronto	11	11	Melbourne	11
22	Vancouver	103	10	Washington, DC	11	21	Barcelona	7
23	Milan	102	28	Madrid	7	22	Montreal	7
27	Dallas	92	29	Montreal	7	23	Vienna	7
28	Montreal	89	30	San Francisco	7	60	Cairo	3
29	Zurich	88	49	Moscow	5	61	Calgary	3
46	Kuala Lumpur	55	50	Calgary	5	63	Dublin	3
47	Calgary	55	51	Genoa	5	67	Tel Aviv	3
48	Oslo	55	85	Tijuana	4	68	Vancouver	3
			86	Vancouver	4	69	Wellington	3

* Number of practitioners employed by KPMG and Coopers & Lybrand in city x.

** Number of offices (Ernst & Young, Arthur Anderson, Price Waterhouse) in city x.

Source: Beaverstock, Smith, and Taylor, 2009. Data is for or slightly after the year 2000.

not the rule. That Montreal, for example, should have an advertising industry about the size of that of Vienna or Barcelona (which is not a bad showing) is not surprising given the size of Canada's French-speaking market, approximately the same size as Austria or Catalonia. For small economies, global success in corporate and financial services is often founded on the safe haven principle—that is, providing an orderly institutional and monetary environment in otherwise volatile and nasty neighbourhoods. This has been the source of success for Switzerland and its banking cities of Zurich and Geneva. By the same token, Singapore and Hong Kong are islands of institutional order in a part of the world where corruption and political meddling are still rife. Miami's role as a regional banking centre for much of Latin America also owes much to the incapacity of South American governments to provide stable monetary and regulatory environments.

Canada's global cities have few such comparative advantages. Their neighbours are US cities, and besides providing access to a much larger market, they also belong to an institutional environment that is no less orderly than Canada's,[6] with the added advantage of a currency less prone to resource-driven fluctuations. Add the border problems, and there is little reason for a global financial institution to headquarter its North American (or world) operations in a Canadian rather than a US city. Exceptions exist. Vancouver has acted as a safe haven for Asian money and probably Toronto as well, but this advantage may disappear as the great Asian economies—notably China and India—become institutionally more stable. Calgary's global standing is almost exclusively driven by the international needs of the Alberta energy economy.

Montreal, because of its size and the French language, occupies a different space in the global economy. Business relationships are sensitive to language and shared culture. Carrol (2007), looking at the interlocking relationships for 350 corporations, finds that Anglo-US corporations headquartered in London, New York, and other English-speaking cities form a relatively closed network, although outwardly global due to their dominant status. Canadian-headquartered corporations are part of this corporate universe but with Montreal as a notable outlier. Large Montreal-headquartered firms are most closely interlocked with firms in Paris and Brussels, in part a reflection of the global ties of Power Corporation (controlled by the Desmarais family). Montreal's European orientation, as well as Vancouver's Asian connection, demonstrates the diffuse nature of urban Canada's international relations, moulded by geography and by language. Abstracting from US links, Canada east of Manitoba exports proportionally more to Europe; west of that, more exports go to Asia (Figure 2.7). However, the relationship with the US remains central and is no less regionally diffuse.[7]

Another indicator of competitiveness at both the North American and international levels is educational achievement. Table 2.9 compares Canadian and American cities and contains both good and bad news. Canadian cities lag behind US cities in the proportion of adults with degrees, suggesting that Canadian urban incomes have been partly sustained by resources rather than by superior know-how. Aside from the national capitals (those lawyers again!), the star performers in North America come as no surprise: San Francisco, Boston, Minneapolis–St Paul, and Seattle.

Some 45 per cent of San Francisco's population between 25 and 65 held a BA degree or higher in 2006, compared to 34 per cent in Toronto, Canada's best performer. The four lowest rungs are occupied by Canadian cities, with Edmonton and Winnipeg at

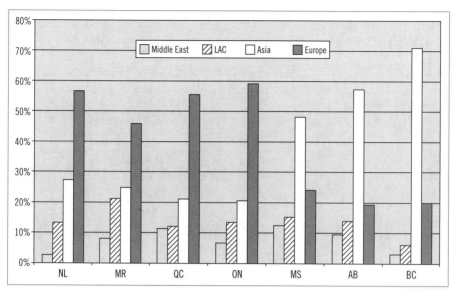

Figure 2.6 Exports by Canadian region to non-US markets.

(Source: Industry Canada. Trade Data Online@strategis.ic.gc.ca)

Note: LAC: Latin American countries.

the bottom end. In the competition for high-tech jobs, these two cities clearly face an uphill battle, in addition to their isolated (and cold) locations in North America. No less damning is the comparison with neighbours. For the same age group, 38 per cent of the population of Seattle held a BA, compared to 31 per cent in Vancouver. In Minneapolis–St Paul, the proportion was 39 per cent, compared to 24 per cent for Winnipeg. One would expect that high-tech firms based in Seattle and Minneapolis would outsource less demanding (and lower-paid) work to their northern neighbour.

Is this to be the future of Canadian cities in the North American knowledge economy? Fortunately, the two other columns in Table 2.9 suggest that such will not be the case. If we look at degrees held by the younger generation (ages 25 to 34), the gap between US and Canadian cities closes noticeably. The star US players—Boston and San Francisco—still have the best educated populations in North America, but all Canadian cities move up at least one or two rungs. Winnipeg and Edmonton are now joined at the bottom by Miami and Dallas. Once we exclude the three top players, Canadian cities are now fairly evenly distributed among the various classes For this younger age group, Toronto's percentage of degree-holders is comparable to New York's, Montreal is comparable to Chicago, and Vancouver is clearly above Los Angeles and close to Seattle. Canadian cities are catching up. All Canadian cities show positive comparative gains in the percentage of degree-holders for the younger cohort (last column), while some US urban areas have witnessed declines, perhaps attributable to the influx of undocumented immigrants to cities such as Miami, Los Angeles, and Dallas.

The future of Canada's cities in the North American knowledge economy lies in the middle range, often founded on niche specializations in particular sectors. Canada

Table 2.9 Percentage of Population (Ages 25 to 65 and 25 to 34) with a Bachelor's Degree or Higher, Selected US and Canadian Metropolitan Areas, 2006

CMA/MSA	% BA Ages 25–65	% BA Ages 25–34	Difference
Washington, DC	48.0	46.9	–1.1
San Francisco	45.2	45.8	0.5
Boston	44.6	49.9	5.3
Minneapolis	38.7	41.1	2.4
Seattle	37.9	38.3	0.4
New York	37.5	41.3	3.7
Denver	37.0	34.4	–2.6
Ottawa	35.4	41.4	6.1
Atlanta	34.8	33.1	–1.7
Chicago	34.2	34.9	0.7
Philadelphia	34.1	38.3	4.2
Toronto	33.6	40.7	7.1
Portland	33.5	31.5	–2.0
Pittsburgh	31.4	38.5	7.1
Vancouver	30.7	36.6	5.9
Calgary	30.6	35.2	4.6
Los Angeles	30.5	29.4	–1.1
Dallas	30.4	27.2	–3.2
Miami	29.7	28.4	–1.3
Halifax	29.1	37.6	8.5
Detroit	28.6	29.0	0.5
Cleveland	28.4	30.0	1.6
Montreal	26.5	33.9	7.3
Quebec City	24.9	31.0	6.1
Winnipeg	24.1	29	4.9
Edmonton	22.9	27.9	5.0

does not have—and is unlikely to develop—a great global high-tech concentration on the scale of the San Francisco Bay Area (which includes Silicon Valley) or Greater Boston. But Canadian cities are significant players. The Milken Institute 2007 ranking of High Tech Metropolitan Areas (Milken Institute 2009) puts Toronto in fifteenth place after Chicago, with Montreal in nineteenth position after Denver, which is not bad, considering that suburban concentrations, part of metropolitan areas, are sometimes also counted.[8] If they are excluded, Toronto and Montreal jump up three and four notches, respectively. Significantly, the ratings of all Canadian cities have increased markedly in recent years, consistent with the improvement in educational attainment. Since 2003, Toronto has moved up 10 notches and Montreal 8, while Ottawa moved from sixty-sixth to thirty-seventh place and Vancouver from sixty-fourth to thirty-sixth.

The middle-range image holds as well for Canadian cities as university centres. Canada's cities will never support giants like Harvard, MIT, Stanford, or Berkeley, systematically rated among the very top universities in the world, but Canada's major cities all house universities that rank in the first 100 or 200, depending on the source consulted—a good performance, considering the overwhelming number of US universities. Of the top 50 in the Webometrics (2009) ranking, 40 schools are in the US, with two of the remaining 10 in Canada and two in Britain. Canada has 16 universities ranked in the top 200, compared to nine for Britain and six for Australia.

The rankings on Table 2.10 should be viewed with caution, given the inherent pitfalls of exercises of this nature, as witnessed by the differences in rankings depending on the source used.[9] The important point is that the same cities rank in approximately the same order: Toronto, Montreal, and Vancouver, followed by Calgary and other metropolitan areas. In Canada, these four cities may deserve the label 'global', whether as financial or corporate service centres or centres of learning and high-tech industry, but probably no more than these four. It is a high number, considering Canada's small population size, and a result of Canada's vast geography, federal structure, and language split, as well as the relatively high inflows of international scholars and professionals through immigration.

Table 2.10 tells two other stories. First, Ottawa's position as a high-tech and producer-service centre rests almost entirely, it seems, on its role as federal capital rather than on the presence of a well-developed university environment. In this respect, Ottawa resembles Washington, DC. Second, the table suggests that some smaller cities host well-rated universities and, by the same token, potential high-tech industries. The four cities identified are all in southern Ontario, with Kitchener-Waterloo the star example as home to Research in Motion, creator of the Blackberry mobile phone and one of Canada's most successful global high-tech corporations. Kitchener-Waterloo is only an hour from Toronto. Thus, size and proximity matter. One may ask whether similar success would have been possible in northern Ontario.

Finally, the North American context can also restrict the development of other international connections, especially given the difficulties at the US border since the 9/11 attack on New York's twin towers. For Canada, the dual presence of a huge economic neighbour and a partially closed border has important consequences for the capacity of Canadian cities to attract global functions. This is exemplified by hub airports, essential features of the global economy. In rankings of global cities, Toronto invariably comes out on top among Canadian cities, as noted earlier, generally followed by Montreal and Vancouver, but although Toronto's Pearson Airport is Canada's largest by far, it is not a major global hub. In 2006, Toronto ranked twenty-ninth internationally in terms of total passenger traffic, after Philadelphia, down from the twenty-sixth spot in 2001 (Airports Council International 2009). Vancouver ranked fifty-ninth, and Montreal was not even in the top 100. Within North America, Canadian cities systematically attract less air traffic than their population size would predict (Discazeaux and Polèse 2007).

Table 2.10 Location of Canadian Universities Ranked in Top 200 in 2009—Three Sources

Urban Area	Ranks given appear in the following order by source: *Webometrics, Shanghai Jiao-Tong, Times-QS.* If no number is given, the city is not ranked in the top 200 by that source*			
Toronto	Toronto 28, 24, 29	York 145, –, –		
Other Ontario	Waterloo 85, –, 113	McMaster 187, 89, 143	Queen's 194, –, 118	Western Ontario 196, –, 151
Montreal	Montréal 72, 126, 107	McGill 86, 60, 18	UQAM 173, –, –	
Vancouver	UBC 49, 35, 40	Simon Fraser 69, –, 196		
Calgary	Calgary 50, –, 149			
Edmonton	Alberta 59, 125, 59			
Quebec City	Laval 161, –, –			
Ottawa	Carleton 163, –, –			
Winnipeg	Manitoba 200, –, –			

Sources: http://www.webometrics.info; http://www.arwu.org; http://www.topuniversities.com/university-rankings.

The principal reason is the relative lack of international transit passengers, the core component of global hubs. International traffic accounts for about 15 per cent of Toronto's passenger traffic once North American cross-border flights are discounted. We may reasonably assume that the overwhelming majority of the incoming international passengers are destined for Canada (and those out-going originate there). Why use Toronto, Vancouver, or Montreal as a stopover if headed for (or coming from) a US city and encounter the double hassle of US and Canadian customs?[10] Canada's internal market is small, distributed linearly along an east–west axis, not an ideal configuration for a hub-and-spoke network. Canada's international airports exist primarily to meet the international needs of its domestic market. Again, the 'global' functions of Canada's cities are essentially driven by the strength of the domestic economy rather than by services rendered to third parties.

The Historical Sequence

In urban studies, the temporal context is at least as important as the spatial. The growth of a city in a particular time period reflects the growth over previous periods and the pattern of nearby spatial competition. Investment decisions and political commitments reflect the contemporary choices available and the economic and political issues at the time. This is as evident in Canada as in any other country. As the colonial and post-colonial settlement pattern expanded from east to west, a succession of cities emerged as regional or national centres, to be gradually replaced by others as the focus of economic activity slowly shifted westward. As Quebec City gave way to Montreal, Montreal gave way to Toronto, and Winnipeg gave way to Calgary and Vancouver. Those are the major stories in Canadian urban history over two centuries, but there are dozens of similar competitions among smaller cities. Figure 2.7 shows the period in which various Canadian cities attained the threshold size of 10,000 population and were added to the urban system. In 1850, there were only eight. Most of them (St John's [not part of Canada at that time], Halifax, Saint John, Quebec City, and Montreal) were the ports of entry for settlers within their regions, while the remainder (Kingston, Toronto, Hamilton) emerged as commercial and political centres at later stages in the settlement process.

Transportation played an important role in shaping urban growth from the very beginning (Apparicio et al., 2007). The ports of entry identified the key water routes to the interior of the continent, to be supplemented by canals at key locations: Montreal emerged around the canal on the St Lawrence; Ottawa and Kingston mark the end-points of the Rideau Canal. St. Catharines anchors the Welland Canal. These nodes were later reinforced by a railroad network, crucial to the nation-building process, which solidified the location advantages of cities in central Canada. At the same time, it stimulated a network of nineteenth-century cities at junctions across the country: the Intercolonial route from Halifax to Montreal, the Grand Trunk across southern Ontario, and the Canadian Pacific to the west coast.

These cities also demonstrate the regional sequence of development. Southern Ontario manufacturing cities dominated the period from 1851 to 1901, and similar cities emerged in Quebec during the following half-century—along with the agricultural settlements of western Canada and the resource centres in the near north. The past 50 years reveal the continuing urban population growth in Alberta and British Columbia, plus the continued growth of favourably located older centres in Ontario and Quebec. The number of urban additions in each half-century period after 1851 is 19, 44, and 75 cities, respectively.

Not Like the Others!

The spatial and temporal context helps to explain the unusual characteristics of Canadian urbanization that will become more evident throughout the later chapters. We can begin with the geography that underlies the urban areas. The third-largest country in the world fronts on three oceans and embraces a wide range of landscapes and climates.

Although the relatively small population of Canada, hence the small number of cities, is huddled along the southern border of the country, there are still enormous

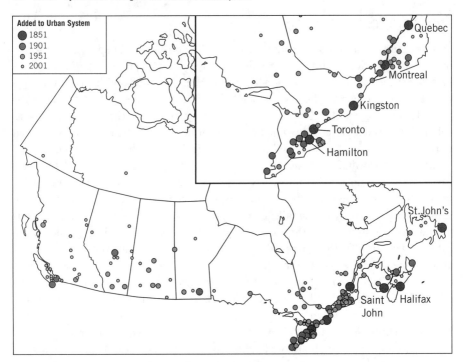

Figure 2.7 The temporal sequence of urbanization.

(Source: Compiled by the authors from Census of Canada, various years: see Simmons 1990)

distances to be covered: for instance, the more than 5000 kilometres between the metropolitan areas of St John's and Victoria. While some cities in the more temperate areas of the country flourish side by side, many urban centres in the North combat long distances and extreme isolation. Inevitably, the low density of cities increases the variations in their physical environments: differences in rainfall, temperatures, altitude, landforms, and economic resources. The geography alone imposes a dramatic pattern of economic specialization.

This dispersed geography with low levels of population and high levels of investment has inevitably been highly sensitive to external events and forces over the years. Colonial empires, coupled with intense economic specialization, determined the points of settlement and shaped the economy to serve foreign markets. For instance, the fur trade outlined the national geography and identified the key locations in a water-based transportation system. Subsequent resource economies and transportation investment extended the patterns of urban settlements, just as the energy sector shapes the current distribution of population growth. Canada's colonial status skewed trade relations and immigration towards France, Britain, and finally, the US.

An extensive and diverse geography, transformed by trade, investment, and immigration over 400 years has produced a system of cities with many dimensions of variation. We have briefly discussed the topography, economy, and history, but the cultural diversity is also unusual. More than 20 per cent of Canadian cities are predominantly francophone, including Montreal, one of the case studies to be discussed, and (in part) Ottawa-Gatineau, the national capital. At the same time, 20 per cent of Canadians are foreign-born, including almost 40 per cent of Vancouver's residents and 45 per cent of Torontonians. Inevitably, this aspect of diversity affects the allocation of political and economic power and the way these cities, their economies and landscapes, evolve. Immigration adds global connections by expanding the 'transnational' communities of interest, but it can also add uncertainty to population growth. Immigration flows can decline dramatically if conditions in source countries change and may even lead to return migration (emigration).

Compared to cities in the US—the cities most similar in terms of age, income, technology, and level of development—Canadian cities generally have higher population densities within their borders, with generally tighter planning regulations (and higher land prices!). The presence of recent immigrants is a significant difference, especially for the largest and fastest-growing cities. Extreme income variations within the metropolitan area are less apparent, however. There are few examples of extended low-income neighbourhoods or abandoned housing in Canadian cities. Canadian cities generally have lower crime rates than US cities of comparable size. By the same token, Canadian metropolitan areas generally house stronger and more vibrant downtowns. Inner-city ghettos are far less prevalent, a reflection again of Canada's different geography and consequent social history, not weighted down by a past legacy of plantation economies and slavery. On the other hand, geography has also meant that Canada has inherited a proportionally much larger Native (First Nation) population, which is now entering its urbanization phase and as such will be an additional factor shaping the social geography of Canadian cities in the future, especially in the west.

References

Airports Council International. 2009. 'Passenger traffic, 2006'. Airports Council International. http://www.airports.org/cda/aci_common/display/main/aci_content07_c.jsp?zn=aci&cp=1-5-54-55-4777_666_2_.

Apparicio, P., G. Dussault, M. Polèse, and R. Shearmur. 2007. 'Transportation infrastructure and local economic development: A study of the relationship between continental accessibility and employment growth in Canadian communities'. http://projetic.ucs.inrs.ca. Montreal: INRS.

Beaverstock, J.V., R.G. Smith, and P.J. Taylor. 2009. As part of their ESRC project 'The geographical scope of London as a world city', Data Set 4 of the Globalization and World Cities (GaWC) Research Network. http://www.lboro.ac.uk/gawc.

Carrol, W.K. 2007. 'Global cities in the global corporate network'. *Environment and Planning A* 39: 2297–2323.

Citizenship and Immigration Canada. 'Permanent residents admitted by country of last permanent residence, Jan. 1980–Oct. 2006'. Ottawa: Citizenship and Immigration Canada.

Discazeaux, C., and M. Polèse. 2007. 'Comment expliquer le déclin de Montréal comme centre de transports aériens : une question de géographie économique'. *The Canadian Geographer* 51 (1): 22–42.

Industry Canada. Trade Data Online. http://www.strategis.ic.gc.ca.

Mainelli, M. 2007. 'Market of markets: The Global Financial Services Index'. *Journal of Risk Finance* 8 (3): 313–19. http://www.zyen.com/Knowledge/Articles/GFCI_V1.htm.

Milken Institute. 2009. '2007 metro ranks'. North America's high-tech economy: The geography of knowledge-based industries. The Milken Institute. http://www.milkeninstitute.org/nahightech/nahightech.taf.

Polèse, Mario. 2009. *The Wealth and Poverty of Nations: Why Cities Matter*. Chicago: University of Chicago Press.

Shanghai Jiao Tong University. 2009. 'Academic ranking of world universities'. http://www.arwu.org/rank2008/EN2008.htm.

Simmons, Jim. 1990. 'The emergence of the urban system'. In D.P. Kerr, ed., *The Historical Atlas of Canada*, vol. III, Plates 10 and 54. Toronto: University of Toronto Press.

———. 1992. 'The reorganization of urban systems: The role and impacts of external events'. Research Paper no. 186. Toronto: Centre for Urban and Community Studies, University of Toronto.

Simmons, J., and S. Kamikihara (with the assistance of L.S. Bourne and I. Escamilla). 2006. 'The North American urban system: The limits to continental integration'. Research Paper no. 205. Toronto: Centre for Urban and Community Studies, University of Toronto.

Statistics Canada. Annual. *Air Passenger Origin and Destination: Canada–United States Report*. Catalogue no. 51-205. Ottawa: Statistics Canada.

———. Annual. *Air Passenger Origin and Destination: Domestic Report*. Catalogue no. 51-204. Ottawa: Statistics Canada.

Taylor, P.J., P. Ni, B. Derruder, M. Hoyler, J. Huang, F. Lu, K. Pain, F. Witlox, X. Yang, D. Bassens, and W. Shen. 2009. 'Commentary. The way we were: Control and command centres in the global space-economy on the eve of the 2008 geo-economic transition'. *Environment and Planning A* 41: 7–12.

Webometrics. 2009. 'Webometrics ranking of world universities'. http://www.webometrics.info/top6000.asp.

World Federation of Exchanges. 2009. 'Annual statistics, World Federation of Exchanges'. http://www.world-exchanges.org/statistics.

Notes

1. Data sources are not always completely compatible. These figures are from the World Bank Development Indicators, http://data.worldbank.org/country/canada.

2. Local public market tenders are not necessarily subject to NAFTA, an issue of current concern in US–Canada relationships.

3. With the exception of Winnipeg and Halifax (author's calculations), these tables are drawn from the Observatoire Grand Montréal of the Montreal Metropolitan Community (CMM, by its French initials): http://observatoire.cmm.qc.ca/swf/index.php. Tables 2.7 and 2.9 are largely based on the same source. We thank Philippe Rivet at the CMM for making these data freely available to us. Employment data for Canada are based on Statistics Canada Labour Force Survey and on the Current Employment Survey of the Bureau of Labor Statistics in the US. Tables 2.5 and 2.9 are also based on the 2006 Canadian census and on the US Census Bureau, American Community Profiles 2006.

4. Formally, two of the biggest banks—the Bank of Montreal and the Royal Bank—have headquarters in Montreal, but this is almost as symbolic as the Bank of Nova Scotia's 'head office' in Halifax. Such interprovincial bank moves would be impossible under the US system.

5. Computer services are included in the 'professional and scientific service' class.

6. However, the 2007–8 sub-prime crisis revealed the (surprising) comparative solidity of Canada's banking system. Unlike in the US, no bank in Canada failed. The US, at the time of writing, was in process of financial reform, which may bring it closer to the Canadian model.

7. For Figures 2.7 and 2.8, as well as Table 2.7, the source is a special tabulation requested by the authors, prepared by Statistics Canada.

8. For example, the San José–Santa Clara area (Silicon Valley), in the number one spot, is counted separately from San Francisco in the number eight spot.

9. For example, Shanghai Jiao Tong University (2009) systematically ranks Canadian cities higher within the first 100 but in the same order. It does not assign precise ranks below 100, however.

10. This impediment may be of less consequence in the future if Canadian (hub) airports introduce transfer facilities—as do many European airports—that allow incoming passengers to transfer to third-party destinations without having to go through local customs and immigration. The Montreal airport has recently taken a step in that direction, allowing international incoming passengers destined for the US to proceed directly to US customs (housed in the airport).

3 Growth and Change in the Canadian Urban System

Larry S. Bourne, Cedric Brunelle, Mario Polèse, and Jim Simmons

Introduction to the Canadian Urban System

Canada's urban population includes some 25 million persons (depending on how it is defined), roughly equivalent to a single one of the world's great global cities such as New York, Tokyo, or Mumbai. In Canada, however, this population is widely dispersed among 144 urban centres scattered across 9 million square kilometres of territory. This chapter introduces the basic facts about the spatial distribution of Canadian cities in terms of city size and region. It also identifies the main processes or determinants of urban growth that affect all cities to varying degrees. It will be evident that a significant aspect of Canadian urban growth over the past two decades has been the continued concentration of growth within the largest cities. There are many factors contributing to this concentration, but we have grouped them into two categories: the economic and demographic processes. Each category is discussed in a separate section to follow.

Defining the Canadian Urban System

As in most countries, urban regions in Canada are composed of diverse municipal units that were created a century ago or earlier and have since evolved in a variety of ways. But at no time have these municipal units been designed for the purpose of either scholarly research or policy. By far the best alternative for that purpose are the formal definitions of urban areas based on functional criteria, imposed by Statistics Canada as part of the five-year census of population. The main spatial units used in the census are laid out in Table 3.1. The list begins with the basic political units: provinces, counties or census divisions, and municipalities (census subdivisions) and then introduces the imposed networks of census spatial units, including census metropolitan areas (CMAs) and the smaller census agglomerations (CAs).

Thus, the cities that we measure, map, and discuss below are all artefacts of the Statistics Canada procedures, and the results will differ from the urban areas based on the definitions developed by statistical agencies in the United States or European countries. The one notable example, discussed in Chapter 10, is Toronto, which as a CMA is clearly under-bounded by Statistics Canada in terms of its current patterns of growth, interaction, and economic integration. The extension of Toronto's CMA boundary to the east and west is constrained by historical precedents that fixed the boundaries of the neighbouring CMAs—Oshawa and Hamilton, respectively—although these two

Table 3.1 Spatial Units: Statistics Canada, 2006

	Number of Units Total	Population Total	Mean Size	Share of Canada Total (%)
Canada	1	31,600,000	31,600,000	100.0
Provinces and Territories	13	31,600,000	2,430,000	100.0
Counties and Census Divisions	288	31,600,000	109,700	100.0
Census Subdivisions	4,606	31,600,000	6,860	100.0
Census Metropolitan Areas (CMAs)	33	21,500,000	652,000	68.0
Census Agglomerations (CAs)	111	4,120,000	37,100	13.0
Total Urban Areas	144	25,620,000	178,000	81.0
Census Tracts (for 48 of Larger Cities)	5,014	22,748,000	4,540	72.0
FSAs (Three-Digit Postal Codes)	1,624	31,600,000	19,500	100.0
Dissemination Areas	52,872	31,600,000	600	100.0

Source: Statistics Canada. *Census of Canada, 2006*. Ottawa: Statistics Canada.

CMAs are closely integrated within the Toronto commuter shed. The three CMAs will be combined within the case study of Chapter 10.

For the purpose of this chapter, the most important units of analysis are the CMAs and CAs. The former are defined as areas of one or more neighbouring municipalities around a major urban centre, with a combined population of 100,000 or more, including at least 50,000 in the core city. A CA must have an urban core of at least 10,000. The inclusion of neighbouring municipalities is determined by the level of commuting to the core city, as a surrogate for economic and social integration.

Canadian Cities in 2006

Canada's urban population in 2006 was mapped in Chapter 1 as the frontispiece and is grouped by city size and region in Table 3.2. The map shows the long-standing concentration of large cities in the industrialized Windsor–Quebec corridor, a concentration that recently has been partly counterbalanced by the accelerated growth of large cities in British Columbia and Alberta. No urban areas with more than 100,000 population are found north of Edmonton. Table 3.2 indicates that there are six metropolitan areas with more than 1 million population, 10 others with more than 300,000, and another 19 places that have more than 100,000 according to the 2006 census. The six largest cities house 14 million people, more than half of the urban population and 45 per cent of the nation's population. Thus, for the first time in Canadian history, almost one-half of Canada's population lives in metropolitan environments with populations of more than 1 million. Regionally, Ontario includes the largest urban population with 10.7 million—roughly one-third of the national total. Ontario and Quebec between them support 16.7 million—equivalent to half of Canada's population. Nonetheless, the big story in this latest census has been the growth of the western cities. Alberta now has twice the urban population of Manitoba and Saskatchewan combined. Both Calgary and Edmonton have populations above the symbolic 1 million mark, while Vancouver has 2.1 million. At the same time, the Atlantic region is barely maintaining its population. The population shares of smaller cities have been declining with each census. Less than one-third of the population now lives in urban areas of less than 100,000 population or in small towns and rural areas.

Urban Growth Patterns

The proportion of Canada's population living in large cities has been growing system-atically over the postwar period (1951–2006), as indicated in Chapter 1 (Figure 1.2). In 1951, the urban population (based on 2001 spatial units) made up only 51 per cent of the total. Fifty-five years later, the proportion has risen to 81 per cent. The urban population has been growing more rapidly than non-urban areas—although both the national and urban growth rates have declined sharply over the study period. Of course, urban growth also depends on the redefinition of urban boundaries as cities expand into the surrounding countryside. The fact that higher growth rates are found in the largest urban areas is a more recent phenomenon, dating from 1981. The causes are many, and we will explore various aspects of this theme in the remainder of the chapter.

The history of Canadian urbanization has been characterized by substantial regional variations in the rate of urban growth. As will be evident in the section to follow, each decade brings a different economic sector to the forefront and thereby defines a new development frontier. This is evident when the growth rates of Canadian cities are examined for the most recent census period, 2001–6. For Canada as a whole, the rate of population growth was 5.35 per cent, and for the 144 census urban areas, the aggregate growth rate was 6.42 per cent (Table 3.3). This latter value is a bit misleading, however, because it includes the very high growth rates for some of the largest cities (Toronto, Calgary, Edmonton) when the average growth rate for all cities was only 3.88 per cent, with a standard deviation of 6.63 per cent. This wide variability in the individual urban growth rates implies that widespread decline is inevitable: 37 of the 144 cities (about one-quarter) lost population over the last census period.

Table 3.3 also disaggregates the national pattern of growth by region and city size. The most striking feature is the enormous concentration of growth in the very largest cities. First, almost all the growth occurred in urban areas—with small towns and rural areas adding only 58,900. Second, two-thirds of this growth took place in the six largest places that have populations greater than 1 million. Third, the smallest urban places have barely maintained their populations. At the same time, there were powerful regional variations, with Ontario and Alberta leading the way, while Quebec (led by Montreal) has been quite successful. Only the Atlantic region—without a

Table 3.2 Urban Population Aggregated by City Size and Region, 2006

City Size/Region	BC	Alberta	MB/SK	Ontario	Quebec	Atlantic	Canada Total
Number of Cities							
Over 1,000,000	1	2	0	2	1	0	6
300,000–1,000,000	1	0	1	6	1	1	10
100,000–300,000	2	0	2	8	3	4	19
30,000–100,000	9	5	3	15	14	5	51
10,000–30,000	15	7	16	12	10	8	58
Total	**28**	**14**	**22**	**43**	**29**	**18**	**144**
Population, 2006 (1000s)							
Over 1,000,000	2116.6	2114.3	–	5959.9	3919.6	–	14,110.3
300,000–1,000,000	330.1	–	649.7	2646.1	715.5	372.9	4714.3
100,000–300,000	321.3	–	428.9	1087.4	480.1	535.9	2853.5
30,000–100,000	575.2	371.3	122.4	826.0	709.7	257.1	2861.6
10,000–30,000	283.8	117.9	97.0	194.3	197.0	156.9	1,046.9
Total	**3627.0**	**2603.5**	**1298.0**	**10,713.7**	**6021.8**	**1332.7**	**25,631.6**
Small Towns and Rural Areas	587.8	686.9	818.6	1446.6	1524.3	952.1	6016.3
Grand Total	**4214.8**	**3290.4**	**2116.5**	**12,160.3**	**7546.1**	**2284.8**	**31,612.9**

Territories cities grouped with BC.

Source: Statistics Canada. *Census of Canada*, 2006. Ottawa: Statistics Canada.

Table 3.3 Population Change, 2001–6

Region/City Size	BC	AB	MB/SK	ON	QC	Atlantic	Canada
	Population Change (1000s)						
Over 1,000,000	129.6	224.9	–	471.0	206.8	–	1032.3
300,000–1,000,000	18.2	–	18.1	152.7	28.9	13.7	231.5
100,000–300,000	26.2	–	10.2	60.4	11.7	12.3	120.8
30,000–100,000	29.4	53.0	1.1	25.3	29.3	5.0	143.1
10,000–30,000	2.1	12.2	–0.8	7.2	–0.7	–0.8	19.2
Total	**205.4**	**290.2**	**28.6**	**716.5**	**276.1**	**30.1**	**1546.9**
Rural Areas	8.8	25.4	–10.5	33.7	32.4	–31.1	58.9
Grand Total	**214.3**	**315.6**	**18.1**	**750.2**	**308.6**	**–1.0**	**1605.8**
	Population Growth Rate %						
Over 1,000,000	6.52	11.90	–	8.58	5.57	–	7.89
300,000–1,000,000	5.83	–	2.87	6.12	4.22	3.81	5.11
100,000–300,000	8.87	–	2.44	5.89	2.50	2.34	4.42
30,000–100,000	5.38	16.65	0.91	3.16	4.31	1.98	5.26
10,000–30,000	0.74	11.54	–0.82	3.84	–0.33	–0.50	1.87
Total	**6.00**	**12.54**	**2.25**	**7.17**	**4.80**	**2.33**	**6.42**
Rural Areas	1.52	3.84	–1.27	2.39	2.18	–4.14	0.99
Grand Total	**5.36**	**10.61**	**0.87**	**6.58**	**4.26**	**–0.004**	**5.35**

Territories cities grouped with BC.

Source: Statistics Canada. *Census of Canada, 2006*. Ottawa: Statistics Canada.

major metropolitan area—and the eastern Prairies were unable to generate sustained population growth. Overall, the aggregate growth rates correlate roughly with the size categories: big cities grew most rapidly, and rural areas barely grew at all. Outside of the areas of metropolitan influence (i.e., within 100 or 150 kilometres of the metropolitan area—see Chapter 5) and excluding Aboriginal communities, most of rural and small-town Canada is losing population.

The table also identifies localities of more intense growth or decline: the most rapidly growing cities are found in southern Ontario around Toronto or in Alberta and southern British Columbia. At the same time, the pattern of population decline is also distinctive: population decline occurs in smaller cities (the largest CMA to decline is Saguenay, population 152,000) that are located in the North or otherwise isolated, especially in the Atlantic region, Quebec, and the eastern Prairies. These are also the communities at risk of further decline.

Table 3.4 provides a sense of the variations in growth environments that Canadians experience in everyday life. The 144 urban areas have been divided into four equal groups of 36 cities based on their relative population growth rates. Cities in the high-growth quartile grew by more than 6.5 per cent, including very high rates for Calgary, Edmonton, and Toronto. In aggregate, the population growth rate is more than 10 per cent for this group. The cities in the group showing moderate growth—ranging from 3.8 to 6.5 per cent— include the Montreal, Ottawa-Gatineau, and Vancouver CMAs. The aggregate growth rate is 5.4 per cent. Cities in the low-growth group have growth rates from zero to 3.8 per cent, while the final quartile experienced negative growth. Fortunately, most Canadians live within the cities of moderate to high growth that account for more than 80 per cent of the total population, while cities in decline include less than five per cent of all urban residents. The table confirms the positive relationship between city size and growth, with the average city size considerably greater in the two higher-growth quartiles.

If the focus shifts to the 36 cities in the declining group, it is apparent that cities decline very slowly—certainly in comparison with the highest growth rates (Simmons and Bourne 2007). In total, these 36 places lost only 27,200 persons, for an average loss of only 757—well within the error of census measurement. Occasionally, the rate of decline can be high, as for Kenora or northern BC communities, and it can be either cyclical or long-term. Growing cities, in contrast, are capable of absorbing huge increases in population in either absolute or relative terms. Consider the addition of 430,000 people to the Toronto CMA between 2001 and 2006, or the growth rate of 46.7 per cent in Okotoks, AB. Clearly, there are powerful forces that attract growth to metropolitan areas and resist the loss of population in small communities. As for decline, a variety of public assistance programs encourage people to stay put, ranging from government transfers to municipalities to the various kinds of social insurance (unemployment benefits) available to individuals and families. There is also the weight of the financial and human capital already invested in a home, business, or community that people (and governments) are reluctant to abandon.

The overall concentration of population growth implied by Table 3.4 is nonetheless impressive. Most of Canada's urban growth problems, such as the provision of infrastructure or new education facilities or services to immigrants, are significant within only half a dozen areas of the country. For the rest of Canada, the problem is the simple

Table 3.4 Variations in the Urban Growth Experience, 2001–6

Quartile	Range (% Growth)	Population, 2006 (1000s)	Population Share (%)	Average City Size (1000s)	Average Change, 2001–6	Average Growth Rate (%)
1 High Growth	> 6.5	10,043.7	39.2	279.0	25.8	10.2
2 Moderate Growth	3.8 to 6.5	11,321.2	44.1	314.5	16.1	5.4
3 Low Growth	0.0 to 3.8	3,113.6	12.1	86.5	1.9	2.2
4 Negative Growth	< 0.0	1,153.1	4.5	32.0	−0.8	−2.3
Total		**25,631.6**	**100.0**	**167.3**	**10.7**	**3.9**

Note: 36 cities in each quartile; n = 144.

Source: Statistics Canada. *Census of Canada, 2006*. Ottawa: Statistics Canada.

lack of growth and the accompanying withdrawal of investment, skills renovation, and innovation that we have come to expect.

Perhaps the most important message to take away from this analysis is the close relationship between urban size and urban growth. This book focuses on Canada's largest cities and by implication the cities that are growing faster, so that it presents only one pole of the two divergent lifestyles and living environments in Canada: big city and high growth contrasting with slow growth or decline in smaller places. The next two parts of the paper explain the reasons for this increasing divergence in growth rates and lifestyle opportunities by examining the economic and demographic bases of growth processes, respectively.

The Urban Economy

Growth is unevenly distributed across Canadian cities, as we have seen. The divergence in urban growth paths is more pronounced than it would be in other smaller, more compact nations because of Canada's size and geography and the diversity of its regional economies. There is little reason why the economy of Sherbrooke should evolve along the same path as that of Thunder Bay or Victoria. The factors that explain growth vary both over time and over space (Shearmur et al. 2007). Specialization in the production of a given resource—oil, wood, hydroelectric power, wheat—may bring growth in one period but decline in the next. By the same token, a given attribute may positively affect growth in certain communities but not in others. A high percentage of university graduates is an advantage for large metropolitan areas competing in the knowledge economy but may have little impact on the economic growth of smaller manufacturing communities that compete mainly on labour costs.

The first decade of the twenty-first century witnessed two surprising upsets in regional fortunes across Canada. The urban areas of southern Ontario were the hardest hit by the 2007–9 global recession, and the traditional powerhouse of the Canadian economy became the laggard. In contrast, Newfoundland and Labrador, a traditionally slow-growth region, has witnessed a period of unprecedented expansion since 2001, in turn fuelling the growth of the St John's economy. Few would have predicted either outcome a decade earlier. Yet these individual trajectories need to be viewed in light of deep-seated structural factors that continue to shape the economic geography of Canada. City size, location, and economic structure remain powerful predictors of growth (Shearmur and Polèse 2007, Shearmur, Polèse, and Apparicio 2007), and the continuing structural transformation of the Canadian economy is a fundamental driver of change, to which we now turn.

The Rise of the New Service Economy

As in Chapter 2, we focus on the sectors of the economy that are (or can be) part of an urban area's economic base: that is, activities whose output—goods or services—can be exported over fairly long distances. Public services, health, education, and retailing are excluded, although these activities can in part be 'exported' to surrounding populations. Other services such as wholesaling and transportation, while important functions of metropolitan economies, largely act as supporting activities for goods and services traded over greater distances. The emphasis is on goods and services on which Canada's cities compete globally as well as nationally.

Canada, like other advanced economies, has witnessed a major shift in the structure of employment, a result of the combined impact of technological change and the growth of the knowledge economy. Figure 3.1 shows the evolution of employment in five industry classes over 20 years, with the year 1989 as benchmark. Knowledge- and information-rich services have shown steady growth, with the most knowledge-intensive class (professional, scientific, and technical; henceforth PST) growing most rapidly, while primary sector employment, agriculture excluded, stagnated, although with growth taking off again since 2001. Manufacturing employment grew steadily after the 1992–3 downturn, in part fuelled by NAFTA and the resultant rising exports to the US. That growth came to an abrupt end in 2004, in part because of the dual impact of a rising Canadian dollar, pushed up by skyrocketing oil prices, and of growing Chinese competition for the US and other markets. Growth was further dampened by the arrival of the US sub-prime crisis in 2007 and the ensuing global recession.

Beyond the cyclical explanations—the recession will end, the oil boom is not eternal, and China will one day cease to be a low-cost economy—the shift away from manufacturing is a sign of more fundamental changes. Manufacturing will not disappear. The decline (or slow growth) in employment is in part a reflection of increased productivity—automation—that allows plants to produce more with less workers. Employment is shifting to higher value-added products, which require fewer workers per unit of output. Manufacturing will continue to provide an essential economic base for many urban areas, especially smaller places. As well, the decline in manufacturing is in part an illusion, a reflection of the increasing fragmentation of production—where 'manual' work is separated out from 'brain' work, often in different locations—with a parallel growth of service inputs into the production processes. Today, almost every

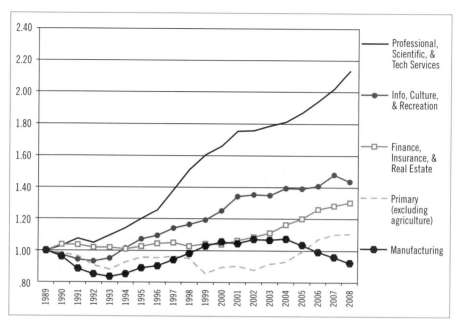

Figure 3.1 Employment growth in five industry classes, Canada, 1989–2008 (1989 = 1.0).

Source: Statistics Canada. 2008. *Labour Force Historical Review 2008*. Catalogue no. 71F004XCB. Ottawa: Statistics Canada.

product is a composite of physical inputs and of numerous unseen 'service' inputs: design, marketing, management advice, engineering, and so on. The pure (brawn) costs of assembling an automobile are only one element—and perhaps not the most costly—of a much more extensive production chain, much of which would be classified as 'services' if carried out by separate firms in the service sector.

It is this shift in the manner in which goods are produced and traded that largely explains the exponential rise in PST services. This has important consequences for Canadian cities, not only because of the link with manufacturing but also because of the equally umbilical relationship with resource industries. As noted in the previous chapter, oil exploration relies on an army of engineers, geologists, and other scientific and technical know-how, not forgetting the managerial, marketing, and financial-packaging talents needed to carry out projects from beginning to end. Brunelle and Polèse (2008) find that the spatial division of labour within the Canadian energy sector is growing over time. Scientific and managerial functions associated with large hydroelectric projects (often located in the North) create more employment in large metropolitan areas than at the project site. By the same token, it is reasonable to assume that oil exploration in northern Alberta is fuelling the growth of the PST and FIRE (finance, insurance, and real estate) sectors in Calgary, just as offshore exploration is fuelling employment growth in these same sectors in St John's.

The mix of activities that fuel the rise of PST services is not necessarily the same in Canada as it is in developed economies less linked to resources, not are the spatial consequences. In a nation such as Britain, the rise in PST services is largely founded on functions such as accounting, advertising, and management consulting, highly concentrated in London, for which the link with 'physical' production or with geography is minimal. This is less true for engineering and associated scientific expertise linked to resource exploitation. PST services cover a broad range of knowledge-intensive services, everything from computer programmers to lawyers. Canadian cities will capture shares of rising knowledge-intensive services for different reasons. The push behind the growth of PST services in Toronto, and probably also Montreal, is not unlike that of London, while in Calgary or St John's the link with resources is more direct. The downside of the close link with resources is the footloose nature of the PST services concerned. Once the resource is no longer profitable, the engineering firms may leave. This, again, is a classic Canadian challenge: to build strong (diversified) urban economies while resource booms last, hoping that the bubble will not burst too soon (see Shearmur and Doloreux 2008; Shearmur 2010).

The growth of the information, culture, and recreation industries is founded both on changing tastes and changing technology. Canadians, like others, consume more and more news and entertainment services, whether by turning on their radio or television, purchasing CDs, or, increasingly, by using the Internet. Modern information technology means that the final product—a song, a document, or a newscast—can now be widely traded at almost no cost to the sender (or seller). This has two implications for urban economies. First, the low—indeed non-existent—transport costs of the final product facilitates centralization; a viewer in Brandon, Manitoba, watching the national news on television will be looking at an anchorman (or woman) seated in a studio in Toronto (or in Montreal if he or she prefers French). Second, the ease of transport increases competition between cities. For many services, competition is global. For international news, our viewer may prefer Atlanta-based CNN.

The FIRE sector represents a more mature industry, less a source of growth than in the past. As noted in Chapter 2, the concentration of financial services in Toronto—banking particularly—is facilitated not only by information technology (IT) but also by Canada's regulatory system, which favours the growth of large nationwide banks. Finally, a word of caution is in order on the evolution of primary employment depicted in Figure 3.1. The annual figures given by the Labour Force Survey (Statistics Canada 2008) include forestry and fishing as well as mining and oil extraction. The first two have been declining over the past two decades and thus hide the impressive upturn in employment in the oil and mining industries since 2001. Between the census years of 2001 and 2006, employment in mining and oil extraction grew at an average annual rate of 7.5 per cent, almost four times the national rate, of which 70 per cent was concentrated in Alberta, also pulling up employment in construction. The years from 2001 to 2006 are thus atypical. This needs to be kept in mind.[2] But then again, booms and busts are quintessentially Canadian.

Economic and Functional Specialization: Effects of City Size

The recent growth history of Canadian cities clearly demonstrates the benefits of population size, and the economic explanations derive from the shift toward services. Information-rich and knowledge-rich service industries are sensitive to city size.[3] For all three service classes, the highest relative concentrations of employment (location quotients[4]) are invariably found in the very largest urban areas, a reflection (on the consumption side) of low transport costs for the final product and (on the production side) of the weight of agglomeration economies. The importance of agglomeration economies for modern high-order services has been demonstrated time and again by studies around the world and in Canada. For Canada, Polèse and Shearmur (2004b; 2006) confirm the positive relationship between city size and high-order services and also the stability of that relationship over time. The glue that ties knowledge-rich service industries to large cities is the need for face-to-face contacts. The more complex the service and the less standardized and predictable the product, the greater is the need for rapid access to a variety of talent, contacts, information, and ideas.

Table 3.5 and Figures 3.2 and 3.3 provide evidence of the hierarchical nature of the relationship between city size and high-order services. The drop from large cities to small places in relative employment concentrations (location quotients) is somewhat less pronounced for financial services (FIRE), which, as noted earlier, represent a fairly mature industry.[5]

Even smaller cities (populations below 100,000) will have banking establishments, although much of the more complex work will be handled at the head office, perhaps in Toronto, or at a regional head office. The most precipitous decline from large to small places is for employment in telecommunications, media, and the performing arts. Clearly, activities in this class need a big-city environment to thrive. The rise of the IT-related entertainment industry—perhaps the most visible manifestation of the modern plugged-in economy—is a powerful force fuelling the continued growth of larger cities. The electronic gadgets that (young) consumers hold to their ears, or in front of which they are glued, may have been assembled in a small city or, more likely, in China, but what they hear or see will have been produced in a studio in some large city.

A warning is in order on the interpretation of location quotients for the smallest size class (small town and rural), which in some cases exhibits quotients and growth higher than for larger size classes, notably for employment in telecom, media, and the arts and in the sciences and engineering occupational class. Two caveats are called for. First, this size class includes places lying just beyond the boundaries of large metropolitan areas in the rural fringe. Second, the census data refer to place of residence, not place of work. If only a small proportion of workers commute into nearby metropolitan areas, the location quotient will rise. The slightly higher quotients for the media and arts class and for occupations in the sciences (which includes university professors, for example) suggest that persons in these fields or occupations are less often tied to nine-to-five jobs requiring a daily presence in the big city and thus enjoy greater residential flexibility. A bank teller or manager certainly enjoys less freedom in this respect than a musician or a journalist.

As long as the largest urban areas exhibit location quotients above unity in rapidly growing industries—meaning higher than proportional employment shares—the simple structural dynamics of the Canadian economy will continue to drive the growth of large urban areas. This is the principal economic factor driving the continued growth of large metropolitan areas in Canada. The slight declines between 2001 and 2006 in location quotients for the largest places for telecom, media, and the arts and professional and scientific services are more indicative of a spatial diffusion process than

Table 3.5 Economic Specialization by City Size: Location Quotients for Five Industry Classes, 2006 and Change Since 2001

CMA	Extraction		Manufacturing		Telecom., Media, Arts		Professional & Scientific		Finance, Insurance, & Real Estate	
	2006	Change since 2001	2006	Change since 2001	2006	Change since 2001	2006	Change since 2001	2006	Change since 2001
Over 1,000,000	0.686	0.078	0.952	-0.020	1.399	-0.045	1.303	-0.057	1.240	-0.016
300,000–1,000,000	0.118	-0.020	1.099	0.001	0.837	0.046	0.927	0.016	1.084	0.024
100,000–300,000	1.230	0.039	0.924	0.023	0.815	0.014	0.837	0.021	0.810	-0.037
30,000–100,000	1.965	-0.018	1.052	0.017	0.514	-0.004	0.661	0.064	0.648	-0.002
10,000–30,000	1.810	-0.115	1.046	0.031	0.504	0.024	0.548	0.028	0.621	0.009
Rural & Small Town	1.947	-0.496	1.079	0.071	0.513	0.045	0.490	0.025	0.582	-0.005

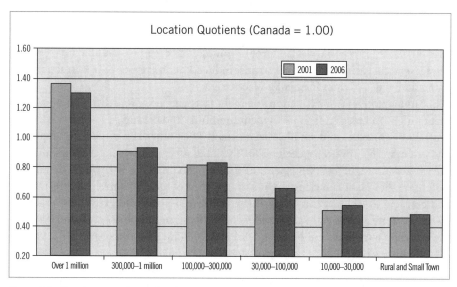

Figure 3.2 Employment in professional and scientific services by city size, 2001–6.

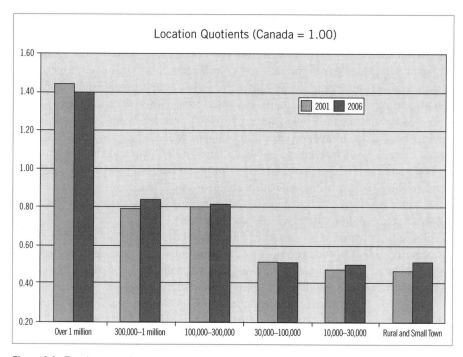

Figure 3.3 Employment in telecommunications, the media, and the arts by city size, 2001–6.

of a decline in the importance of agglomeration economies. For the professional and scientific service class, the gains of smaller and mid-sized cities are in part linked to the input–output relationship with extraction and manufacturing for services such as engineering, industrial design, programming, bookkeeping, and payroll services. These 'service' gains are, in other words, induced by the presence of extraction and manufacturing activities in smaller places.

Location quotients for extraction and manufacturing tell a very different story from that of high-order services. Employment in mining and oil and gas extraction is concentrated in middle- and smaller-sized places. The main drivers here are geography and geology. The important declines for the smallest size classes suggest that much of the work is increasingly taking place in larger cities, consistent with our earlier observations on the rise of knowledge inputs in all sectors of the economy, including mining and extraction. The growth of 'fly-in/fly-out' exploration and extraction practices also contributes to the centralization of employment in large centres. For extraction, the increase in the quotient between 2001 and 2006 for the largest urban size class is almost entirely attributable to Calgary and Edmonton.

For manufacturing, the highest quotients are for urban areas in the 300,000 to 1 million population range, a reflection of the weight of southern Ontario cities such as Windsor, Hamilton, and Oshawa (see Table 3.7). The principal gains (excluding the urban/rural fringe) are found among urban areas in the 10,000 to 100,000 range. The quotient for the largest metropolitan areas has declined. This is consistent with trends observed in all advanced economies: mid- and low-tech manufacturing is deserting large metropolitan areas for smaller places where land prices and wages are lower. This industrial exodus, also referred to as crowding-out, is the corollary of the concentration of rapidly growing information- and knowledge-rich services in large metropolitan areas, diverting less knowledge-intensive activities to other locations. Typically, manufacturing plants seek out smaller cities well-located for trade (near highways and rail links) and not too far from a major urban area, precisely because of the weight of service inputs in the production chain. Consultants, bankers, programmers, managers, and so on constantly travel back and forth between the plant and the neighbouring big city, and there are also (material) input–output relationships with customers and suppliers. Mid-sized cities located within about an hour's drive from a large metro area are as a rule most successful in attracting manufacturing (Polèse and Shearmur 2004b; 2006), and this pattern is evident in Figure 3.4.

The resulting division of labour among different-sized urban places produces a centre–periphery relationship, with knowledge-rich, white-collar jobs disproportionately concentrated in large metropolitan centres and blue-collar jobs in smaller places. The inverse relationship between city size and blue-collar employment is almost perfectly symmetrical (Figure 3.5). In other words, the occupational mix of Canadian cities is very much a function of size (Table 3.6). The symmetry of the relationship—not simply a replication of the manufacturing/service split—suggests that it is being driven as much if not more by the division of functions within industries as by industry-mix differences. In Chapter 2, we noted the social divide between cities, generally large, with important immigrant populations and those without, generally smaller cities. Here, we encounter another social divide related to city size. Small-town Canada is different from metropolitan Canada on

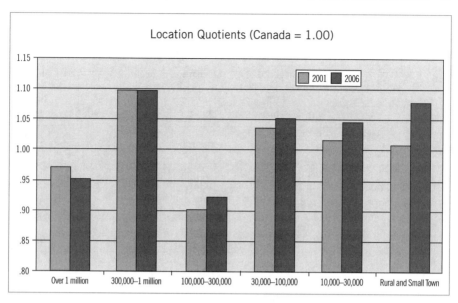

Figure 3.4 Employment in manufacturing by city size, 2001–6.

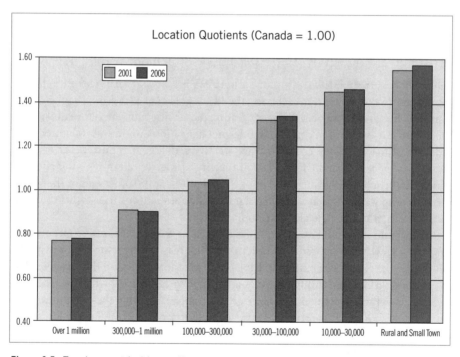

Figure 3.5 Employment in blue-collar occupations by city size, 2001–6.

Table 3.6 Functional Specialization by City Size: Location Quotients for Three Occupational Classes, 2006 and Change Since 2001

CMA	Managerial		Sciences & Engineering		Blue-Collar	
	2006	Change since 2001	2006	Change since 2001	2006	Change since 2001
Over 1,000,000	1.168	−0.010	1.210	−0.019	0.774	0.007
300,000–1,000,000	0.980	0.003	0.999	−0.003	0.903	−0.009
100,000–300,000	0.892	0.007	0.861	0.003	1.046	0.012
30,000–100,000	0.788	−0.002	0.742	0.010	1.339	0.019
10,000–30,000	0.770	−0.002	0.665	0.009	1.462	0.009
Rural/Small Town	0.734	−0.026	0.739	0.018	1.576	0.021

several dimensions. The chances of meeting a manager, professional, or scientist are much lower in the former than in the latter. This appears to be a surprisingly stable feature of Canada's urban landscape, since locations quotients for the three occupational classes barely changed between 2001 and 2006.

Economic and Functional Specialization: Principal Metropolitan Areas

Tables 3.7 and 3.8 show the same information as Tables 3.5 and 3.6 but identify specific census metropolitan areas. In each case, the 14 CMAs (out of 33) with the highest location quotients are shown, with a line separating those with quotients above unity from those below. It is common practice to interpret a location quotient above unity as an indicator of an export—or economic base—activity: that is, an industry that 'exports' goods or services beyond the borders of the city or metropolitan region identified, which can be to a neighbouring town or a destination at the other end of the world. We shall simply assume that a quotient significantly above unity suggests that the given industry contributes to the city's economic base.

Employment in mining and oil and gas extraction is by far the most concentrated, limited to a few metropolitan areas. Metropolitan areas that benefited most from the 2001–6 oil and (to a lesser extent) mining boom are those closest to the resource. The five metropolitan areas with location quotients above unity are—with the exception of Sudbury—linked to the oil economy, with St John's showing the greatest relative gain. The high location quotients, again with the exception of Sudbury, reflect high concentrations of management and support functions within the industry rather than actual mining or oil and gas extraction activity within the metropolitan boundaries. As such, the economic fortunes of these cities are to a large extent tied to the resource endowments of their home province and surrounding areas (we might also include fish and agriculture) and world fluctuations in demand for those resources. The upsurge in demand for grain-derived ethanol fuel, for example, gave a boost to the Saskatchewan

Table 3.7 Location Quotient by Industry: 14 CMAs with the Highest Values in Descending Order, 2006

	Extraction			Manufacturing			Telecom., Media, Arts	
CMA	Location Quotient 2006	Change since 2001	CMA	Location Quotient 2006	Change since 2001	CMA	Location Quotient 2006	Change since 2001
Sudbury	5.29	−0.90	Windsor	1.95	−0.07	Regina	2.06	0.43
Calgary	4.66	0.00	Kitchener	1.84	0.02	Montreal	1.61	0.06
Edmonton	2.28	0.33	Sherbrooke	1.41	−0.13	Vancouver	1.52	−0.14
St John's	1.72	0.35	Trois-Rivières	1.29	0.05	Toronto	1.48	−0.01
Saskatoon	1.68	−0.11	Hamilton	1.28	−0.06	Saint John	1.41	0.08
Saint John	0.72	0.51	London	1.25	0.11	St John's	1.40	0.03
Regina	0.53	0.01	Oshawa	1.23	−0.09	Halifax	1.38	0.07
Thunder Bay	0.48	−0.20	Toronto	1.14	0.01	Calgary	1.15	−0.11
Saguenay	0.37	0.02	Montreal	1.14	−0.08	Winnipeg	1.11	0.07
Halifax	0.31	−0.12	Saguenay	1.13	−0.04	Ottawa	1.02	−0.15
Vancouver	0.27	0.07	St Catharines–Niagara	1.09	−0.13	Saskatoon	0.98	−0.01
Abbotsford	0.26	0.10	Abbotsford	1.01	0.12	Victoria	0.93	0.03
Hamilton	0.15	0.02	Winnipeg	0.94	−0.01	Oshawa	0.92	0.16
Windsor	0.14	−0.08	Quebec City	0.76	0.09	Hamilton	0.86	0.07

	Professional and Scientific Services			Finance, Insurance, and Real Estate (FIRE)	
CMA	Location Quotient 2006	Change since 2001	CMA	Location Quotient 2006	Change since 2001
Calgary	1.52	−0.02	Toronto	1.59	−0.01
Toronto	1.40	−0.08	Regina	1.44	−0.16
Ottawa	1.32	−0.20	Kitchener	1.32	0.05
Saint John	1.29	0.01	Vancouver	1.25	−0.05
Vancouver	1.27	0.01	Quebec City	1.19	0.07
Montreal	1.17	−0.03	London	1.18	−0.05
Halifax	1.15	−0.09	Oshawa	1.12	0.02
St John's	1.11	0.12	Halifax	1.11	−0.03
Victoria	1.08	0.02	Hamilton	1.10	0.00
Edmonton	1.07	−0.05	Montreal	1.07	0.03
Oshawa	0.96	0.04	Winnipeg	1.07	0.05
Quebec City	0.94	−0.04	Calgary	1.00	−0.09
London	0.93	0.05	Victoria	0.96	−0.02
Kitchener	0.92	0.05	Edmonton	0.88	−0.04

Table 3.8 Location Quotient by Occupational Class: 14 CMAs with the Highest Values in Descending Order, 2006

CMA	Managerial			CMA	Science & Engineering			CMA	Blue-Collar	
	Location Quotient 2006	Change since 2001			Location Quotient 2006	Change since 2001			Location Quotient 2006	Change since 2001
Ottawa	1.25	−0.04		Ottawa	1.80	−0.14		Abbotsford	1.45	0.14
Toronto	1.25	−0.03		Calgary	1.47	0.03		Windsor	1.20	−0.09
Calgary	1.23	0.00		Quebec City	1.26	−0.07		Kitchener	1.15	−0.04
Vancouver	1.18	−0.01		St John's	1.23	0.03		Oshawa	1.04	−0.04
Victoria	1.12	−0.02		Victoria	1.19	−0.05		St Catharines	1.03	−0.08
Halifax	1.10	−0.01		Toronto	1.18	−0.01		London	1.02	0.03
Regina	1.08	0.02		Vancouver	1.09	−0.01		Trois-Rivières	1.01	−0.02
Hamilton	1.05	0.01		Montreal	1.09	−0.01		Edmonton	0.99	0.07
Montreal	1.05	0.03		Halifax	1.08	−0.05		Hamilton	0.98	−0.03
St John's	1.03	0.01		Regina	1.07	0.00		Saguenay	0.98	−0.04
Oshawa	1.03	0.04		Edmonton	1.06	−0.03		Sherbrooke	0.97	−0.11
Edmonton	1.00	−0.01		Kingston	1.02	0.03		Saskatoon	0.95	0.05
Winnipeg	0.98	0.04		Kitchener	1.02	0.04		Sudbury	0.95	0.04
Kitchener	0.97	−0.06		Saint John	1.00	0.12		Thunder Bay	0.89	−0.06

economy, which—added to rising oil prices—fuelled the growth of the Saskatoon and (to a lesser extent) Regina economies.

Manufacturing cities are an entirely different lot. All metropolitan areas with manufacturing quotients above unity are in either southern Ontario or Quebec, except for Abbotsford, which matches the pattern of the middle-sized city within an hour's range of a large metropolis, Vancouver in this case, close to the US border. The seven cities with the highest quotients (above 1.20) are all close to Montreal or Toronto, or Detroit in the case of Windsor. They are well located for trade with the US industrial midwest and with the great cities of the US northeast. This remains Canada's industrial heartland, with Toronto and Montreal as the two anchors. However, Ontario and Quebec's industrial bases are not the same, tied to different regional markets (recall Table 2.4 in Chapter 2). Southern Ontario's manufacturing sector is heavily weighted toward the automobile industry, while manufacturing in southern Quebec is more diversified. In both provinces, the manufacturing in northern cities—outside the industrial heartland—is largely resource-dependent, generally in pulp and paper or smelting.

Manufacturing in the city of Saguenay (former Chicoutimi-Jonquière) is almost entirely concentrated in aluminum smelting—which, de facto, is a resource industry, founded on the area's hydroelectric potential. Smelting and logging towns are also common in other provinces—the Prairies excepted—but concentrated in urban areas below the CMA population threshold.

For the three high-order services, comparative urban specialization, as measured by location quotients, yields a more diversified picture, mainly because urban areas have developed niches in specific sub-sectors. Recalling an example from Chapter 2, the insurance industry is less sensitive to agglomeration economies than other financial services, allowing smaller urban areas such as London, Ontario, and Quebec City to register fairly high location quotients within the broader FIRE sector, although neither city comes remotely close to Toronto's mass (in terms of total employment) as a financial centre.

Professional and scientific services are, arguably, the most diverse class. The three top cities for this class—Calgary, Toronto, and Ottawa—are founded on different mixes of knowledge-intensive service functions. Half in jest, one is tempted to say that the three urban service sectors are respectively peopled by geologists, accountants, and lawyers, an oversimplification that conveys the essential message that the knowledge economy covers a wide array of possible urban specializations. Six of the 10 metropolitan areas with location quotients above unity are capital cities, and three of the remaining four have populations above 1 million (Montreal, Vancouver, and Calgary). Specialization in knowledge-intensive services is clearly helped by size and political links, with the latter often complemented by a large provincially funded university. On this score, the high quotient for Saint John is somewhat of a mystery.[6] The sharp decline in Ottawa's location quotient between 2001 and 2006 for professional and scientific services (−0.20) likely reflects the bursting IT bubble and subsequent fall of Nortel, the Ottawa-based telecommunications giant.

The telecommunications, media, and arts group illustrates yet again the possibility of niche specializations not irrevocably tied to size. The surprisingly high quotients for Regina and Saint John are partly attributable to the fact that each is home to the regional telecommunication company, respectively SaskTel and Bell Aliant.[7] Once this essentially technological function is discounted, the role of size, location, and language is again visible. The remaining metropolitan areas with quotients above 1.10 are Canada's six regional capitals, each one the cultural and media centre for its market, with its 'regional' newspapers, television studios, and entertainment industry. St John's is the cultural capital for Newfoundland and Labrador, as Vancouver is for BC. The role of language is brought home by Montreal's high quotient. Montreal, even more than Toronto, is the centre of a captive market—some 7 million French-speakers—for information, entertainment, and related services. The language split also ensures that the information and entertainment industries are not monopolized by a single dominant metropolis.

Cities also specialize in functions within industries (Table 3.8). For managerial and professional/scientific-type functions, being a capital city again is a definite advantage. With the exception of Winnipeg, all capital CMAs exhibit high quotients on the second category and most on the first as well (the two remaining provincial capitals, Fredericton and Charlottetown, are not CMAs). Winnipeg has a lower share of knowledge/white-collar-type workers than its position as a capital would suggest. Recall as well Winnipeg's low score on educational attainment (Table 2.9, Chapter 2) and its absence in Table 3.7 in the top 14 professional/scientific services class. In part, this explains its low growth rate among CMAs. Edmonton also scores lower on the same three attributes than one would normally expect of a large capital city that is also, like Winnipeg, home to a major university. Neither attribute, it appears, is sufficient to deliver sustained growth. Edmonton has the good fortune of being the capital of an oil-rich province, but its employment growth rate has nonetheless remained consistently below Calgary's (Figure 3.6).

Once the capital city factor is discounted, population size largely explains a city's propensity to attract managerial and professional/scientific-type functions. The big four—Toronto, Montreal, Vancouver, and Calgary—are the control/ corporate/ professional/ white-collar centres of the nation. It is also these four that most frequently show up in global rankings, generally in that order (see Chapter 2). Toronto's high rank needs little further explanation. Calgary's strength depends, it must be repeated, almost exclusively on the energy sector, which raises the question of its long-term role as a corporate service centre. Perhaps the most surprising finding is Montreal's comparatively low score both for management and professional/scientific functions. Part of the explanation rests with its historical legacy, going back to the early twentieth century, as a low-wage industrial city, rooted at the time in an abundant (francophone) labour force, fuelled by high birth rates. A more recent explanation is the flight of much of the old Anglo business elite from Montreal to Toronto during the 1960s, 1970s, and 1980s, taking their capital and head offices with them, from which Montreal has yet to totally recover. In sum, Montreal has an inherited urban size beyond its role as a corporate and business service centre, which in part explains its consistently slower growth since the mid-1960s (Polèse and Shearmur 2004a) and contributes to Toronto's unparalleled rapid growth over the same time period.

Blue-collar CMAs are generally mid-sized manufacturing cities, many of them in southern Ontario (Table 3.8). Following Abbotsford, which seems to be evolving into a working-class outlier of Vancouver, Windsor is Canada's most strongly blue-collar

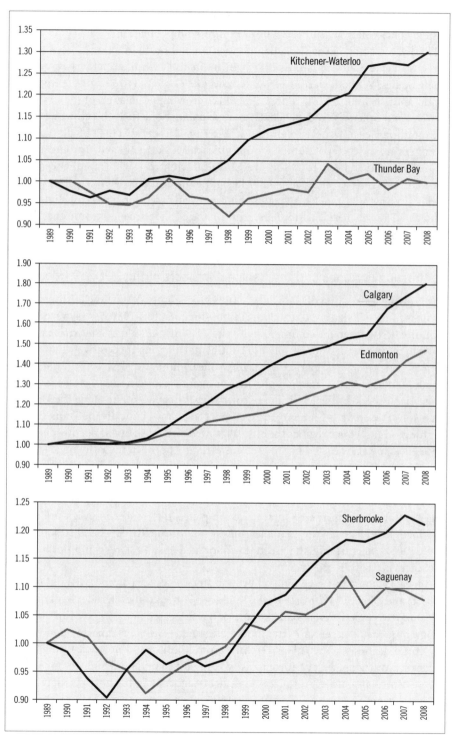

Figure 3.6 Employment 1989–2008: Six paired CMAs (1989 = 1.00).

CMA, based on the auto industry. Numerous mining and resource towns are also strongly blue-collar but lie below the CMA threshold population.

Finally, the aging of the Canadian population and resultant migration of retirees has introduced a new economic base, which affects CMAs like Victoria and several smaller cities. Without warm weather retreats like Florida, Arizona, or the Riviera to attract tourists and well-off retirees, Canada's growing population of seniors, semi-retirees, and otherwise independently mobile populations fuels the growth of small communities with natural amenities within easy driving distance of a large CMA. Examples are the cottage country communities in Muskoka north of Toronto and the Laurentian and Eastern Townships regions near Montreal, with rapidly expanding communities such as Ste Agathe and Magog. Kelowna also owes much of its growth to purely residential choices, helped in this case by an exceptionally warm climate (for Canada). The weather/ isolation/ residential choice factor accentuates another divide that affects growth—namely the north–south axis, to which we now turn.

Contrasting Fortunes and the Regional Impacts of the 2007–9 Recession

The growing integration of Canada into the US economy (despite the less-than-open border) since the signing of the Free Trade Agreement in 1989 has favoured communities best located for commerce with the US. Various studies confirm that proximity to the border and accessibility to continental markets are positively linked with employment growth (Apparicio et al. 2007; Shearmur and Polèse 2007; Shearmur et al. 2007). Within each Canadian province or region, some cities are better located for trade than others. In the Maritimes, growth tends to concentrate along the Halifax–Moncton–Fredericton–Edmundston transport corridor—basically the Trans-Canada Highway—but in most regions the critical growth differentiation is north–south. The differences are accentuated by the sectoral variation between the vulnerable logging, smelting, and mining towns—often found in northerly locations—and the more diversified urban economies in the south. The rift is most evident in Quebec and Ontario, but also in BC and to some extent Alberta (Fort McMurray notwithstanding). Southern and northern Ontario are completely different economic worlds, with the north exhibiting all the characteristics of a traditionally depressed region (Slack et al. 2003). Increasingly, the greatest economic disparities are found not between provinces but within provinces.

Figure 3.6 compares employment growth since 1989 for paired CMAs in Ontario, Alberta and Quebec.[8] Within Ontario, Kitchener-Waterloo and Thunder Bay are evolving along two very different paths. Thunder Bay is located in the heart of an immense geographic area—essentially the Canadian Shield—where stagnating and declining communities are the rule, stretching from Manitoba to northeastern Quebec. Kitchener-Waterloo, west of Toronto, is evolving into a mid- and high-tech manufacturing centre. In Quebec, resource-dependent Saguenay in the north is increasingly diverging from Sherbrooke. The latter typifies the advantages of a well-located mid-sized 'southern' city. Sherbrooke is close to the border with a direct highway link to Boston (about a four-hour drive), 130 kilometres from Montreal, and enjoys pleasant natural surroundings. Saguenay is not only further north—as far from Sherbrooke as the latter is from Boston and colder—but it is a high-wage economy, driven by the large, highly capitalized plants (smelters) typical of many resource-dependent communities, thus blocking economic diversification, a fundamental dilemma for many of Canada's northern communities.

The north–south divide is not a perfect predictor, however, which underlines the difficulty of explaining urban growth (and decline) in a country as geographically diverse, economically open, and vast as Canada. Within Saskatchewan, Saskatoon has been growing consistently faster than Regina (the capital) to the south. The explanation lies in part in different geological and soil conditions and in the overall distance of the province (whichever CMA is considered) from major markets in the US.

The global recession—still effective at the time of writing—unleashed by the 2007 subprime crisis in the US, provides a particularly salutary lesson of the dangers of predictions based on past trends and generalizations (Table 3.9). For the first time since the publication of annual unemployment data (1966), Toronto's unemployment rate is higher than Montreal's. Between 2006 and July 2009, the former gained 3.4 points, compared to 1.2 points in Montreal. Unemployment climbed 4.7 points in Kitchener-Waterloo but only 0.6 points in Sherbrooke. The hardest-hit cities are in southern Ontario, followed by cities in Alberta and BC, while those least affected are in Quebec, Atlantic Canada, Manitoba, and Saskatchewan, all historically slower-growing regions. The cities of southern Ontario

Table 3.9 Unemployment Rate, July 2009* and Change Since 2006** by CMA

Most Affected by Recession			Least Affected		
	Rate	Change		Rate	Change
Windsor	15.2	6.2	Montreal	9.6	1.2
London	10.9	4.7	Halifax	6.1	1.1
Kitchener-Waterloo	9.9	4.7	Saguenay	9.8	1.0
Abbotsford	9.0	4.5	Kingston	7.2	1.0
St Catharines	10.5	4.1	Thunder Bay	8.5	1.0
Calgary	6.9	3.7	Ottawa	5.9	0.7
Toronto	10.0	3.4	Winnipeg	5.3	0.7
Oshawa	9.7	3.2	Sherbrooke	8.5	0.6
Edmonton	7.0	3.1	Saskatoon	4.7	0.3
Vancouver	7.0	2.6	Trois-Rivières	8.3	0.2
Victoria	6.1	2.4	St John's	8.1	0.0
Sudbury	9.5	2.3	Quebec City	4.8	−0.4
Hamilton	8.2	2.3	Saint John	5.0	−1.1
			Regina	3.2	−1.7

*3 month rolling average/deseasonalized.

** Annual average.

are paying the price of their dependence on one industry (automobiles) and the US market. Does this mean that the manufacturing cities of southern Ontario are doomed to stagnate on the model of the cities of the US Rustbelt? Such a prediction is surely premature. Toronto is unlikely to become another Cleveland.

The fundamental forces driving Canada's economic geography have not been revoked by the 2008–10 recession. The structural transformation of the Canadian economy and the weight of agglomeration economies will continue to fuel the growth of large metropolitan areas to the benefit of Toronto, Montreal, and others. Vancouver's favourable location on the Pacific—opposite the world's most dynamic economies—will continue to bolster its growth. Mid-sized cities, well-located both for international trade and for interaction with large metropolitan centres, will continue to attract manufacturing employment. Colder and more distant communities, tied to the 'wrong' resources, will continue to face difficult challenges. However, the vulnerability of Canada's urban economies to outside forces—plus the volatility and diversity of those forces—means that even the most solid predictions can be overturned by unforeseen events.

The Demographics of Urban Growth

In Canada, demographic processes have long played an important role in directly shaping the amount and location of urban development. Waves of immigrants (and emigrants, during the 1930s) have produced significant fluctuations in the national rate of population growth and urbanization from decade to decade. But these processes generate important regional variations as well. Since recent immigrants may have little attachment to place, they provide a particularly volatile source of population growth, responding quickly to new economic opportunities in frontier communities or large cities. But Canadians themselves are also continually on the move, and in certain times and under certain conditions as many as 400,000 persons (gross) or 80,000 (net) per year may relocate from one province to another. And while the regional variations in the rate of natural increase have subsided in recent years, 50 years ago the difference in birth rates between the big urban centres and the rural areas in the Atlantic region or Quebec was substantial.

The strategy here is to describe separately the three components of population change—natural increase, internal migration, and immigration/emigration—and then to discuss the interaction among the three processes and their relationships to cities and their economies. It will be seen that there is a basic congruence between the economic and demographic processes that operate at this spatial scale, notably in the relation between urban growth and population size, but it should be emphasized that the temporal patterns of growth are very different for the two kinds of process, at least at this scale. Urban populations change gradually, so an increase of 10 per cent over a decade is dramatic, but city employment can change by 10 per cent within a year. Economic phenomena are much more volatile (Simmons, Bourne, and Cantos 2005).

Over the past 35 years, Canada's annual growth through natural increase of population has remained within a fairly narrow range—from 200,000 declining to 110,000—with a maximum value in about 1990 and gradual reduction thereafter. Net immigration fluctuates more widely, from 25,000 in the years 1978 and 1984 to 220,000 in 1993, with higher growth and greater consistency over the past 15 years. The joint effect of the two processes is an overall national growth in population of about 400,000 per year in recent years, equivalent to about 1.3 per cent of the population annually.

These demographic growth processes are closely linked to the age of individuals (Figure 3.7). The births and deaths that determine natural increase are linked to the reproductive cycle, with highly predictable results. Migration, too, is highly age-dependent, with mobility concentrated between ages 15 to 35, coinciding with the major lifecycle changes: e.g., the pursuit of higher education, employment, marriage, child-raising. A long period of stability follows until retirement age introduces another lifecycle sequence that may include retirement, sickness, or spousal loss, which, in turn, triggers more moves—but often over short distances. Immigration is less markedly age-dependent, but the peak rate of movement at age 30 is three to four times higher than it is in later years.

Fortunately, the abruptness of these age-dependent processes is mitigated by the relative lack of variability in the age structure among Canada's urban areas. The city with the oldest average age is the retirement community of Parksville, BC (51.0 years), while the youngest is Thompson, MB (30.0 years), so the spatial variation in rates of natural increase or migration is less than one might assume. As well, age structure primarily influences the out-migrant rate, so migration may be more important in determining age structure than the reverse. In-migrants are more spatially concentrated as they seek jobs or housing or lifestyle amenities, thus supporting the highly structured economic

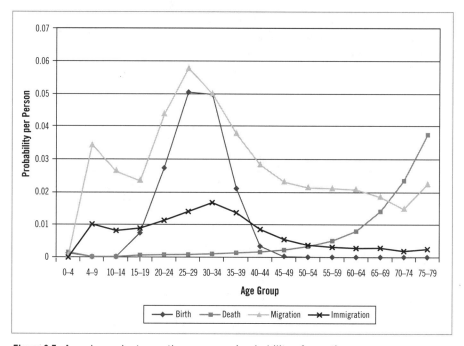

Figure 3.7 Age-dependent growth processes (probability of event).

(Source: Statistics Canada. *Annual Demographic Statistics*. Catalogue no. 91-513. Ottawa: Statistics Canada)

growth patterns observed earlier. The overall effect of age-dependent demographic processes adds a modest level of natural increase to most cities while generating a continuing stream of out-migrants that must be replaced if the city is to maintain its population, economy, and level of services.

Natural Increase

As indicated above, the rate of natural increase in Canada is in decline, with lower birth rates and higher death rates as the average age of the population continues to rise. Natural increase contributes smaller and smaller shares to Canada's population and labour force growth so that in most parts of the country, natural increase, in the absence of either net migration or immigration, generates very low or negative levels of population growth.

The geographical distribution of natural increase is inexorably linked to the age distribution of the population and thus to the recent demographic history of each community. Cities with high levels of recent growth, such as cities in Alberta (high economic growth), attract younger people and thus display the highest rates of natural increase (see Table 3.10). At the same time, the very largest cities are characterized by lower rates of natural increase, presumably because of lifestyle factors and partly because of the loss of young households to nearby urban areas, those overflow cities located close enough to larger urban centres to attract commuter households—places like Barrie, ON, Okotoks, AB, and Abbotsford, BC. Retirement communities in British Columbia and across the country may have zero or negative rates of natural increase. Of the 144 urban areas in Canada, 22 (15 per cent) registered negative natural increase. The range of values is significant—from −7 per cent (Elliot Lake, ON) to +20 per cent (Okotoks, AB) over five years—especially for the provincial governments that have the responsibility to provide education and health care.

Table 3.10 Natural Increase by City Size and Region, 2001–6

City Size	BC	AB	EP	ON	QC	ATL	Total
	(Per Cent of Initial Population)						
Over 1,000,000	−0.47	4.55	–	1.29	0.85	–	**1.38**
300,000–1,000,000	1.71	–	0.91	1.74	2.74	3.08	**1.87**
100,000–300,000	2.02	–	3.90	2.92	2.64	2.72	**2.89**
30,000–100,000	1.84	12.37	3.25	2.10	3.33	2.44	**3.64**
10,000–30,000	1.40	8.35	2.23	1.87	2.79	0.97	**2.48**
Total	**0.47**	**5.80**	**2.18**	**1.64**	**1.62**	**2.55**	**2.87**

BC indicates British Columbia; AB, Alberta; EP, Saskatchewan and Manitoba; ON, Ontario; QC, Quebec; and ATL, the Atlantic region.

Source: Statistics Canada. *Census of Canada, 2006*. Ottawa: Statistics Canada.

Internal Migration

While natural increase is largely a demographic phenomenon, the population changes resulting from net migration within Canada are most closely linked to changes in economic opportunity. Domestic migration accelerates during times of prosperity (e.g., 1975, 1997) and slows down during recessions (1984). Depending on the annual fluctuations in the geography of job creation, the spatial impacts of migration can be almost neutral if the flows into a province offset the outflows, or they may become spectacularly redistributive when the net flows are large and link different regions of the country, as in the case of flows to Alberta in the 1980s or during the most recent five years.

While the level of interprovincial net migration is the migration process that is most widely analyzed, the continuing flows of individuals up the urban hierarchy within each province or region are more substantial and have been of greater significance for the urban system in recent years (Table 3.11 and Figure 3.8). The range of values and the impacts of net migration are much greater than those for natural increase. For example, Okotoks, near Calgary, attracted net migration equivalent to 22.8 per cent of the initial population over the past five years, while isolated Kitimat, BC, lost 12.2 per cent of the 2001 population through net out-migration. Both regional and city-size variations are apparent in Table 3.11, but first note the substantially higher levels of migration—both in and out—for smaller communities.

The economic problems behind the north–south differences were discussed earlier, but these places also lose migrants because they cannot provide the variety of lifestyle choices with respect to education, jobs, or partners that are found in larger centres, even though the out-migrants may return at a later date. Still, the potential for decline is substantial. In the absence of in-migration, a small urban centre can expect to lose as much as 10 per cent of its population over five years.

Fortunately for smaller places, in-migration shows a similar pattern of variation with city size so that inflows largely compensate for the out-movement. The place-to-place variation is greater for smaller urban places, however, because migrants are attracted to a small number of high opportunity locations. Note the attraction of smaller centres in British Columbia or Alberta compared to places in the eastern Prairies or the Atlantic region. At the same time, there is a strong regional pattern to all migration rates in that urban areas to the east of Montreal have lower turnover rates in general.

The variation among migration destinations accounts for most of the differences in net migration, as shown in the final panel of the table and as mapped in Figure 3.8. The result is widespread population loss due to out-migration, while the gains are concentrated in a small number of regional centres within the growing regions. Alberta's substantial migration growth contrasts with the declines for cities of all sizes in the eastern Prairies and the Atlantic region (although recent migration flows have reversed some of this decline). For the most part, cities across the northern periphery are losing migrants. British Columbia, Ontario, and Quebec display diverse results, based on events in particular cities and on the growth of regional economies. Although larger cities generally demonstrate more growth through net migration than the smallest places, the very largest metropolitan centres—Toronto,

Table 3.11 Net Internal Migration by City Size and Region, 2001–6

| | **(Per Cent of Initial Population)** | | | | | | |
| | **Out-Migration to CMAS, CAS** | | | | | | |
City Size	**BC**	**AB**	**EP**	**ON**	**QC**	**ATL**	**Total**
Over 1,000,000	5.30	6.24	–	5.05	3.09	–	**4.75**
300,000–1,000,000	9.20	–	5.07	6.93	4.97	8.08	**6.60**
100,000–300,000	11.15	–	8.94	8.96	7.60	6.30	**8.43**
30,000–100,000	11.33	11.71	10.49	8.26	7.37	8.00	**9.13**
10,000–30,000	13.17	16.10	10.79	11.76	8.48	6.77	**11.11**
Total	**7.77**	**7.44**	**7.20**	**6.27**	**4.94**	**7.18**	**9.63**
	In-migration to CMAS, CAS						
City Size	**BC**	**AB**	**EP**	**ON**	**QC**	**ATL**	**Total**
Over 1,000,000	4.66	8.84	–	3.94	2.80	–	**4.46**
300,000–1,000,000	11.40	–	3.26	8.25	5.22	7.40	**7.20**
100,000–300,000	15.21	–	5.95	10.68	6.26	5.37	**8.67**
30,000–100,000	13.93	14.04	7.01	8.51	7.68	6.70	**9.80**
10,000–30,000	11.41	17.37	7.11	13.29	5.16	4.94	**9.76**
Total	**8.22**	**9.94**	**4.75**	**6.19**	**4.09**	**6.14**	**9.48**
	Net Migration to CMAS, CAS						
City Size	**BC**	**AB**	**EP**	**ON**	**QC**	**ATL**	**Total**
Over 1,000,000	–0.64	2.60	–	–1.11	–0.29	–	**–0.29**
300,000–1,000,000	2.20	–	–1.81	1.32	0.24	–0.68	**0.59**
100,000–300,000	4.06	–	–2.89	1.73	–1.34	–0.93	**0.24**
30,000–100,000	2.59	2.33	–3.48	0.25	0.31	–1.30	**0.67**
10,000–30,000	–1.77	1.27	–3.67	1.53	–3.32	–1.84	**–1.34**
Total	**0.45**	**2.50**	**–2.45**	**–0.08**	**–0.35**	**–1.04**	**–0.21**

BC indicates British Columbia; AB, Alberta; EP, Saskatchewan and Manitoba; ON, Ontario; QC, Quebec; and ATL, the Atlantic region.

Source: Statistics Canada. *Census of Canada, 2006*. Ottawa: Statistics Canada.

Montreal, and Vancouver—lose population to nearby smaller cities through suburban overspill.

The systematic flow of migrants from smaller cities to larger centres provides useful insights into the organization of the urban system in the sense of defining regional relationships among the various cities. These linkages are shown in Figure 3.9 as the 'largest outflow' from each urban centre. What is the single most likely destination for an out-migrant from a particular city? The map reveals a dozen or more

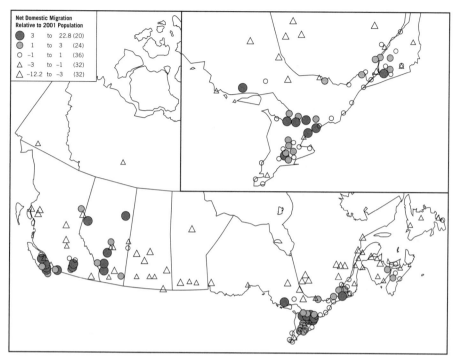

Figure 3.8 Net domestic migration, 2001–6.

(Source: Statistics Canada. *Census of Canada, 2006*. Ottawa: Statistics Canada)

regional centres that serve as the primary destinations for smaller places nearby. By implication, the regions defined by these flows represent the trade or service areas of the regional centres, while zones of competition separate the trade areas. Toronto, Montreal, and Vancouver are the most important regional centres, with Toronto playing a national role as the destination choice for residents moving from smaller regional centres. Calgary and Edmonton share Alberta, and Saskatoon/Regina serve Saskatchewan as Winnipeg serves Manitoba. Ottawa is the regional centre for eastern Ontario; Quebec City plays the same role in eastern Quebec. Halifax and Moncton serve the Maritimes, and St John's is the destination of choice for most migrants in Newfoundland. These patterns reflect the city size and distance tradeoffs that are still relevant within the urban system. Exceptions occur in the two western provinces of Alberta and Saskatchewan, where the provincial governments maintain an uneasy balance between two competing and growing urban centres of roughly the same size. The structure of the national urban hierarchy is evident in the relationships between the regional centres and the largest places such as Toronto, Montreal, and Calgary.

Immigration

Net population growth through natural increase in Canada is currently about 120,000 per year, or 600,000 over five years, while migration within Canada redistributed

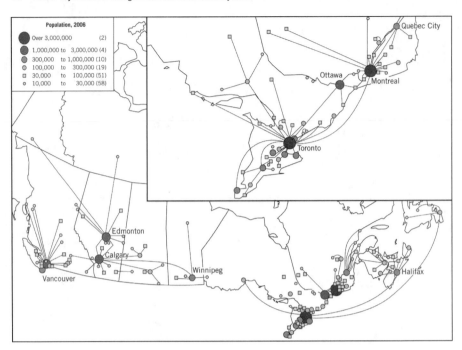

Figure 3.9 Migration: Largest outflows, 2001–6.

(Source: Statistics Canada. *Census of Canada, 2006*. Ottawa: Statistics Canada)

about 190,000 people (from cities losing to cities gaining) for all urban areas combined over the same five years. Over the same period, however, net immigration added some 200,000 new residents of Canada each year for a total of almost 1 million, thus becoming the most important demographic growth process for Canadian cities and the source of most of the growth in labour force. Thus, immigration, the most policy-driven and politicized kind of population growth, has become the main source of Canadian urban growth. The federal government sets targets and defines immigration selection criteria and staffs Canadian immigration offices around the world. Quebec has chosen to operate its own immigration agency in order to focus on francophone immigrants. Manitoba and some other provinces have also become involved in sponsoring immigration. Provincial and municipal agencies are also responsible for providing a variety of services, housing, and employment for immigrants. As pointed out in Chapter 2, part of the Canadian enthusiasm for immigrants derives from the presence of the 6 million foreign-born Canadians who have a strong interest in the welfare of their compatriots.

The annual level of immigration has varied widely, ranging from 100,000 to 250,000 per year until the intake began to stabilize in the 1990s. With emigration steady at 50,000 per year, the net contribution is now about 200,000 per year. Although the number of immigrants is smaller than the total number of domestic migrants annually, the impact of the movement is much greater because the reverse flows (the emigrants) are relatively few. Immigrants tend to stay in Canada, especially in the 'gateway' cities.

Domestic migration, in contrast, exhibits substantial reverse flows so that the amount of 'net' migration is relatively small.

Table 3.12 and Figure 3.10 show the demographic effect of immigration on the urban system. The table suggests that the impact of immigration is further intensified by the concentration in a small number of destinations—Canada's largest cities—adding some 6.8 per cent to their growth rates in aggregate while contributing less than 1 per cent to the smallest centres. Large cities have a long tradition as immigrant destinations, with institutions and compatriots to assist in the transition, and they provide a wide range of economic and educational opportunities that attract native-born Canadians as well. Immigrants with less education and fewer language skills are especially attracted to communities with previous immigrants, while expectations of economic opportunity may be more important for professionals.

The regional variation in the scale of immigration is also substantial, ranging from less than 1 per cent in the Atlantic region to more than 5 per cent in British Columbia. It reflects differences in economic opportunity, the presence of earlier migrants, and the historical traditions of the region. British Columbia has become a magnet for Asian immigrants, while the Atlantic region and Quebec (outside Montreal) have relatively few foreign-born concentrations. The map in Figure 3.10 indicates that many smaller places in Atlantic Canada and the northern parts of Ontario and Quebec attract almost no immigrants, while in Vancouver, Toronto, and Calgary (and other Alberta cities), immigrants add the equivalent of 5 per cent of population over five years.

The Interrelationships

The analysis reported above underlines the variety of ways in which a city's population can grow (or decline). Growth may vary by city size and region, as well as in response to changes in the local economy or age structure. The growth of the very largest cities is driven by massive immigration, although the levels of net domestic migration and natural increase are only modest, since these cities tend to lose retirees to small towns and young families to adjacent urban areas. Medium-sized cities, in contrast, attract fewer immigrants but absorb net migrants from the larger places and from smaller communities nearby, while the influx of young migrants stimulates reproduction and natural increase. The smallest cities lose population through out-migration (with a few exceptions), but they retain high rates of natural increase.

Regionally, cities in the Atlantic region have difficulty attracting and holding immigrants and lose domestic migrants to Ontario and the west but maintain relatively high rates of natural increase. Aside from Montreal, immigration to Quebec cities is weak, domestic migration tends to be negative, and even the rates of natural increase are below average, despite recent gains. Ontario cities gain substantial numbers of immigrants and are approximately neutral in terms of domestic migration, but natural increase rates are relatively low despite the immigration. The eastern Prairies lag in every aspect of demographic growth: low immigration, substantial out-migration, and relatively low rates of natural increase. Alberta, in contrast, scores highly on each measure. Recent economic growth in Saskatchewan may bring the two regions into balance. British Columbia is a study in contrasts, including both the highest- and lowest-growth

Table 3.12 Net Immigration by City Size and Region, 2001–6

City Size	BC	AB	EP	ON	QC	ATL	Total
(Per Cent of Initial Population)							
Over 1,000,000	7.63	4.76	–	8.40	4.79	–	6.80
300,000–1,000,000	1.92	–	3.56	3.06	1.23	1.41	2.65
100,000–300,000	2.79	–	1.42	1.23	1.20	0.56	1.29
30,000–100,000	0.943	1.95	1.17	0.81	0.68	0.84	0.95
10,000–30,000	1.11	1.99	0.54	0.44	0.19	0.36	0.73
Total	5.09	4.24	2.44	5.65	3.36	0.82	4.48

BC indicates British Columbia; AB, Alberta; EP, Saskatchewan and Manitoba; ON, Ontario; QC, Quebec; and ATL, the Atlantic region.

Source: Statistics Canada. *Census of Canada, 2006*. Ottawa: Statistics Canada.

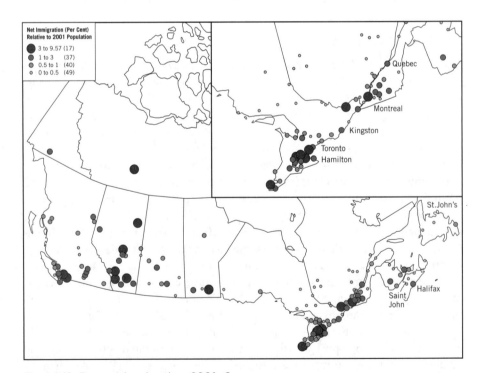

Figure 3.10 Per cent immigration, 2001–6.

(Source: Statistics Canada. *Census of Canada, 2006*. Ottawa: Statistics Canada)

communities, some cities that grow through immigration (Vancouver) and others that grow through retirement (Parksville), while others are economic boomtowns (Fort St John and Kelowna) or declining resource communities (Kitimat).

The correlations shown in Table 3.13 reflect the complexity of the growth processes. The overall growth rate is modestly related to city size and dependent on all three growth processes, most closely with net migration and natural increase and to a lesser extent with immigration. Although the first two processes are widespread across the country, immigration is largely concentrated in a small number of cities. The link between natural increase and net migration occurs through the age structure: migration, on balance, brings in young people of reproductive age. The age structure of immigrants, however, is dispersed over a wider range of age groups. Note that net immigration is only modestly related to net migration. These processes affect different populations with different origins, information, and objectives. Finally, net migration is primarily dependent on the in-migration decisions. Moving out of a community is a universal process, but the choice of destination is more selective.

Table 3.14 lists the urban areas that rank highest and lowest with respect to each demographic growth process. Except for immigration, almost all of the places with extreme values are smaller centres, which tend to be more susceptible to rapid change. Each growth process identifies a different set of locations that gain or lose population; thus, each map is different. Despite these variations over time and space, demographic growth processes are far more stable over time than economic growth and mitigate much of the volatility in levels of production, employment, and income. Once settled, urban residents do not relocate easily, especially as they age and/or if they live in big cities.

Reconciling Demographic and Economic Growth

It is difficult to generalize about the relationships between demographic and economic growth, since so many of the old verities have been overwhelmed. As noted above, overall economic growth no longer depends on an identifiable economic base in the sense of jobs within primary or secondary sectors. Instead, analysts deal with the place-to-place variations in the growth of various business services and the role of consumer and government spending within the various market areas. At the same time, immigration now dominates the demographic growth processes, while domestic migration—traditionally driven by employment opportunities—affects fewer and fewer persons each year. Local economic forces have less impact on immigration now that the selection process is largely bureaucratic. While these relationships—or lack of relationships—are true for Canadian cities in general, it is possible to disaggregate the patterns for large and small cities and observe different results.

The Big Cities

Immigration has now overtaken natural increase as the principal source of big-city growth, and while neither process is particularly responsive to local economic conditions, the impact of the former is widely dispersed in space, while the latter is highly concentrated. The demographic growth process accumulates large

Table 3.13 Growth Parameters, 2001–6*

Variable	Mean	Standard Deviation	Correlation Coefficients						
			1.	2.	3.	4.	5.	6.	7.
1. LogPop01	4.70	0.59	1.000						
2. Growth Rate	3.89	6.63	0.211	1.000					
3. Nat. Incr. (%)	2.87	3.68	-0.039	0.755	1.000				
4. Net Imm.(%)	1.22	1.46	0.625	0.440	0.190	1.000			
5. Net Migr.(%)	-0.21	4.04	0.156	0.795	0.261	0.187	1.000		
6. Urban In (%)	9.48	5.45	-0.227	0.694	0.422	0.117	0.712	1.000	
7. Urban Out (%)	9.63	3.90	-0.487	0.149	0.326	-0.030	-0.040	0.673	1.000
8. GR Employ't	8.36	6.34	0.224	0.841	0.613	0.239	0.634	0.548	0.549
9. GR Market	4.52	9.29	0.136	0.807	0.699	0.275	0.485	0.546	0.797
10. GR Inccap	0.87	5.42	-0.116	0.290	0.412	-0.001	0.063	0.298	0.318

1. Log 10 of the 2001 Population.
2. Population Growth Rate, 2001–6
3. Natural Increase Rate as (#2—#4—#5)
4. Immigration Rate, 2001–6
5. Domestic In-migration Rate—Domestic Out-migration Rate
6. Domestic In-migration Rate
7. Domestic Out-migration Rate
8. Growth Rate Employment, 2001–6
9. Growth Rate in Market Income, 2001–6
10. Growth Rate of Income per Capita
*144 CMAs and CAs

Source: Statistics Canada. *Census of Canada, 2006.* Ottawa: Statistics Canada.

Table 3.14 Demographic Winners and Losers, 2001–6

Cities	Growth Rate	Natural Increase	Net Domestic Migration	Net Immigration
1. Okotoks	46.7	Okotoks	Okotoks	Toronto
2. Wood Buffalo	23.6	Grande Prairie	Parksville	Vancouver
3. Grande Prairie	22.3	Red Deer	Barrie	Calgary
4. Red Deer	22.0	Wood Buffalo	Chilliwack	Montreal
5. Barrie	19.3	Lloydminster	Port Hope	Brooks
6. Calgary	13.4	Yellowknife	Oshawa	Wood Buffalo
7. Yellowknife	13.1	Lethbridge	Collingwood	Windsor
8. Lloydminster	12.8	Camrose	Kelowna	Kitchener
9. Oshawa	11.6	Rivière-de-Loup	Kawartha Lakes	Squamish
10. Canmore	11.6	Medicine Hat	Courtenay	Abbotsford
. . .				
135. Bathurst	–3.4	Edmundston	Elliot Lake	Valleyfield
136. Elliot Lake	–3.4	Campbellton	Rouyn	Cowansville
137. Kenora	–4.2	Kitimat	Terrace	Amos
138. North Battleford	–4.4	Quesnel	Yellowknife	Port Alberni
139. Campbellton	–5.0	Port Alberni	Gander	Elliot Lake
140. Williams Lake	–5.1	Port Hope	Thompson	Temiskaming
141. Terrace	–7.0	Parksville	Baie-Comeau	La Tuque
142. Quesnel	–8.1	Williams Lake	Kitimat	Kenora
143. Prince Rupert	–12.5	Prince Rupert	Labrador City	Shawinigan
144. Kitimat	–12.6	Elliot Lake	Prince Rupert	Bay Roberts

numbers of people in a small set of urban places. As Table 3.15 suggests, the six largest urban areas that include at least 1 million people, with 55 per cent of the population, attract two-thirds of the population growth, including 82.5 per cent of the immigrants, while the aggregate level of natural increase is modest and the net domestic migration is negative. Clearly, the traditional relationship between income and job creation and domestic in-migration is no longer the central theme in this growth story at present. Urban growth in these places largely depends on the jobs and income that have been created for and by immigrants. The share of growth in employment and aggregate income in these cities is less than their share of population growth—despite the contribution of Calgary. At the same time, the largest cities attract disproportionate numbers of skilled professionals and degree-holders in general, as demonstrated in recent studies (Beckstead et al. 2008). As noted earlier in this chapter, it is the quality of the jobs and the workforce that drives the growth of the large cities.

Table 3.15 The Six Largest Cities

	Population, 2006	Change, 2001–6	Natural Increase	Net Immigration	Net Migration	Change in Jobs	Change in Market ($m)
Toronto	5113	430	55	448	–73	214	6206
Montreal	3636	185	29	165	–10	145	2917
Vancouver	2117	130	–9	152	–13	109	1742
Ottawa	1131	63	19	35	9	38	328
Calgary	1079	128	41	58	29	92	8745
Edmonton	1035	97	45	32	20	80	5242
Total	14,110	1032	180	890	–38	678	25,252
Per Cent of Total	55.0	66.7	38.5	82.5	–	58.7	63.5
The 138 Remaining Cities	11,521	515	288	189	38 (232*)	477	14,528
Per Cent of Total	45.0	33.3	61.5	17.5	–	41.3	37.5
All 144 Cities	25,632	1547	468	890	0	1155	39,750

All values in 1000s.

* The absolute value of net migration for all cities.

Source: Statistics Canada. *Census of Population, 2006*. Ottawa: Statistics Canada.

Simmons and Bourne (2007) suggest that an expanded definition of the largest cities (the mega urban regions) would include some of the most rapidly growing smaller centres and could partially redress the apparent patterns of net out-migration and low rates of natural increase in those regions. As an example, most of the out-migration from Toronto goes to nearby CMAs and smaller cities.

The Remaining Cities

At the same time, the 138 Canadian cities with fewer than 1 million inhabitants retain many of the earlier patterns of interdependence between economic change and demographic growth. Natural increase is still more important than immigration, and net migration continues to play a substantial role in population growth. The correlations for this set of cities (and for the full set of cities, see Table 3.13) indicate that both net migration and natural increase are more important than immigration in determining population growth. Net migration, immigration, and natural increase are only weakly correlated, and net migration depends largely on the level of in-migration, since out-migration is so universal. From an economic point of view, population growth is closely correlated with the growth of employment, the labour market, and the market (total income) but less so with the growth of income per capita. Each of the demographic growth processes has slightly differing relationships with parallel economic processes: natural increase is most closely associated with the growth of the market. Immigration is only weakly linked to any of these measures. Net migration is closely tied to employment growth but with no relation to changes in income per capita.

References

Apparicio P., G. Dussault, M. Polèse, and R. Shearmur. 2007. 'Transport infrastructures and local economic development: A study of the relationship between continental accessibility and employment growth in Canadian communities'. Study funded by Infrastructure Canada. INRS-UCS. http://www.ucs.inrs.ca.

Beckstead, D.M., M. Brown, and B. Newbold. 2008. *Cities and Growth: In Situ versus Migratory Human Capital Growth.* Canadian Economy in Transition Series no. 019. Ottawa: Statistics Canada.

Brunelle, C., and M. Polèse. 2008. 'Functional specialization across space: A case study of the Canadian electricity industry, 1971–2001'. *The Canadian Geographer* 52 (4): 486–504.

Newbold, K. Bruce. 1996. 'Internal migration of the foreign-born in Canada'. *International Migration Review* 30 (3): 728–47.

Newbold, K. Bruce, and M. Bell. 2001. 'Return and onwards: Migration in Canada and Australia: Evidence from fixed interval data'. *International Migration Review* 35 (4): 1157–84.

Polèse, M., and R. Shearmur. 2004a. 'Culture, language and the location of high-order service functions: The case of Montreal and Toronto'. *Economic Geography* 80 (4): 329–50.

———. 2004b. 'Is distance really dead? Comparing industrial location patterns over time in Canada'. *International Regional Science Review* 27 (4): 1–27.

———. 2006. 'Growth and location of economic activity: The spatial dynamics of industries in Canada, 1971–2001'. *Growth and Change* 37 (3): 362–95.

Shaw, R. Paul. 1985. *Intermetropolitan Migration in Canada: Changing Determinants over Three Decades.* Toronto: New Canada Publications.

Shearmur, R. 2010. 'Scale distance and embeddedness: Knowledge intensive business service location and growth in Canada'. In D. Doloreux, M. Freel, and R. Shearmur, eds, *Knowledge Intensive Business Services: Geography and Innovation.* Farnham, UK: Ashgate.

Shearmur, R., P. Apparicio, P. Lizion, and M. Polèse. 2007. 'Space, time and local employment growth: An application of spatial regression analysis'. *Growth and Change* 38 (4): 691–717.

Shearmur, R., and D. Doloreux. 2008. 'Urban hierarchy or local milieu? High order producer service and (or) knowledge-intensive business service location in Canada, 1991–2001'. *Professional Geographer* 60 (3): 333–55.

Shearmur, R., and M. Polèse. 2007. 'Do local factors explain local employment growth? Evidence from Canada, 1971–2001'. *Regional Studies* 41 (4): 453–71.

Shearmur, R., M. Polèse, and P. Apparicio. 2007. 'The evolving impact of continental accessibility on local employment growth, 1971–2001'. Working Paper no. 2007-04, INRS-UCS. http://www.ucs.inrs.ca.

Simmons, J., and L.S. Bourne. 2003. 'The Canadian urban system, 1971–2001: Responses to a changing world'. Research Paper no. 200. Toronto: Centre for Urban and Community Studies, University of Toronto.

———. 2004. 'Urban growth in Canada, 1971–2001: Explanations and interpretations'. Research Paper no. 201. Toronto: Centre for Urban and Community Studies, University of Toronto.

———. 2007. 'Living with population growth and decline'. *Plan Canada* 47 (2): 13–21.

Simmons, J., L.S. Bourne, and J. Cantos. 2005. 'How cities grow: A study of time series data for Canadian cities'. Research Paper no. 202. Toronto: Centre for Urban and Community Studies, University of Toronto.

Slack, E., L. Bourne, and M. Gertler. 2003. 'Small, rural, and remote communities: The anatomy of risk'. Research Paper Series, RP (18), Panel on the Role of Government. Toronto: Government of Ontario.

Statistics Canada. Various years. *Census of Canada*. Ottawa: Statistics Canada.

———. 2008. *Labour Force Historical Review 2008*. Catalogue no. 71F004XCB. CD Format.

Notes

1. Source: Statistics Canada. 2008. *Labour Force Historical Review 2008*. Catalogue no. 71F004XCB. Ottawa: Statistics Canada.

2. Detailed analysis by industry prior to 2001 is more complicated because of the introduction in 1997 of NAICS (North American Industrial Classification System).

3. The five industry groups in Table 3.5 are composites of NAICS classes. The extraction group covers mining and oil and gas extraction. The other groups are largely self-explanatory. Detail is available upon request from the authors. Tables 3.5 to 3.8 and Figures 3.2 to 3.5 are based on special Statistics Canada tabulations from the 2001 and 2006 censuses for the Spatial Analysis and Regional Economics Laboratory housed at INRS in Montreal (SAREL: http://laser.ucs.inrs.ca/EN/Home.html).

4. The location quotient is a standard measure of relative concentration. In this case, the per cent

of employment in the given industry class (say, manufacturing) in a place is compared to the national average. A result above 1.0 means that the place has a higher share of employment in the industry than the national average.

5. For reasons of space, not all the information in Tables 3.5 and 3.6 is reproduced in graph form. We chose four cases to illustrate typical relationships.

6. A possible explanation is the Point Lepreau nuclear generating station, which like all such facilities relies on engineering and scientific services. Within a relatively small economy such as that of Saint John, a single firm or establishment can significantly affect the value of the location quotient, one of the drawbacks of this measure when applied to small economies.

7. We have no explanation, however, for the sharp increase in Regina's quotient for this industry class between 2001 and 2006.

8. Source: Statistics Canada 2008.

4 Political Economy, Governance, and Urban Policy in Canada

Jim Simmons, Larry S. Bourne, Tom Hutton, and Richard Shearmur

At this point in Canadian history, there can be no doubt that the progress and future of our cities is closely linked with the fortunes and actions of government. Cities are in one sense the creation of governments, or at least of collective action. Once created, the growth of cities in turn shapes governments, although often with a considerable time lag. Specifically, the immense population size and density of modern cities, and the disparities of wealth within them, require an extraordinary variety of regulations, interventions, and public investments: these are necessary to impose order, manage development, build roads and pipes and other infrastructure, maintain schools and hospitals, provide social services (including fire and police), and develop parks, other amenities, and convention centres. The character of both cities and governments reflect the overall level of economic growth and technological change at any given time, but the intensity of human contact in daily urban life also leads to a variety of conflicts and externalities (or spillover effects) that invite further government intervention.

A particular problem has arisen in recent years as urban economies in most developed countries, Canada included, have become more specialized while the demographics of cities have become more dependent on immigration. If the temporal and spatial fluctuations of the country's economy and its demography are out of phase with each other, the consequence may be insufficient jobs in some regions and insufficient workers at other locations. For all the reasons above, the increasing level of urbanization in developed countries has been accompanied by the increased role of governments, directly or indirectly, in urban development, and regardless of differences in political ideology.

But it is also apparent from earlier chapters that the development paths of major cities are often the accidental by-products of national (or provincial) policies concerning other issues, such as trade or immigration or transportation—as will be demonstrated in the following case studies. For example, Montreal's economy is in part the result of political decisions by the provincial government. Ontario's (and Toronto's) recent boom and subsequent slowing growth is closely linked to the sequence of decisions that led to further integration of the markets of Canada with the United States. Calgary's economy was battered by government intervention in the energy sector during the 1970s, while Vancouver has been transformed by Asian immigration and investment.

This chapter revisits the role of governments at all levels in the urban process in Canada. This discussion serves as a framework for analysis and interpretation that is important in its own right, but it also serves as a platform for extracting lessons learned from the case studies and for ordering suggestions for new policy initiatives in later chapters. The relevance of these initiatives cannot be identified in a political vacuum. The first section examines the country's political economy using data on expenditures as a guide to the changing roles of different levels of government. The second section tracks the impacts of policy in creating differential urban growth within the broader urban system, the third section looks at the federal government's varied roles, while the fourth section examines the contribution of planning and local public policy to the spatial form and structure of governance in Canada's cities and urban regions. The latter section also illustrates the diversity of approaches to land-use planning, economic development, and sustainability evident among cities and in so doing anticipates many of the policy issues to be addressed in the five case studies.

Introduction: Canada's Political Economy

In Canada as a federal state, government activities and the responsibility for public policy formulation are usually assigned to one of three levels: federal, provincial, or local. Relationships among the federal government and the provinces are defined by the Constitution and arbitrated by the Supreme Court, but local governments are entirely controlled by the province in which they are located, with designated obligations but without any inherent rights. The federal government deals with foreign affairs, trade, and commerce and over the years has managed to take over most of the management of financial affairs, taxes, social benefits, and the like. The provinces deliver more local and regional services: health and education, infrastructure investment, and regulated municipal government. In the case of the latter, the provinces define the responsibilities and powers of cities and may provide much of their financial base. The entire governmental system involves substantial transfers of resources (monies) among levels of government: from federal to provincial and from provincial to local.[1]

Figure 4.1 summarizes the evolving roles of government in Canada since 1961, using direct public expenditures as an index of activity. The magnitude of expenditures is, of course, not the whole story of government participation, but it does indicate relative shifts in the level of activity and has implications for the perceived risks of economic uncertainty. The upper graph compares the expenditures of the various levels of government to the country's gross domestic product (GDP). Total government spending (and budget deficits) peaked in 1992–3 just before the government of the time enforced drastic cuts to reduce the size of the deficit—cuts that affected all levels of government through changes in the transfer systems. It is shown in Chapter 6 that this same period coincided with a substantial increase in income inequality. Since the 1990s, the level of expenditures by the federal government has declined, and so has that of local government, while the level of provincial spending has been maintained, primarily because health care has absorbed a larger and larger share of GDP. Note that the federal government spent almost twice as much as the provinces in 1961, but half a century later it now spends considerably less. Local government's role, at least in terms of expenditures, is largely unchanged over this period. These expenditure levels, especially of the senior levels of government, again increased dramatically in the form of stimulus spending after the onset of the 2008 recession.

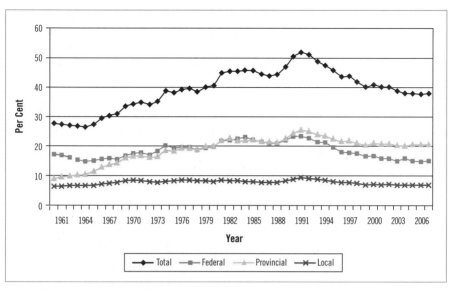

Figure 4.1 Government spending/GDP.

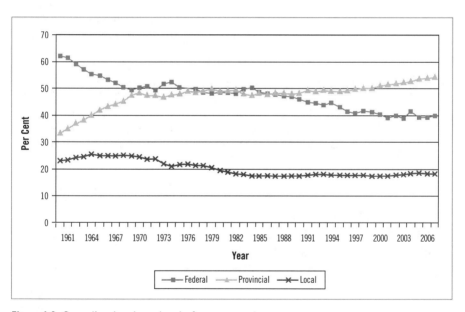

Figure 4.2 Spending by share level of government.

The second graph makes this shift in role explicit by showing the changes in the proportional shares of government expenditures by different levels of government. The share spent by the federal government has undergone a continuous decline over the study period, while the collective role of the provinces has gradually increased until they now control more than half of the country's public economy.

This control is only partial, however, since various transfers from the federal government are negotiated in order to impose restrictions on how the money is spent. The share of spending by local governments peaked in 1965 at the height of the suburban expansion and is currently quite stable at about 18 per cent of the total. Note that this level of expenditure combines all municipalities in the country and there are marked variations from province to province and place to place.

The expanded financial role of the provincial governments, and their absolute control over the actions of local governments, means that cities are highly dependent on the provincial economies in which they are located. Table 4.1 shows the variation in the financial resources that are made available to cities by the provinces. The first point to make is the extraordinary range in the size (and implicitly, the administrative capacity) of the provincial governments. Prince Edward Island is obviously small and largely rural. Five other provinces range in size from half a million to 2 million population, but the four that house the five case study cities are very large indeed, varying from 3.5 million (Alberta) to almost 13 million (Ontario) residents. This creates an interesting political situation in which the prosperity of the major city is a significant factor within the provincial economy but its priorities can be offset by political pressure from extensive rural and resource-based economies that may have disproportionate political influence because of differences in constituency size. Cities and urban regions, as a result, are often under-represented in the national political arena. This in turn has made it difficult to develop strategic urban policies

Table 4.1 Provincial Economies and the Support of Local Government, 2001–8

Provinces	Population 2008	GPP/Capita	GPP Growth Rate 2001–8	Local Expenditure/ Capita 2008
NL	506	$61.8	89.1%	$2.44
PE	139	33.1	15.5	2.00
NS	937	36.5	13.1	2.71
NB	747	36.6	13.4	1.25
QC	7753	39.0	11.8	3.50
ON	12,936	45.4	11.1	4.14
MB	1206	42.1	23.9	3.17
SK	1014	62.7	64.3	3.65
AB	3596	81.0	65.0	4.34
BC	4384	45.2	27.1	3.13
Canada	**33,327**	**48.0**	**23.8**	**3.69**

All values in 1000s
GPP = Gross Provincial Product (2008)

Source: Statistics Canada. CANSIM database.

at the national scale and has led to frequent calls for more resources and political power for cities.

Consider the variation in size among the provincial economies. In 2008, Alberta (hence Calgary and Edmonton) generated more than twice the output per person as Quebec (Montreal) and substantially more than the provinces containing the other case study cities. This implies that more resources are available for schools, hospitals, transportation, and economic development projects of all kinds. Without this kind of economic base and/or the institutional support for municipal initiatives such as airports, transit, or regional government, the most aggressive and far-thinking local government is almost powerless. Cities, then, are hostage to provincial governments and their economies, and in Canada many of these regional economies are resource-based and highly volatile—no matter how much the city tries to build its own techno-logical base. One result of this variability is evident in the differing growth rates of the provincial economies over the past eight years: from 11 per cent in Ontario and Quebec to 60 to 80 per cent in energy-rich Saskatchewan, Alberta, and Newfoundland.

Table 4.2 further explores the variation in the sources of provincial revenue. It shows the five main revenue sources: three of them are based on aspects of the provincial econ-omy—the taxes on personal and corporate income and the taxes on level of production (including agriculture), while the federal transfers depend (inversely) on the level of provincial income—hence the ability to provide services—relative to the rest of the country. Resource revenues include royalties and licences relating to oil and gas, hydro-electricity, mining, and forest industries. The largest provinces (Ontario and Quebec)

Table 4.2 The Fiscal Basis of Canadian Provinces, 2008

				(Share of Total Revenue, %)	
	Economic Taxes			**Federal Transfers**	**Resource Revenue**
Provinces	**Personal**	**Corporate**	**Production**		
NL	12.9	2.3	21.8	53.2	9.8
PE	17.0	3.2	33.2	43.3	3.3
NS	22.4	4.4	27.0	41.9	4.3
NB	17.7	3.0	30.3	38.6	10.3
QC	32.0	5.3	37.4	19.1	10.3
ON	32.3	9.4	37.4	17.2	3.7
MB	21.5	3.3	30.9	33.4	10.9
SK	16.4	5.4	32.5	21.8	23.8
AB	19.5	9.9	18.7	9.4	42.5
BC	22.1	4.5	39.2	17.5	16.7
Canada	**27.0**	**6.9**	**32.5**	**20.4**	**13.2**

Source: Statistics Canada. Bulletin 13-016, 'Provincial and territorial economic accounts'.

depend largely on their own economic taxes, but the smaller and poorer Atlantic provinces receive substantial federal transfers, while the wealthier western provinces rely more heavily on their resource base. These latter revenues, driven by commodity prices, tend to be more variable over time so that other revenue sources must compensate.

These provincial revenues, however, need not benefit local governments. The level of transfers to municipalities reflects provincial traditions (such as the allocation of governmental activity between the provinces and the cities in the Atlantic provinces) and long-standing budget patterns (e.g., Ontario), as well as the recent prosperity of Alberta and Saskatchewan. The overall message is clear: Canadian cities do not have carte blanche to initiate new policies or projects—even the almost costless ones. They operate within a complex network of legislation, budgeting, and bargaining practices that have been created by their political masters, the provinces.

This discussion continues below, following the multi-scalar approach outlined in Chapter 1, by examining the interactions of policy and urban development in Canada at various geographical scales: national, regional, and within the city. As the scale changes, the policy formulation process becomes more regional and/or local, and local decision-makers play a more active part in that process.

The Urban System

In Chapter 3, the concept of the urban system was used to describe the overall distribution of urban population within the nation—including the size and number of cities, their distribution in space, the patterns of demographic and economic variation, and the dynamics of growth and change. The latter operate largely though the complex networks of interaction that link the cities, such as the production cycle and supply chains, population migration and immigration, as well as the political mechanisms that transfer funds from one location to another. Simmons (1986) has summarized the variety of ways by which government interventions shape the future of cities within an urban system (Table 4.3)—often as the by-products of policy decisions rather than by design. Four types of intervention are discussed here: (1) bounding, (2) connecting, (3) homogenizing, and (4) stabilizing.

Bounding refers to government activities that control flows and connections across the national border—for example, policies concerning immigration, trade, or climate change. As a result of such policies, the national market may become either more intensely integrated and isolated or more fragmented and thus weakened. Through trade, the relative prices of goods may be altered, and the differential growth of sectors and regions may increase. In the past, trade policies in Canada were intensely debated, but over time a number of international agreements have eroded tariff structures and weakened the impact of significant events such as the NAFTA treaty of the early 1990s. The results of the latter treaty were more generalized in the sense that the effects were spread across the country rather than location-specific. Trade with the US and the level of US investment increased in many different sectors in all parts of the country so that the two economies became much more interdependent. Courchene and Telmer (1998), among others, have argued that one side-effect has been the widespread adoption of American economic strategies and policies by Canadian firms and governments.

Table 4.3 Government Intervention and the Urban System

Intervention	Recent and Current Issues
Bounding Immigration Tariffs and other controls Monetary policy	1. Immigration and the growth of big cities 2. NAFTA and the dependence on trade with US 3. Copenhagen Agreement on carbon emissions
Connecting Transportation investments Rates, subsidies, restrictions Communications and utilities	1. Quebec loses corporate and financial links 2. Hydro Quebec attempts to take over NB Power to control access to US market
Homogenizing Government services Spatial transfers (e.g., equalization) Regulation: wages, prices, environment Change in subsystem autonomy	1. The future of health care and pensions as the baby-boomers retire 2. The funding of public transit in large cities. 3. How special are asymmetrical federalism and Quebec?
Stabilizing Income tax Transfers to individuals Employment insurance Equalization Public employment	1. The current recession: stimulus package 2. Debates about employment benefits 3. Support of innovation

Immigration policy, as highlighted in earlier chapters, has arguably had a greater impact on growth and change within the urban system because of the concentration of recent immigrant populations in the very largest cities and their immediate neighbours, especially in and around Toronto and Vancouver where the presence of pre-existing immigrant communities has attracted newcomers. Because the federal government has raised immigration targets and modified the criteria for admission, largely for domestic political reasons, Canada's urban population growth has become even more concentrated in a small number of cities such that immigration has become the country's dominant urban policy. The case studies suggest that problems have emerged in the destination cities where the influx of newcomers has overloaded housing submarkets and services and in some cases exceeded the absorption capacity of local labour markets.

Another major challenge comes from the pressure on Canada to join the world community in a climate change agreement. While such an agreement will affect the consumption patterns for all Canadians, the major impacts on the production side will be felt in Alberta as home to the oil sands developments. A recent study of the economic impact of the required adjustments (Pembina Institute 2009) suggests that Alberta and Saskatchewan could jointly suffer a reduction of economic growth of about 10 per cent from the currently projected levels by 2020—which would still leave them better off than the rest of the country. Given that Alberta is the home and political base of the current prime minister, Canada's participation in any such agreement is uncertain.

The second category of governmental activity is *connecting*—that is, the creation of networks that link cities together (e.g., new highways) or exclude them. Historically, the examples would include the great canal and railway construction projects of the nineteenth century that benefited Montreal and Ottawa, but the most significant recent example—thoroughly discussed in Chapter 8—is the partial withdrawal (disconnection?) of Quebec from Canada's financial and corporate leadership as part of the fall-out from political events in the province during the 1970s. A number of leading firms that were unwilling to participate in the francophone community moved from Montreal to Toronto. Toronto became the largest city and leading financial centre. While one can debate the degree to which this shift was already underway much earlier, the immediate impact was quite dramatic.

In contrast, most current connecting events have relatively little impact within an urban system such as Canada's that is already intensely interconnected. New airports, pipelines, or media regulations do not much change our world in the short term—although future projects such as a national electrical grid or expanded alternative energy projects may become significant. Newfoundland's campaign against the expansion of Hydro Québec into New Brunswick in order to control access to the US market suggests the potential for debate.

Homogenizing, the third category, uses government activity, and especially public expenditures, to impose a common way of life and uniform living standards throughout the spatial system. This includes the distribution of such services as health care, education, and pensions and regulations concerning labour organizations, wages and prices, and environmental controls. The mechanisms of homogenization include both the direct funding of services and the elaboration of transfer payments among governments. For example, Canada's system of equalization payments to provinces is expressly designed to overcome the differences in financial resources and fiscal capacity among provinces and to help them provide the same level of services for all Canadians, as identified in Table 4.2.

The entire transfer system will, however, be under increasing stress within the next decade as a result of the current recession and rising debt loads. It may be further tested as Canada absorbs the demographic shock of increased numbers of retirees with fewer new workers coming online. From the urban system viewpoint, the larger urban areas typically have higher incomes (and expenses) than rural areas so that the processes of homogenization typically result in massive transfers of money down the urban size hierarchy into smaller centres, as Simmons (1984) and Kneebone (2007) have demonstrated. Neither federal nor provincial governments have shown any inclination to repay the urban centres for their generosity with special privileges or subsidies or to respond to the pleas for assistance, for example, to build public transit systems and reduce automobile use.

A significant illustration of the potential for reversal of homogenization has occurred with respect to the province of Quebec. A complex political situation in that province has given a succession of Quebec premiers the opportunity to negotiate special relationships with the federal government that include such issues as taxation, transfer payments, control of immigration, and certain language rights. The result is a province with institutions that are increasingly different from those of the rest of Canada.

Stabilization, the final form of government intervention under discussion, is especially relevant during the current economic crisis. Since the Great Depression, Canadian governments have taken on the responsibility for stabilizing the space-economy—both nationally and at the urban scale—in order to avoid the catastrophic economic events that might destroy productive capacity and increase social and regional inequalities. Many stability mechanisms are simply routine transfers of funds among governments and individuals that on balance redistribute money from areas of growth to areas of decline, from rich to poor, from urban to rural, from employed to unemployed, and from young to old. In this sense, the public economy itself tends to be more stable than the private sector, and an expanded public sector tends to reduce the variability of economic activity over space and time—but may also reduce the overall level of economic growth, as many economists will argue.

In Canada, the geographical variance in economic growth has always been a major problem for governments. The country's population—and thus its labour force—grows slowly and regularly, but job creation or destruction can be massive, abrupt, and highly erratic, especially in resource towns. An elaborate system of employment insurance schemes, industrial subsidies, labour market incentives, and complex regional development schemes has been the result. Only recently in Canada have governments begun to support the innovation processes in the big cities that now drive the economy, as argued in Chapter 1 and illustrated in the case studies to follow.

Canada's urban history, and the case studies themselves, suggest that the choice of which cities will grow or decline is seldom the result of systematic analysis or policy design and intervention. Political decisions are taken, sometimes in crisis, to introduce a tax, alter a tariff, or build a highway or airport, and as a result of these decisions a nearby regional economy succeeds or does not. The logic of political cause and effect in determining the economic dimensions of the urbanization process, and in accounting for the differences in the economic development of individual cities, is poorly understood at this scale.

National Urban Policy?

It is frequently said that Canada is one of the few developed countries (aside perhaps from Australia) that has no national urban policy. Nor, as previous chapters made clear, does it have a set of strategies targeted to individual cities and city-regions. These assertions are both true in the sense that federal governments of all political stripes have not taken on the task of developing either an explicit national policy for cities or a national economic development strategy that gives prominence to the role of the country's cities, not even the larger cities. There is, for example, no formal recognition that these cities are the engines of the national economy or that these cities serve as the 'hubs' of regional economies, as the Conference Board (2006 and 2007) and other research institutes (Bradford 2005) have argued. Unlike many European countries, with France and the UK as obvious examples, Canada has no policies designed to facilitate and encourage economic growth in the capital cities in response to an increasingly competitive global economy. Most of these other countries recognize that their large cities are potential instruments of economic growth and competition policies and that the prosperity of these cities not only drives their respective national economies but offers that economy certain competitive advantages.

The review above suggests why an urban agenda is largely absent in Canada: cities, at least as political entities (municipalities), are seen as the constitutional responsibility of the provinces. The federal government in effect has no statutory role in cities and no (or few) direct routes or institutional mechanisms to work with individual municipalities. Although there are a number of tri-level public partnership agreements operating with particular cities and for particular projects, each agreement has to be constructed for the purpose at hand. A second factor, cited above, is the political under-representation of cities in the federal Parliament relative to rural areas and less-urbanized parts of the country. This situation tends to dilute the influence of cities in the national political arena and thus the emphasis given to urban as opposed to, say, agricultural issues in political discourse. A similar situation of representational imbalance between rural and urban areas exists in most provincial legislatures. Looking ahead, it is also the case that the emergence of 'city-regions' as described in this volume—urbanized regions that in many situations extend well beyond local municipal boundaries—are not represented by any political constituency.

Historians of the city and urban policy will remember the experience of the Ministry of State for Urban Affairs (MSUA) during the 1970s, the first serious effort by the federal government to become a player in managing urban development. MSUA was established with the precise purpose of co-ordinating federal actions and programs in cities and as a second-order objective to establish a research base that would provide the evidence for more informed policy-making generally (Lithwick 1971). The MSUA experiment did not last long and was dismantled within a decade. It was successful in raising awareness of the rapid growth of urban Canada through the urbanization process, and it left a legacy of an informative literature and an impressive research inventory. It failed to survive because of the combination of a lack of political commitment at the federal level and pressures from provincial governments who saw the federal initiative as invading areas of traditional provincial jurisdiction and responsibility.

This situation is still relevant and is symptomatic of the challenges of Canadian federalism and of the pitfalls facing public policy formulation (Gibbins, Maioni, and Stein 2006). The latest attempt to define a new tri-level relationship with cities was the 'new deal for cities' initiative proposed by the Liberal government in the mid-1990s (Bradford 2005; Infrastructure Canada 2006). That initiative essentially died with the subsequent federal election, although elements of the new deal, such as the transfer of part of the gas tax to municipalities and GST exemption for local governments, did survive. One consequence of this continuing impasse has been even louder voices calling for new fiscal arrangements between the three levels of government and new revenue sources for cities (FCM 2006). In parallel, others have called for greater political autonomy for cities, particularly the large cities that are the subject of the case studies in this book (Rowe 2000; Bradford 2008). We pick up the theme of multi-level governance again later in this chapter.

Although there is no explicit national urban policy in Canada, the preceding review suggests that the actions of the federal government do have substantial impacts on cities.

Policies regarding trade, taxation, housing, income security, transfer payments, resource development, transportation, and immigration, to name but a few, do have significant, long-term, and often uneven effects on growth and change in cities and regions across the country. These effects, in combination, constitute what might be called an *implicit* national urban policy. The problem is that such policies seldom have urban objectives, and as argued above, the result is that the effects are uncoordinated, sometimes contradictory, and often unintended. In the absence of a coherent national or provincial policy context for cities and city-regions, governments at the local level are largely left to their own devices and their own source revenues, as we discuss in the following section.

Urban Planning and Policy in Canada: Traditions and Innovations

The division of powers in the Canadian Constitution recognizes only the federal and provincial levels as true orders of *government.* Local (municipal) governments derive such powers as they exercise from their respective provincial governments and for the most part exercise legal (i.e., executive) powers primarily through the regulation of land use via zoning, building by-laws, and other discretionary policies. Many local governments in Canada are heavily reliant on the property tax base to fund local infrastructure and the provision of services, with various fees and levies representing ancillary sources of local revenue. But transfers to local government from senior government levels—for housing, infrastructure, and immigrant settlement, for example—have become more important over the years. That said, the decision of the Conservative federal government of Brian Mulroney in the 1980s to substantially cut back funding for new housing administered through the Canada Mortgage and Housing Corporation (CMHC) has left a gap that the combined efforts of local and provincial government, non-governmental organizations (NGOs), and the private sector have struggled to make up.

Policies for managing growth and regulating land use in Canada, as in other jurisdictions, have evolved in part in response to long-run changes in the dominant economic regime within the city: hence the introduction of *zoning* a century ago as a means of managing the negative externalities of the industrial city and to separate incompatible land uses within the city's territory by means of exclusionary zoning ordinances. With the decline of inner-city manufacturing and allied industries over the 1970s and 1980s and the associated diminution of negative externalities such as pollution, excessive noise, and congestion linked to goods movement, this exclusionary zoning model came under attack from numerous quarters, including civic activists and reformist city councils. This in turn has led to the increased use of mixed-use zoning as a means of encouraging diversity and economic renewal as well as facilitating social interaction and overall urban vitality.

With the rapid spatial expansion of cites and urban communities in the postwar period and with growing problems of sprawl, the encroachment of urban activities into the countryside, and associated challenges of financing such growth, Canadian cities have experimented with a wide variety of planning and policy programs (Grant 2008; Hodge and Gordon 2008). The case studies to follow illustrate this diversity of policy initiatives and their relative success or lack of success. Many of these plans, however, did not appear out of nowhere. Many have their intellectual and practical roots in the *urban containment* model (see Hall et al. 1973 for the classic treatment of this model), including the allocation of green spaces (parks, agriculture, and in some cases wilderness areas), establishing new towns and regional town centres in suburban and exurban territories, and (again in a few cases) planning controls on the growth of the central city. The Livable Region 1976–1986 plan developed for Greater Vancouver represents a Canadian example of this containment model, while future chapters will discuss policies for the Ottawa Greenbelt (Chapter 9) and Ontario's recent Places to Grow (Chapter 10).

As another demonstration of the complex linkages between economic change and local policy responses, programs of *growth management* emerging during the late twentieth century sought to contain the impacts of both suburban expansion and central-city office growth associated with the service-based (or post-industrial) city, notably the costs of long-distance commuting and associated local dislocations in inner-city neighbourhoods. Growth management programs and policies typically included the imposition of metropolitan growth boundaries, as well as controls on central-city commercial office growth and the introduction of market pricing as a means of allocating scarce resources. The latter, although still limited in scope in most Canadian cities, is perhaps most evident with respect to transportation (roads) and energy (electrical distribution) infrastructure and also in the growing use of development cost charges levied on developers to help cover the costs of servicing new growth.

Constitutional limitations on local governments notwithstanding, the more recent sequences of economic restructuring have stimulated experimentation in new forms of local–regional governance, institutional innovation, and the identification of interrelated clusters of policies, as disclosed in the case studies presented in Part II. Broadly, the incentives for policy experimentation can be traced to the dislocations of industrial decline in many cities, the pressures (and opportunities) of globalization, shifts in senior government policies and programs (including the retrenchment in commitments to social assistance and public housing since the 1980s and 1990s), and the associated pursuit of competitive advantage, each of which requires a rethinking of governance structures and the scope of local policy reach.

A review of the experience of Canadian cities in the realm of economic development also reveals a range of policies and strategies designed to address the challenges of economic change: to redeploy marginalized labour and to adapt 'redundant spaces' in the city for new industries; to mobilize other city resources more fully in order to attract new investment, enterprise, and talent; and to rebrand and market the city and its spaces and places to accommodate the 'new economy'. These efforts have typically involved a mix of urban design initiatives, civic improvements, and adaptive re-use of the built environment. Such programs have become commonplace among both advanced and transitional urban societies; indeed, they are part and parcel of what David Harvey described more than 20 years ago as a strategic shift in local government visions and approaches from the 'managerial' to the 'entrepreneurial' city (Harvey 1989).

These broad program commonalities aside, the case studies of Canada's largest city-regions disclose further contrasts in the range, depth, and degree of commitment to economic development, growth management, and social innovation in the public economy. Possibly the polar extremes are represented by Montreal, which embraces a rich program of policies for a wide range of economic initiatives including cultural industries, film and video production, and various technology-based industries, on the one hand, and Vancouver, which lacks a region-wide economic program per se, although there are indirect policies such as those for compact development, transit-oriented development, and the preservation of land for agricultural industries. In the Toronto case, detailed in Chapter 10, the policy constraints on local government include the uniquely under-bounded nature of the city-region and the extra responsibilities for (notably) housing and social assistance programs devolved from the government of Ontario—without a commensurate expansion of powers and resources to adequately discharge these responsibilities. Ottawa, of course, represents a special case, governed in part by the National Capital Commission, which provides investment and expertise in the development of institutions and infrastructure designed to showcase Canada's federal capital—a privilege and a responsibility exclusive to the Ottawa-Gatineau region.

Planning and the Reshaping of Urban Structure and Form in Canadian Cities

One aspect of local planning and policy that merits special consideration here is *spatial planning*, which concerns a suite of local policies and programs (comprising regulation, investments, and incentives) designed to shape the overall structure and morphology of city-regions. Traditional spatial planning has its origins in concerns for physical and environmental public policy goals, but there has always been at least a subsumed social component as well, and we can also readily identify implications for urban economic development.

In the postwar period, the spatial planning regime for city-regions in Canada, as in the UK, Europe, and Australia (among other states and jurisdictions) has incorporated policies for promoting a balance of growth within the broader metropolitan region, for reducing congestion in the urban core, and, especially, for reining in 'sprawl'—the phenomenon of near-constant encroachment of low-density suburban development into the countryside. In the early stages, the primary sources of planning ideas and legislation in Canada were of British and European origin, not North American. The provenance of spatial planning can thus be traced in part to the garden city movement of a century ago, but it was more fully realized in the classic regional plans for world cities such as London (including the wartime Barlow and postwar Abercrombie plans for Greater London) and Paris (notably the Delouvrier plan for Paris and the Île-de-France), which included programs for new and expanding towns, green belts or agricultural zones, transportation networks, and housing.

Following the relative success of these influential plans, a second generation of spatial plans was developed for many more city-regions, including some Canadian cities as well as urban areas in Europe, the US, and Australia, over the 1970s. In this period, the growth of long-distance commuting, a product of an over-concentration of (primarily office) jobs in the core, automobile ownership, and the expansion of suburban housing, was widely acknowledged as the cardinal planning question for many metropolitan areas, notably those with a growing service economy (Daniels 1975). New and expanding towns (the latter designated as centres for new public and private investments) and regional town centres were proposed as means of diverting growth pressure from the core, while providing—at least in theory—a spatial focus for suburban development. At the same time, the uneven pattern of office growth within the broader urban system, notably the concentration in larger cities, was perceived as a spatial planning issue for national governments, including Canada, as discussed in the Ottawa chapter, as well as in Britain and other states.

Since the 1980s, the relevance of spatial planning has significantly increased in importance, partly in response to the ecological impacts of urbanization—including the unsustainable dimensions of the urban ecological 'footprint' (Rees 1992) and high levels of energy consumption, as well as the more localized consumption of environmental assets—which have in part stimulated the growth of the global environmental movement. Public acceptance that spatial planning is useful has also been a response to the rising social costs of sprawl and low-density suburban/exurban development, the latter including the increasing costs and debt burden of local/regional infrastructure and service provision. In addition, this shift in thinking is at least in part due to the awareness of the role spatial planning can play in improving the efficiency of the urban–regional economy.

But while there is a more or less clear policy logic to the exercise of spatial planning in city-regions, the experience in individual jurisdictions in Canada tends to be highly variable over time, if not untidy. This may be attributable to a relatively weak political commitment and often to within-region jurisdictional conflicts and boundary issues (and more particularly, under-boundedness—notably in the Toronto case), the inadequacy of local–regional regulatory powers to overcome the power of the market, and policy disagreements between upper- and lower-level local governments and/or between the local and provincial governments. There is also a question of how local political culture and tradition may come into play. The wide variation in the postwar experience of spatial planning in Canada has undoubtedly been shaped by contrasts in inherited governance structures, planning styles, and practices, not to mention by the personalities of those in leadership positions in both public and private sectors.

To illustrate the point, the contrasting aspects of spatial planning in Canada and their impacts on regional structure are examined in a major study of urban growth patterns and development policies published by researchers working under the auspices of the Neptis Foundation. The Neptis study focuses its analysis on the three fastest-growing metropolitan regions in Canada—Toronto, Calgary, and Vancouver—each of which is included in our city case studies presented in Part II (Chapters 10, 11, and 12, respectively). The study, co-authored by Zack Taylor and Marcy Burchfield, identifies Vancouver as the Canadian success story in terms of efforts to achieve a more compact and in theory more sustainable urban form through intensification of urban development since the 1970s. They attribute this relative success to Vancouver's (1) consistency of regional planning institutions and principles, (2) strong regional governance model, (3) sustained commitment to intensification policies and programs, and (4) promotion of regional nodes (i.e., 'regional towns' or 'town centres') (Taylor and Burchfield 2010, 22). In contrast, Calgary is characterized as having achieved some consistency in regional institutions and governance but also by a relatively low commitment to intensification and the promotion of nodal development, resulting in a long record of suburban sprawl and recurrent annexations. On this scale, Toronto falls in between. Municipalities in the Toronto region have striven (especially since the

1990s) to promote intensification and nodal centres, but the region suffers from 'intermittency' of regional planning institutions and principles and 'weak to nil' regional governance (Taylor and Burchfield 2010, 22).

Vancouver's apparent success notwithstanding, the regional authority (initially the Greater Vancouver Regional District and now Metro Vancouver) has struggled to impose coherence on the form and space-economy of the metropolis, as evidenced by the rapid growth of low-density employment centres in the region beyond the officially designated regional town centres (RTCs). Local (municipal) authorities—*not* the regional authority—constitute the 'basic units' of local government and enjoy stronger legal powers over urban structure and land use, which they may exercise in opposition to (or only in partial compliance with) the regional plan. And all of Canada's large cities have encountered difficulties in coping with the spatial impacts of accelerated rates of economic restructuring, a theme to be picked up in each of the city case studies presented in Part II.

Although land-use planning processes in Canadian cities are often viewed by critics as inconsistent and intermittent, there is nonetheless more emphasis on (and acceptance of) regulation and on developing and maintaining nodes of higher-density development than is common in American cities. The direct involvement of the provincial governments in urban development tends to be stronger than that of most state governments in the US. At the same time, metropolitan government is more prominent (although not regional governments for the expanding city-regions discussed here), and competition among municipalities for assessment is more muted. As a result, population densities, at least on average, tend to be higher (along with land values and transit use), the central areas and downtown cores are stronger, and there are more planned employment clusters in the suburbs.

Multilevel Governance and Urban Development in Canada

One novel feature of economic development and governance in Canadian cities is the continuing interest in multi-level governance experiments, including metropolitan governments, in which the limitations of local (or any single level of) government are at least partially overcome by the combination of resources (including intellectual resources and specialized knowledge), policy, and regulatory powers. The very different conditions of the early twenty-first century presented challenges that have stretched the capacity of local authorities. These challenges, to reiterate, include (1) recurrent industrial restructuring, including structural unemployment and socioeconomic polarization; (2) the volatility of globalization as it affects international trade and capital markets; (3) increasing levels of international immigration and associated issues of settlement and social integration; (4) gaps in the financing of infrastructure, due to the physical deterioration of municipal stocks, in an era of uncertain public revenues and limited borrowing capacity; and (5) new commitments to achieving *sustainable* urban development, which require major changes in transportation, land use and urban structure, and energy consumption, in order to reduce the urban ecological 'footprint'. As we will see later in the volume, the degree of progress in meeting each of these challenges is rather mixed.

At the same time, local governments in Canada have had to adapt to changes in federal–provincial relations resulting from 'asymmetrical federalism', largely but not exclusively linked to Quebec's political aspirations, and to the implications of Aboriginal treaty claims and prospective settlements of land claims and other resource issues, which increasingly affect the roles of local government. Moreover, cities and towns in Canada have been thrust forcefully into arenas of globalization and the pursuit of competitive advantage, implying a major enlargement of policy *scope*, while simultaneously contending with the constraints imposed by domestic politics and the country's political economy.

Despite these challenges, the pressures (and uncertainties) of change have stimulated the search for policy innovation throughout the Canadian urban system, including collaboration with senior governments and engagement with social actors (including NGOs, community-based organizations [CBOs], and business interests) as well as partnerships with Aboriginal groups. For their part, federal and provincial agencies and actors have experimented with new forms of inter- and multilevel governance across a spectrum of policy fields and program areas, while social organizations have encouraged wider participation in innovative policy formulation, development, implementation, and monitoring. These experiments in collaborative forms of policy-making can be examined within the rubric of *multilevel governance*.

Most forms of multilevel governance share a number of attributes: (1) they bring in a broader representation of social and economic constituencies and actors for more equitable policy formation and effective program delivery; (2) they combine the resources (financial, intellectual, and regulatory) of two or more levels of government; (3) they reflect the multi-scalar nature of development processes and policy issues in a modern federal state and society; and (4) they in theory increase the potential for experimentation and innovation. The refinement of policy through multilevel, interagency experimentation is particularly relevant in the context of this volume during an era of increasing economic volatility and uncertainty due to intensified globalization (Young and Leuprecht 2008; Lazar and Leuprecht 2007).

In Quebec, the provincial government and the Montreal municipal authority collaborate on a wide range of initiatives—part of the 'Quebec Inc.' model of statist economic development derived in part from the experience of the European Union. In Vancouver, in the absence of a convincing regional economic development authority or program, the city and metro areas have substantially benefited from an array of specific multilevel governance programs over the past quarter-century. These programs have included the construction of the three fixed-rail transit projects since the mid-1980s (including the Canada Line in August 2009); significant investments in 'hallmark events' (Expo 86 and the 2010 Vancouver Olympic Games), including both capital investment and marketing; and at a smaller spatial scale, the Vancouver Agreement for the revitalization of the Downtown Eastside (DTES), incorporating multi-level expenditures and management for social housing, economic development, and substance abuse treatment programs (Hutton 2009).

To conclude, although Canada's cities and specifically its local governments are captives of a constitutional division of powers established a century and a half ago when the nation was still largely rural, the past several decades have seen considerable experimentation in governance in response to increased urbanization. These experiments include the changing relations between local and provincial governments (see Sancton and Young 2009; Bradford 2005) and the inclusion of social actors (e.g., business and community-based organizations) in policy development and implementation, as well as recurrent restructuring of urban–regional boundaries, especially as experienced in Toronto and Montreal. It seems fair to say, though, that governments are difficult to change and that innovation is slow and often painful. There is still a very large gap between the daunting scope of problems confronted by cities and local governments and the limited powers and resources of those governments in Canada in the twenty-first century. This gap contains significant implications for encouraging economic growth and renewal in Canadian cities, as demonstrated in subsequent chapters.

References

Bennett, R.J. 1980. *The Geography of Public Finance: Welfare under Fiscal Federalism and Local Government Finance*. London: Methuen.

Bradford, N. 2005. *Place-Based Public Policy: Towards a New Urban and Community Agenda for Canada*. Ottawa: Canadian Policy Research Networks.

———. 2008. *Multi-level Governance: Challenges and Opportunities for the Toronto City-Region*. Ottawa: Canadian Policy Research Networks.

Conference Board of Canada. 2006. *Canada's Hub Cities: A Driving Force of the National Economy*. Toronto: Conference Board of Canada.

———. 2007. *Mission Possible: Successful Canadian Cities*, vol. III, *Canada Project*. Ottawa: Conference Board of Canada.

Courchene, T.J., and C. Telmer. 1998. *From Heartland to North American Region State: The Social, Fiscal and Federal Evolution of Ontario*. Toronto: Centre for Public Management, University of Toronto.

Daniels, P.W. 1975. *Office Location: An Urban and Regional Study*. London: Bell.

Dobuzinskis, L., M. Howlett, and D. Laycock. 2007. *Policy Analysis in Canada: The State of the Art*. Toronto: University of Toronto Press.

FCM (Federation of Canadian Municipalities). 2006. *Our Cities, Our Future: Addressing the Fiscal Imbalance in Canada's Cities Today*. Ottawa: FCM.

Gibbins, R., A. Maioni, and J.G. Stein. 2006. *Canada by Picasso: The Faces of Federalism*. Ottawa: Conference Board of Canada.

Grant, J., ed. 2008. *A Reader in Canadian Planning: Linking Theory and Practice*. Toronto: Thompson/Nelson.

Hall, P.G., R. Thomas, H. Gracey, and R. Drewett. 1973. *The Containment of Urban England*. 2 vols. London: George Allen and Unwin and Sage Publications.

Harvey, D. 1989. 'From managerialism to entrepreneurialism: Transformation in urban governance in late capitalism'. *Geografiska Annaler Series B— Human Geography* 88B: 145–58.

Hodge, G., and D. Gordon. 2008. *Planning Canadian Communities*. 5th edn. Toronto: Thompson/Nelson.

Hutton, T.A. 2009. *Multilevel Governance and Urban Development: A Vancouver Case Study*. Report prepared for the Multilevel Governance MCRI Project. Vancouver: Centre for Human Settlements, University of British Columbia.

Infrastructure Canada. 2006. *From Restless Communities to Resilient Places: Final Report of the External Advisory Committee on Cities and Communities*. Ottawa: Infrastructure Canada.

Kneebone, R.D. 2007. *Following the Money: Federal and Provincial Budget Balances with Canada's Major Cities*. Publication no. 249. Toronto: C.D. Howe Institute.

Lazar, H., and C. Leuprecht, eds. 2007. *Spheres of Governance: Comparative Study of Cities in Multilevel Governance Systems*. Montreal: McGill-Queen's University Press.

Lithwick, H. 1971. *Urban Canada: Problems and Prospects*. Ottawa: Ministry of State for Urban Affairs.

Pembina Institute. 2009. *Climate Leadership, Economic Prosperity: Final Report on an Economic Study of Greenhouse Gas Targets and Policies for Canada*. Calgary: Pembina Institute.

Rees, W. 1992. 'Ecological footprints and appropriated carrying capacity: What urban economics leaves out'. *Environment and Urbanisation* 4 (2): 121–30.

Rowe, M., ed. 2000. *Toronto: Considering Self-Government*. Toronto: The Ginger Press.

Sancton, A., and R. Young. 2009. *Foundations of Governance: Municipal Government in Canada's Provinces*. Toronto: University of Toronto Press.

Simmons, J.W. 1984. 'Government and the Canadian urban system: Income tax, transfer payments and employment'. *The Canadian Geographer* 28 (1): 18–45.

———. 1986. 'The impact of the public sector on the Canadian urban system'. In A.A. Artibise and G. Stelter, eds, *Power and Place: Canadian Urban Development in the North American Context*, 21–50. Vancouver: University of British Columbia Press.

Taylor, Z., and M. Burchfield (with B. Moldofsky and J. Ashley). 2010. *Growing Cities: Comparing Urban Growth Patterns and Regional Growth Policies in Calgary, Toronto and Vancouver*. Toronto: Neptis Foundation.

Young, R., and C. Leuprecht, eds. 2008. *Canada: The State of the Federation: Municipal–Federal Relations in Canada*. Montreal and Kingston: McGill-Queen's University Press and the Institution of Intergovernmental Affairs, University of Western Ontario.

5 Canada's Changing City-Regions: The Expanding Metropolis

Richard Shearmur and Tom Hutton

Introduction

In this chapter we set out to describe the way in which the space-economy is changing in and around Canada's city-regions—i.e., cities and their surrounding peri-metropolitan areas. The first part of the chapter focuses on intra-metropolitan changes. These changes are first described at the level of the CMA (census metropolitan area), detailing the transformations in urban form and employment location that are occurring across the case-study cities. The nature of the changes occurring in the metropolitan core is then discussed in more detail, providing an opportunity to describe the connections that exist between local urban restructuring and wider processes of industrial, workforce, and social change. Although such connections between local change and global processes play out across the entire metropolitan area, our focus on the metropolitan core—a key area for both symbolic and functional reasons—allows these connections to be illustrated in a particular context.

The second part of the chapter focuses on peri-metropolitan space: as metropolitan areas expand, and as communication and transport technology enable the geographic co-location of workplace and residence to relax (although not disappear), the surrounding areas are increasingly drawn into the city-region. These areas, often comprised of small towns and rural districts, have diverse economic and socio-demographic structures that are rapidly evolving. These changes can give rise to conflict and to a certain degree of polarization between 'gentrifying' localities and those that retain a more traditional peri-metropolitan role, a problem more commonly associated with urban neighbourhoods. Each of the cities under study is thus located within, and is strongly interconnected with, a peri-metropolitan zone.

In the following sections, a variety of concepts are used to classify spaces within and around metropolitan areas. Indeed, depending on the scale and nature of the analysis, different taxonomies and concepts are applied: although necessarily simplifications of the complex spatial structures that exist across the various scales, they nevertheless provide some idea of the difficulty inherent in discussing a city-region. For instance, a discussion of centrality requires each point to be located in relation to the centre, whereas a discussion of peri-metropolitan space requires spaces within a metropolitan area to be distinguished from spaces that surround it. While each concept and spatial taxonomy is briefly described when it arises, Figure 5.1 presents key concepts used in this chapter and illustrates how each relates to the others.

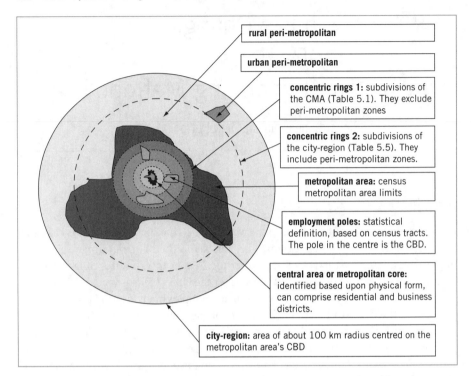

rural peri-metropolitan

urban peri-metropolitan

concentric rings 1: subdivisions of the CMA (Table 5.1). They exclude peri-metropolitan zones

concentric rings 2: subdivisions of the city-region (Table 5.5). They include peri-metropolitan zones.

metropolitan area: census metropolitan area limits

employment poles: statistical definition, based on census tracts. The pole in the centre is the CBD.

central area or metropolitan core: identified based upon physical form, can comprise residential and business districts.

city-region: area of about 100 km radius centred on the metropolitan area's CBD

Figure 5.1 Concepts used to categorize and describe spaces across the city-region.

Note: In Table 5.5, the entire city-region is divided into concentric zones that extend 15, 30, and 60 kilometres from the centre: only one is shown for illustrative purposes ('concentric rings 2' in the figure).

Finally, it is perhaps useful to emphasize that many of the structures and changes described in this chapter are influenced by public policy and by governments, which are active (or not) at various scales (Shearmur and Alvergne 2003; Keil and Young 2008; Desfor et al. 2006). In particular, the articulation between city-region integration, economic growth, and transport networks is a key dimension that shapes metropolitan areas and their surrounding regions and which is as much an outcome of policy and political decision-making as it is of economic and demographic forces (see Chapter 4). However, the political dimensions of metropolitan expansion will only be alluded to in this chapter, our main focus being on the description rather than the explanation of Canada's changing city-regions.

The Changing Intra-metropolitan Space-Economy

The Intra-metropolitan Distribution of Jobs

Between 1981 and 1996, employment growth in all of Canada's major metropolitan areas was faster outside of the central business district (CBD) (Shearmur and Coffey 2002), and this trend has continued over the period 1996–2006[1] (Tables 5.1 and 5.2). The commonly accepted narrative is therefore one of weakening CBDs, of burgeoning suburban employment centres, and of employment sprawling across space, tempered somewhat by the rehabilitation of selected central neighbourhoods in recent years, to which we turn in the next section (this story is one often told about US cities and informs the perception of Canadian metropolitan areas).

A closer look at Canada's five largest cities reveals, however, that each has a very distinctive spatial pattern of employment and that suburban growth is not occurring in the same way in all five. Furthermore, faster growth in the suburbs does not necessarily imply a weakening CBD, as will be argued below.

There are a variety of ways to analyze the distribution of employment in urban areas, and it is useful to employ at least two in order to capture different dimensions of employment distribution and provide a certain 'parallax' to the analysis. Each approach informs the other, and it is by combining them that some degree of understanding can be achieved. Despite debates about the relevance of the spatial patterns first described and theorized by researchers of the Chicago School (Park, Burgess, and McKenzie 1925; Shearmur and Charron 2004), a good starting point for analyzing intra-metropolitan employment is basic patterns such as concentricity, polarization, and spatial segmentation; given the constraints of a book chapter, only the polarization and concentricity of employment will be studied below.

The polarization of employment can be assessed by determining whether or not employment is concentrated in employment centres (defined in this case as aggregations of contiguous census tracts,[2] each of which contains more than 5000 jobs [Coffey and Shearmur 2001]). They can be further classified as the CBD, primary (with a nucleus containing more than 12,500 jobs), secondary (with a nucleus containing more than 5000 jobs), and isolated (single census tracts with between 5000 and 12,500 jobs) centres. As in all taxonomies, these thresholds are somewhat arbitrary[3] but have proved useful in analyzing Canadian cities and represent concentrations of employment that are significant at a metropolitan scale. A second approach to analyzing the distribution of employment is to examine its concentric distribution around the CBD, and it is to this that we now turn.

The Concentric Distribution of Employment in Canadian Cities

With about 28 per cent of all employment within a five-kilometre radius of downtown, economic activities in Montreal and Vancouver are more central than they are in Toronto (Table 5.1). This can, of course, be partly explained by Toronto's geography—its centre is close to the shore of Lake Ontario—but since Vancouver's centre abuts the sea to one side and water (then mountainous terrain) to another, local physical features are not the only determinants. At the other extreme, the share of employment found more than 20 kilometres from the centre is 46 per cent in Toronto and 30 per cent in Vancouver but only 19 per cent in Montreal. There are thus three different concentric patterns to employment in these three cities:

1.   In Montreal, a strong employment centre is surrounded by rings with progressively fewer jobs. This is a 'classic' concentric pattern.
2.   In Vancouver, a strong employment centre is surrounded by rings where employment decreases up to about 20 kilometres away but then increases in the more distant suburbs.

3. In Toronto, a moderately strong employment centre immediately abuts an employment-poor zone, and the rings then progressively house higher numbers of jobs as one moves out to a distance of 30 kilometres; employment levels remain high beyond this distance. This pattern is commensurate with the idea that suburbs are taking on key economic roles and that the centre's centrality (i.e., its capacity to attract activity and to organize the surrounding territory) may be weakening.

Employment growth in all three cities is faster in the outer suburban rings (Table 5.1), but differences in employment distribution across rings mean that these rates should be interpreted differently. In Montreal, about 88,000 jobs were added to the outer suburbs (beyond 20 kilometres) from 1996 to 2006. This is considerably less than was added in the inner parts of the metropolitan area (186,000). In Vancouver, the strength of the outer suburbs is more evident (77,000 new jobs), since fewer jobs were added to the central areas of the city (72,000). In Toronto, about 350,000 jobs were added to the outer suburbs during the decade under consideration; this is three times more than were added closer to the centre. Furthermore, Toronto is the only city to have witnessed a *decline* in the number of jobs within the two innermost rings (during the 2001–6 period).

A final difference of note among these three cities is the relationship between employment and resident labour force, which summarizes the degree to which people commute into or out of an area.[4] In Montreal, workers living in the outer suburbs have a strong tendency to commute toward central areas. In Vancouver, this tendency is weaker than in Montreal, and the city core houses relatively more resident workers: downtown Vancouver has a more residential flavour than either Montreal or Toronto. In Toronto, the central ring (five kilometres around the downtown core) has a ratio of jobs to people similar to that of Montreal, but the ring surrounding it has a more residential character than Montreal's. However, Toronto's outer suburbs (especially 20 to 30 kilometres from downtown) are major attractors of in-commuters, highlighting the economic importance of these places (Shearmur and Coffey 2002).

Calgary and Ottawa are discussed separately because their smaller size means that they are organized at a different geographic scale. In both of these cities, about 45 per cent of all employment is within five kilometres of the centre, and about 90 per cent is within 20 kilometres (Table 5.1). Although employment is growing at a slower rate in the centre than in the suburbs, this is partly due to the low initial numbers of jobs in suburban rings. Having said this, the areas beyond 10 kilometres from Calgary's CBD have been growing at between 5 and 10 per cent per year for 10 years, and this is clearly extending the city.

Notwithstanding the very rapid suburban growth, 47,000 jobs have been added in Calgary's centre, which shows no sign of weakening. Although growth rates are slower, much the same can be said of Ottawa, where 29,000 jobs have been added in the centre. However, in Ottawa's centre there has been a more rapid increase of residents than of jobs, while the increase of both has been identical in Calgary. In both cities, jobs have grown substantially faster than population in the second (five- to 10-kilometre) ring, especially in Calgary.

Although only one dimension of job distribution is captured here, a number of important points can be made.

1. High suburban growth rates do not necessarily imply a weak downtown: in Montreal, the centre, 0–5 kilometres, is growing (adding 60,000 jobs over 1996–2006), and the suburbs, while they are also growing quickly, are growing in tandem with the centre. Much the same can be said of Calgary and Ottawa. Vancouver is in a somewhat similar situation but with a weaker centre. Toronto's centre, after fast growth in the late 1990s, actually declined in the early 2000s even as the suburbs, and particularly the outer suburbs, continue to grow at a higher rate in relative *and* absolute terms.

2. Second, the three larger cities seem to have a ring, built before World War II and located about five to 10 kilometres from the centre, that has a residential rather than an economic vocation. On the whole, the relative weight of employment and population in this ring has remained stable over 1996–2006. In each large city, the 10- to 15-kilometre rings (and, except in Vancouver, the 15- to 20-kilometre rings) have at least as much employment as people, whereas the ratio of jobs to people declines toward the outer suburbs in Montreal, it remains high in the Toronto suburbs. In Calgary and Ottawa, out-commuting tends to rise monotonically up to 20 kilometres from the CBD, then stabilizes or decreases slightly.

3. Finally, the most general point is the following: A straightforward analysis of these five metropolitan areas reveals quite different patterns of employment distribution and growth. Some of these differences may be due to geographic features (e.g., water, mountains), but they primarily reflect different spatial organizations of the city. Toronto, for instance, has a number of very strong employment sub-centres located at some distance from the centre (Mississauga, Vaughan, Markham, Brampton), Montreal only has one major sub-centre located closer in (Ville Saint-Laurent–Dorval), and Vancouver has a major sub-centre close to the airport and a multiplicity of smaller ones, particularly in Burnaby and Surrey (which themselves house more than one sub-centre). Both Ottawa and Calgary have strong CBDs—possibly a consequence of their smaller size (and thus of their lower level of polynucleation) and sizeable employment concentrations close to their airports, which in all cities play a key role as access points to the global economy (Keil and Young 2008). It is to these sub-centres that we now turn.

The Concentration and Diffusion of Economic Activity in Canada's Metropolitan Areas

The geographic distribution of employment can also be assessed by identifying the degree to which it is concentrated in space. To this end, employment centres (or poles) are identified, and the criteria used have been mentioned above. Such an analysis can tell us about the strength of agglomeration economies internal to the city (assuming that economic activities concentrate in order to benefit from local urbanization and localization economies) but may also reveal something about planning regulations and the decisions that have been made by successive administrations regarding the location of economic development (Shearmur 2007b; Williams 1999). In this section,

Table 5.1 Concentric Distribution of Employment in Five Metropolitan Areas

	Toronto		Montreal		Vancouver		Ottawa		Calgary	
	1996	2006	1996	2006	1996	2006	1996	2006	1996	2006
Total Employment and Concentric Distribution										
0–5 km (%)	22	20	30	28	32	29	48	45	48	43
5–10 km (%)	10	8	25	23	16	15	27	26	33	30
10–15 km (%)	13	12	19	18	18	18	10	10	13	18
15–20 km (%)	15	13	11	12	8	7	6	8	2	4
20–30 km (%)	27	30	11	12	16	19	6	8	2	3
over 30 km (%)	13	16	5	7	10	11	3	3	1	2
0–20 km (%)	60	53	84	81	74	70	91	89	96	95
over 20 km (%_	40	47	16	19	26	30	9	11	4	5
Total	2,034,340	2,503,905	1,459,905	1,734,440	828,285	977,560	486,010	579,245	404,170	561,430
Employment to Labour Ratios										
0–5 km	2.35	2.45	2.41	2.32	2.00	1.83	2.31	2.20	2.45	2.45
5–10 km	0.69	0.66	0.93	0.89	0.72	0.72	0.81	0.90	0.86	1.08
10–15 km	1.00	0.98	1.01	1.05	1.48	1.47	0.60	0.58	0.46	0.55
15–20 km	0.88	0.90	0.93	1.01	0.62	0.61	0.49	0.47	0.36	0.36
20–30 km	1.09	1.17	0.51	0.54	0.71	0.78	0.72	0.83	0.45	0.41
over 30 km	0.63	0.61	0.57	0.63	0.69	0.76	0.37	0.43	0.54	0.60
0–20 km	1.11	1.13	1.21	1.21	1.17	1.14	1.10	1.07	1.04	1.07
over 20 km	0.87	0.89	0.53	0.57	0.71	0.77	0.54	0.65	0.48	0.46

Annualized Employment Growth %

	1996-2001	2001-2006	1996-2001	2001-2006	1996-2001	2001-2006	1996-2006	1996-2006
0–5 km	3.3	-0.2	1.9	0.7	0.6	1.2	1.2	2.2
5–10 km	0.9	-0.8	0.6	1.2	0.9	1.4	1.4	2.4
10–15 km	0.9	1.0	2.6	0.4	2.2	0.8	1.8	6.5
15–20 km	0.8	0.4	3.9	1.7	1.3	0.8	4.3	9.7
20–30 km	5.0	2.0	2.2	2.6	3.5	2.7	4.8	8.4
over 30 km	4.5	3.5	5.0	4.3	2.9	3.1	2.0	5.5
0–20 km	1.8	0.1	1.9	0.9	1.1	1.1	1.5	3.2
over 20 km	4.8	2.5	3.2	3.2	3.3	2.9	3.9	7.3
Total	**3.0**	**1.2**	**2.1**	**1.3**	**1.7**	**1.6**	**1.8**	**3.3**

Employment Growth, Absolute

	1996-2001	2001-2006	1996-2001	2001-2006	1996-2001	2001-2006	1996-2006	1996-2006
0–5 km	76,740	-3,935	42,565	18,015	7,610	17,210	28,915	47,620
5–10 km	9,640	-8,815	10,715	23,460	6,290	10,060	19,175	35,495
10–15 km	12,465	13,890	36,990	5,625	17,285	6,965	8,855	46,365
15–20 km	12,355	5,895	32,280	16,755	4,160	2,890	15,400	12,685
20–30 km	148,300	72,070	18,510	24,455	25,555	22,865	17,365	10,885
over 30 km	67,645	63,315	21,765	23,400	12,630	15,755	3,525	4,210
0–20 km	111,200	7,035	122,550	63,855	35,345	37,125	72,345	142,165
over 20 km	215,945	135,385	40,275	47,855	38,185	38,620	20,890	15,095
Total	**327,145**	**142,420**	**162,825**	**111,710**	**73,530**	**75,745**	**93,235**	**157,260**

Distance = distance from census tract centroid to CBD (the tract with the highest employment total in the CMA).

the patterns will be described, but some of the processes behind them will be discussed in chapters that deal more specifically with each city.

Overall, employment has tended to grow within employment centres (identified on the basis of 2006 employment levels): this does not necessarily mean that employment growth is focused in pre-existing employment centres, although other studies have shown that this tends to be the case (Shearmur et al. 2007). However, it does show that employment has not dispersed across the entire metropolitan area but has agglomerated in certain key zones—namely, census tracts with more than 5000 jobs. In 2006, the polarization of jobs was lower in Montreal (43 per cent of jobs were in employment centres; Table 5.2) than in Toronto (52.5 per cent), Ottawa (50.1 per cent), and Vancouver (56 per cent), and this is an enduring feature of each city (Shearmur and Coffey 2002). Calgary stands out as having 70.2 per cent of its employment in centres; although this polarization may be partly attributable to the large census tracts being used, this does not appear to be the only reason. Indeed, the proportion of tracts selected as poles in Calgary is not dissimilar from that in Vancouver, Toronto, or Ottawa; whereas jobs are quite dispersed in Montreal and moderately so in Vancouver, Toronto, and Ottawa, they are highly focused in sub-centres in Calgary.

The weight of each type of employment centre is remarkably similar in Montreal and Toronto (with Toronto's centres systematically gathering a few extra percentage points than Montreal's). About 15 per cent of total employment is located in the CBD, 17 to 20 per cent in primary centres, 6 to 7 per cent in secondary centres, and 3 to 4 per cent in isolated centres (Table 5.2).

The main difference between these two cities lies not so much in their degree and type of polarization as in the number and location of these poles. Taking only primary poles, four of Toronto's contain more than 80,000 jobs (Brampton, Central Mississauga, Markham, and Vaughan), and these poles form an arc about 20 to 30 kilometres from downtown: this reflects the 'edge city'/polycentric phenomenon, and it is further corroborated by the fact that certain high-order functions locate in these sub-centres (Shearmur and Coffey 2002). Montreal, on the other hand, only has one major sub-centre (Ville Saint-Laurent–Dorval), containing more than 200,000 jobs—almost as large as, but far more sprawling than, Montreal's CBD. It is primarily a light-industrial, warehousing, and transport hub with little high-order tertiary activity. Montreal's other sub-centres are far smaller (central Laval, 60,000 jobs; then Anjou, 40,000; and Chabanel, 37,000). Thus, Toronto is a polycentric city with strong and fairly diversified sub-centres, whereas Montreal is a monocentric city with one, strong, industrial sub-centre and some smaller industrial and/or lower-order service-oriented poles.

Vancouver is quite different. Its CBD gathers 20 per cent of all the agglomeration's employment. This may be partly attributable to the more sprawling nature of its CBD: there is a stronger mix of residential and economic activities there (as witnessed by its low employment to labour force ratios), and it physically covers a wider area than either Montreal's or Toronto's. While Vancouver's primary poles contain about the same proportion of jobs as those in Toronto and Montreal, the high proportion of jobs in very small, isolated centres is unique: more than 8 per cent of jobs are in this type of pole, against only 4 per cent in Toronto and less then 3 per cent in Montreal. There is thus both a macro (strong CBD) and a micro (many smaller poles) polarization in

Vancouver, whereas in Toronto, and especially in Montreal, the CBD gathers proportionately fewer jobs and outside of the major poles, jobs are more dispersed. The Ottawa pattern resembles Vancouver's except that the CBD does not contain as many residents.

Calgary's CBD is particularly strong, gathering 26 per cent of all employment. Its primary poles—to the northwest (airport) and southwest of the CBD—gather a further 31 per cent of all jobs and have been growing fast. Thus, more than 57 per cent of all jobs in Calgary are in the CBD and its primary poles—yet its secondary and isolated employment centres gather as many jobs as Toronto's: as already noted, the areas outside of employment centres in Calgary house substantially fewer jobs than they do in the other cities.

Montreal stands out as having the highest employment to labour force ratios in its CBD, about twice Toronto's and four times Vancouver's. Given that these ratios are similar across the central cities more broadly defined (0- to 5-kilometre rings; Table 5.1), this shows that there is a considerable degree of segregation between economic activity and residential areas *within* Montreal's city centre. This may reflect its industrial heritage (renovated industrial buildings are generally not located in residential areas) and the fact that many areas close to its core (such as the Plateau, Outremont, Mile End, Westmount) retain their nineteenth- and early twentieth-century residential architecture; there are thus only a limited number of locations in central Montreal where new commercial real estate development has occurred, and it is in these places that much of the CBD employment concentrates.

The detailed study of the geography of these employment poles, their industrial structure, their evolution as they spread out along highways and around airports, and indeed their internal spatial structure could easily take over the entire book. However, a key point, already emphasized in the previous sections, re-emerges when employment poles are considered: there is no single template that will summarize and explain the spatial distribution of economic activity in Canada's major metropolitan areas. Furthermore, beyond the self-evident observation that in growing cities, growth will tend to be faster in the suburbs, it is unlikely that any single model (polycentricity, scattering of jobs, sprawl) can adequately sum up what is occurring in other Western cities. Indeed, the internal economic geography of cities often results from the interaction of a series of general processes (such as the trade-off between costs and amenities, global processes, technological change) with local industrial mix, policies, geography, and preferences (Shearmur 2007b; Shearmur et al. 2007; Keil and Young 2008). The end result of this interaction is not predictable—except in its broadest outlines—although the inertia inherent in the urban fabric and the transport infrastructure, combined with each city's economic and social history, ensures that changes are usually not abrupt, taking decades to alter existing urban form.

The Restructured Central Area and the New Inner City

The previous section highlighted the fact that Canada's inner cities—except perhaps Toronto's in the early 2000s—have been generating jobs and all have been attracting population. Thus, in Canada high levels of suburban growth and lower central growth *rates* are not incompatible with dynamic downtowns. In this section, we look below these surface figures to describe some of the processes that are currently shaping

Table 5.2 Polarization of Employment in Five Metropolitan Areas

	Toronto		Montreal		Vancouver		Ottawa		Calgary	
	1996	2006	1996	2006	1996	2006	1996	2006	1996	2006
Total employment and distribution across employment centres										
CBD	15.3%	15.1%	14.0%	14.0%	22.7%	20.8%	20.4%	19.5%	27.3%	26.0%
Primary	20.0%	23.5%	16.7%	17.5%	18.7%	20.1%	20.7%	22.9%	28.2%	31.3%
Secondary	8.6%	8.2%	6.4%	7.0%	4.4%	4.7%	1.4%	2.4%	7.6%	7.3%
Isolated	4.2%	4.7%	2.8%	3.5%	8.1%	10.5%	5.3%	5.3%	4.5%	5.7%
Not in a centre	52.0%	48.5%	60.1%	57.9%	46.1%	44.0%	52.2%	49.9%	32.4%	29.8%
% tracts not in centre		*87.0%*		*92.0%*		*64.0%*		*88.0%*		*83.0%*
Total	2,034,340	2,503,905	1,459,905	1,734,440	828,285	977,560	486,010	579,245	404,170	561,430
Employment to labour force ratios										
CBD	9.15	7.81	19.60	19.00	5.25	3.70	10.76	7.44	7.56	7.10
Primary	3.31	2.32	4.22	4.25	3.32	3.29	2.96	2.21	3.21	2.29
Secondary	2.15	1.96	2.95	2.54	2.37	2.70	1.66	0.95	3.08	3.98
Isolated	1.74	1.85	1.18	1.24	1.28	1.34	3.04	3.47	1.01	0.88
Not in a centre	0.61	0.60	0.66	0.66	0.56	0.56	0.59	0.60	0.40	0.40

Annualized employment growth %

	96–01	01–06	96–01	01–06	96–01	01–06	96–01	96–01
CBD	3.8%	0.1%	2.4%	1.2%	0.3%	1.2%	1.3%	2.8%
Primary	4.8%	2.6%	4.1%	0.4%	3.6%	1.2%	2.8%	4.4%
Secondary	2.7%	0.6%	2.5%	2.6%	1.5%	3.0%	7.4%	3.0%
Isolated	4.7%	2.0%	4.3%	3.8%	4.2%	4.4%	1.8%	5.9%
Not in a centre	2.0%	0.8%	1.4%	1.4%	1.1%	1.3%	1.3%	2.5%
Total	**3.0%**	**1.2%**	**2.1%**	**1.3%**	**1.7%**	**1.6%**	**1.8%**	**3.3%**
Employment growth, absolute								
CBD	63,420	2,745	25,435	13,900	3,295	11,635	14,055	35,250
Primary	107,110	75,959	53,860	6,695	30,240	11,160	31,950	61,650
Secondary	24,990	5,555	12,595	14,320	2,745	6,325	7,190	10,610
Isolated	21,900	11,225	9,620	10,210	15,315	19,760	5,120	13,685
Not in a centre	109,725	47,300	61,315	66,585	21,935	26,865	34,920	36,065
Total	**327,145**	**142,420**	**162,825**	**111,710**	**73,530**	**75,745**	**93,235**	**157,260**

central cities. In doing so, we move from a strictly statistical definition of the CBD to a more loosely defined concept of 'central area' (or 'metropolitan core')—i.e., the area of downtown generally recognized, either by the local population, local businesses, or local planning authorities, as the central business area. Needless to say, the two concepts refer to the same general location within the metropolitan area, whereas the statistical definition allows quantitative comparisons to be made between CBDs on the basis of a reproducible entity, the more loosely defined concept of central area is finer-grained and allows for residential as well as employment zones to be considered.

Why focus on the central area? This area, while including, as we have just seen, a diminishing fraction of the metropolitan economy as a whole, nonetheless contains a disproportionate share of the most specialized and dynamic economic functions: it bears the imprint of successive rounds of industrial innovation and restructuring. It has been the focus of much research (see for example Gottmann 1974; Gad 1979; Edgington 1982; Bourne 1982; Ley 1996; Hall 1998; Hutton 2004 and 2008; Pratt 2009) that establishes a connection between the metropolitan core and the development of the wider metropolitan area.

Historically, as *regional central place*, the metropolitan core was characterized by a mix of administrative and market functions. To these were later added a manufacturing function that largely defined the emergence of the classic *industrial city*, which was then succeeded by the remarkable expansion of the corporate office complex, acknowledged as a hallmark of the *post-industrial city* of the late twentieth century but also as a reassertion of its central-place function. The collapse of manufacturing in the central city from the 1960s onwards generated major changes in land use and in the built environment and undercut the viability of working-class communities: these major changes are preconditions to the current restructuring of the inner city.

Processes of Restructuring in the Contemporary Metropolitan Core

Although the basic contours of the post-industrial city remain, the past two decades have seen new episodes of innovation and restructuring in the central area, as well as new experiences of transition, succession, and displacement (Hutton 2008; 2009). The late twentieth-century platform of finance and business services continues to dominate, notably in the larger globally connected metropolitan regions, although the speculative office booms of the 1970s and 1980s have given way to the more selective construction of 'designer' buildings, intent on proclaiming corporate status and identity in the global marketplace: this is part and parcel of the place-making and marketing of the globalizing city (Kaika and Thielen 2005).

The past two decades have also wrought significant changes in the stratified office workforce, seen in pressures on specific occupational groups that comprised the bulk of office workers. First, corporate mergers, takeovers, and acquisitions, coupled with the effects of deep recessions, have cut into the ranks of middle managers in many companies, producing a leaner employment profile. Second, the clerical workforce, one of the largest occupational groups in the urban economy of the 1980s, has been severely affected by the continuing substitution of capital for labour as increasingly powerful computers have transformed such office tasks as manuscript preparation and production, accounting, and data storage. The consequences for clerical workers include large-scale labour shedding, coupled with upskilling and upgrading for para-professional staff. Third, intermediate services are increasingly outsourced (as are other functions), with consultants and other service providers used as a means of controlling staff costs and of capitalizing on external sources of expertise.

Canadian cities have been subject to these pressures of competition and technology, but the effects have been unevenly spread throughout the national urban system. Toronto boasts well over 100 million square feet of office space in its CBD, a reflection of its status as Canada's global city, although new office construction has fallen off from its peak in the 1970s and 1980s. Toronto's central area contains the head offices of Canada's major banks and the majority of multinationals operating in Canada, while many back-office functions have been decentralized to Mississauga and other regional business centres in the Greater Toronto Area. In international financial and business service markets, Toronto competes not so much against the first-tier world cities such as New York and London but rather against centres such as Chicago and Boston.

Montreal lost its national banking and corporate head office primacy within Canada to Toronto during the 1970s (Polèse and Shearmur 2004a) and has since experienced slow growth of office space relative to other cities of its scale and economic importance. But a number of its intermediate services firms (business consultancies, engineering, project development services) have found new clients within international markets, in *la francophonie* and elsewhere, and constitute an important segment of Montreal's economic base. Montreal's downtown also hosts major Quebec utility corporations, resource company headquarters, and domestic financial agencies, notably the Caisse des Dépôts, as well as lively consumption landscapes, part of the commingling of social and economic worlds characteristic of knowledge-based industries (Thrift and Olds 1996). Paradoxically, it may be Montreal's relatively slow growth that has ensured its central area's continued strength, since it has limited the pressure on lower-tier professional services and back-office functions to move to suburban locations.

Ottawa represents a special case, since more than one-fifth of the metropolitan labour force in Ottawa-Gatineau is employed in public administration, notably in federal agencies. Correspondingly, Ottawa's urban core is heavily given over to public administration but now also houses a significant business services sector, including many firms that consult to government. Other important central-city economy elements in Ottawa's core include tourism, culture (including major national galleries and museums), and other forms of consumption.

But the big story in the downtown office sector among Canadian cities concerns the dramatic expansion of Calgary's corporate office complex, specializing in management, finance, and business services catering to Alberta's oil and gas industry. Compared to the more balanced downtown economies and land-use mix of Canada's big three metropolitan cities, Toronto, Montreal, and Vancouver, each characterized by strong representations of cultural industries, higher education, housing, and amenities, Calgary, the leading management centre of the 'New West', projects a more mono-industrial profile based on corporate control and related commercial functions (see Chapter 11). The growing importance of Calgary as a corporate centre is underscored by the construction of the million-square-foot EnCana building, the tallest office tower west of Toronto and a flagship project for Alberta's largest metropolis.

Vancouver, on the other hand, projects an almost 'post-corporate' central area profile, since its base of regional resource corporation head offices was largely stripped by the growing concentration of control in first-tier cities, shaped by deregulation and globalization. What remains (in a central area comprising approximately 25 million square feet of floor space) is a platform of provincial utilities, Crown corporation head offices, and a dynamic SME-scale services economy with (like Montreal's) a significant if sectorally selective presence in international markets, notably within the Asia-Pacific.

From Post-industrial to 'New Industrial' in the Inner City?

Within the central area, a sequence of innovations has transformed the industrial mix, labour force, social morphology, and built environment. The tentative recolonization of the inner city by artists and designers over the 1970s and 1980s was a precursor to the innovation and restructuring that followed, including the technology-driven 'new economy' phenomenon composed of new media industries such as video game production, computer graphics and imaging, digital arts and design, Internet services, and software design. The labour force engaged in these new economy industries includes 'neo-artisanal' workers—cohorts combining advanced technological skills with creative talent in the arts and applied design (Norcliffe and Eberts 1999), including freelancers who associate with firms on a project-by-project basis (Grabher 2001).

Many of the dot.coms were swept away in the tech-crash of 2000, which followed a period of oversupply and inflated stock values. What has emerged from the ashes is an apparently more robust 'new cultural economy' and an urban 'creative class'. Indeed, this cultural makeover of the central city has its roots in the 1980s, particularly in the establishment (often community- or government-led) of precincts of galleries, museums, exhibition space, and interactive leisure and recreation areas: it is typified by the well-known prototype Yerba Buena project in San Francisco's South of Market Area (SOMA), followed by a succession of such projects in London, Berlin, and Barcelona, as well as Canadian cities, notably Toronto and Montreal (Evans 2001).

The extent to which the cultural economy penetrates beyond these rather superficial manifestations to modify the functions of the central area's economy is an open question. The central city has historically been associated not only with markets but with the accommodation, leisure, and cultural activities that often accompany markets. Amid the 'new' economy of amenity and spectacle there operates a specialized production sector—a new generation of high-order services (a traditional central-area function) comprising new media industries as well as established, 'technologically retooled' industries such as architecture, fashion design, and industrial design. As is the case for each new generation of economic actors, many of these firms are small, operate within thin profit margins, and experience the volatility generated by competition and shifting market demands and tastes.

These firms therefore operate within the normal economic space of markets, competition, and production networks, but it is argued that they derive particular benefit from some specific attributes of the central city, including the micro-scale spatiality of inner-city precincts, the distinctive built environment and amenity package of the inner city, and the resultant 'milieu effects' of these precincts, as demonstrated by a number of cases studies (see Indergaard 2004 for a study of new media in Manhattan's

'Silicon Alley', Lloyd 2006 for an exposition on connections between art and commerce in the 'neo-Bohemian' neighbourhood of Wicker Park, Chicago, and Pratt 2009 for a saga of the Hoxton [London] new economy site). The distinctive 'social density' of inner-city districts, combined with a mixture of human, social, and cultural capital in the urban core, attract cultural production and creative industries and labour in a manner reminiscent of Marshall's (1927) inner-city manufacturing districts of the late nineteenth century.[5]

Despite the volatility and turnover of firms and enterprises associated with the new economy of the central area, the cultural sector has attracted the attention of city officials and planners, as well as the natural constituencies of interest in the arts community. Public policies and programs have been instrumental in the development of the cultural economy of the city; they include infrastructure provision (e.g., galleries, museums, and exhibition and performance space), public realm improvements and 'liveability' measures, heritage programs, institutional and educational investments, and urban regeneration and business improvement area programs. In recent years, these initiatives have entered the mainstream of urban economic development strategies, vigorously promoted by Richard Florida (see Florida 2002; 2008), although it remains unclear to what extent the cultural economy is a driver or a consequence of wider economic changes.

The cultural, leisure, and incubator functions of the central city are not new phenomena, and neither are the milieu effects that such dense concentrations of activity allow. Indeed, Florida's model of the creative class as successor to the service elites of the 'new middle class' has been actively contested on ideological, empirical, and policy grounds. The universal application of the type of cultural program he prescribes across a very broad spectrum of urban typologies has been vigorously questioned (Evans 2009; Scott and Storper 2009; Shearmur 2007a), while the apparent underestimation of creative industry volatility, labour market mismatches, and localized dislocations has been seen as a serious blind spot (Peck 2005; Catungal, Leslie, and Hii 2009).

The Social Reconstruction of the Central Area and Its Economic Implications

The growth of residential development in the urban core since the 1990s, evidenced by the increase in population relative to jobs in the CBDs of Canadian cities (Table 5.2), has been portrayed as the social correlate of the core's revitalization. Here, we can point to the preference of new economy and creative class workers for 'loft living' (after Zukin 1989), enabled by the adaptive re-use of heritage buildings, and to the proliferation of live–work lifestyles within the landscapes of the city core. The creative ideal consists in part of a co-location of work, residence, and recreation within the renovated central city (Ley 2003): certainly the contemporary (re)development of central areas in Montreal, Toronto, Vancouver, and Halifax incorporates a juxtaposition of economic activity and residential uses. The development of new condominium neighbourhoods has even been advanced as an instrument of economic development, as well as a social program, as evidenced in Vancouver's 'Living First' program in the post–Central Area Plan (1991) era (see Chapter 13).

However, the social reconstruction of the central city carries with it far more complex and problematic aspects for the renewed economy of the urban core, and they must be acknowledged in our profile of restructuring experiences. First, within the heart of the metropolitan core we observe a forest of new condominiums, both one-off developments and larger, more integrated mega-projects (Olds 2001). Conzen's (2006) depiction of the infiltration of condos and other forms of new housing in Chicago's famous Loop, the world's first central high-rise office district, represents a case in point, and there are important Canadian examples in Toronto, Vancouver, and other cities. Second, the past decade or so has seen examples of adaptive re-use of older commercial buildings for condominium development and apartments in many of these cities. Third, many of the new high-rise towers in the CBD (such as the Shangri-La and Shaw towers and the Wall Centre in Vancouver) have been designed, built, and marketed as mixed-use structures, with a layering of office, hotel, residential, and retail uses. Together, these forms of development in the central city have significantly altered the balance of land use for employment and housing, diluting to an extent the office functions in this critical zone of the metropolis but also diversifying them. Naud, Apparicio, and Shearmur (2009) describe a process of co-gentrification (i.e., social and economic gentrification) in Montreal, whereby high-order services and their professional employees are co-locating in central districts around the traditional core, with older industrial buildings and workers' houses being converted for these new occupants.

The redevelopment just described can sometimes compromise the viability of new industries and firms. Rezoning and land-use change in the inner city have tended to favour new housing and public realm enhancements over employment-generating uses. Further, the inflation of land values and rents produced by the renewed attraction of inner-city property drives a filtering process for economic activity, which Pratt (2009) has described as a form of 'industrial gentrification' in the London case, similar to Montreal's 'co-gentrification'. Together, these contemporary forces—technological change, constant experimentation in design and creativity, new divisions of production labour and social class reformation, the behaviour of the property market, and policy factors—all contribute to a context of ongoing restructuring in the central city, underscoring its role as a unique zone of innovation. And while the way in which new activities are overtaking the old is particular to each specific cycle of development, the process by which older activities are expelled as newer ones develop in the central area is (almost) as old as cities themselves (Bourne 1971).

The Evolution of Peri-metropolitan Spaces

The spatial organization of Canada's metropolitan areas is evolving, and the first section of this chapter describes what is happening and discusses some of the key processes that are affecting change and renewal within metropolitan areas. In this section, we turn to an examination, in general terms, of the territory close to—but outside of—these metropolitan areas. These areas are variously referred to as peri-metropolitan (Holmes 2008) or central (Polèse and Shearmur 2004b) and correspond to areas within about an hour's drive of a major city.[6] This distance criterion, although arbitrary (why not 110 kilometres? or 98.5 kilometres?), illustrates an important attribute of the areas being discussed: they are close enough to a major city for individuals—whether professionals, retirees, or manufacturers—who require regular but not necessarily daily access to its services, to its international airport, to its business networks, or to its markets. They can live there, often in semi-rural surroundings (Bruegmann 2005), and remain connected to the global space of flows (Castells 1996).

Peri-metropolitan spaces are increasingly sites of tension. First, and most recently, they have become an area of tourism, rural living, and leisure consumption for (ex-) city dwellers (Bruegmann 2005; Holmes 2008; Costello 2007, Desfor et al. 2006). In this respect, they have come to the attention of a variety of scholars who may previously have considered them beyond the remit of urban studies (Bruegmann 2005). Second, and more traditionally, peri-metropolitan spaces gather considerable primary industries and resource-based communities. Agriculture surrounds large cities: even in an era when much food is shipped across the world, dairy products and fresh fruit and vegetables are often produced locally. But metropolitan areas also require large quantities of building materials—bulk products such as sand, gravel, and other basic construction materials. They are usually produced and transported from local sites: again, the 100-kilometre radius is important. It allows for the delivery of produce and materials early in the day without forfeiting a whole day's work. A third function of these areas is as sites for manufacturing and wholesale activities: manufacturing industries have been priced out of urban locations and, if they haven't relocated to low-cost countries, now tend to locate on cheaper sites within easy reach of markets and major transport networks. Similarly, land-intensive wholesale and storage activities need to be located close to markets but can afford to be on their periphery. Here too, the 100-kilometre distance—or about an hour's drive at most—ensures easy face-to-face access to clients and business services in the metro area and the possibility of delivering products within a short time frame.

Needless to say, these various uses are incompatible (Holmes 2008; Costello 2007; Desfor et al. 2006)—or, to state things a little less bluntly, tension can emerge between rural idylls, agricultural odours, and tractor trailers. These conflicts occur in the economic as well as in the social spheres. Philips (1993; 2005) has coined the term 'rural gentrification': as people closely connected with metropolitan economies encroach on communities that rely on low-paid and traditional activities, they have attempted to impose their vision of the countryside and, indirectly, have priced local people out of the housing market (Desfor et al. 2006). This gentrification process, however, is not only occurring in rural areas but also in some small towns and cities within 100 kilometres of major metropolitan areas (Costello 2007).

It is important to emphasize that although this issue has been treated by a number of Canadian scholars, it has not (yet?) become a key theme in the development literature[7]: this is partly because, unlike in Europe, the vast spaces and high availability of land surrounding Canada's cities have tended to minimize conflict. It can also be attributed to the fact that until recently, sprawl has been occurring *within* the boundaries of Canada's large metropolitan areas. However, as Canada's population has grown from 21 million in 1970 to 33 million in 2009, and as its cities have grown commensurately (Toronto from 2.7 to 5.5 million, Montreal 2.7 to 3.5, Vancouver 1 to 2.2, Calgary 0.4 to 1.1, and Ottawa, 0.6 to 1.2), so pressures in peri-metropolitan settings have increased (Desfor et al. 2006).

Peri-metropolitan areas have, since the 1970s, represented about 27 per cent of

Canada's population and 25 per cent of its employment (Table 5.3). This stable proportion has been maintained in a context of rapid absolute growth. These areas have higher dependency ratios than both major metropolitan areas and remote peripheral regions. This reflects three possible trends: the first is an aging population, with retirees downsizing their urban residences and moving to smaller towns or rural areas. The second is families choosing to live there because of the lower residential costs.

Table 5.3 Canada's Peri-Metropolitan Areas: Some Basic Statistics

	1971	1981	1991	1996	2001	2006
Population						
Metro	42.6%	43.3%	45.6%	46.3%	47.7%	48.6%
Peri-metro	26.9%	26.7%	26.7%	26.7%	26.7%	26.8%
Peripheral	30.4%	30.0%	27.7%	27.0%	25.6%	24.6%
Canada total	21,568,080	24,083,210	26,993,515	28,523,875	29,639,340	31,241,049
Employment						
Metro	45.7%	47.6%	48.5%	48.3%	49.4%	50.3%
Peri-metro	25.8%	24.6%	25.0%	25.2%	26.0%	25.3%
Peripheral	28.5%	27.8%	26.5%	26.4%	24.6%	24.4%
Canada total	7,752,098	10,929,222	12,814,564	13,278,595	14,984,185	16,006,439
Dependency Ratio						
Metro	2.59	2.01	1.98	2.06	1.91	1.88
Peri-metro	2.91	2.39	2.24	2.27	2.03	2.07
Peripheral	2.97	2.37	2.21	2.20	2.06	1.97
Canada total	2.78	2.20	2.11	2.15	1.98	1.95

Population Growth Rates	1970s	1980s	1991–1996	1996–2001	2001–2006
Metro	1.28%	1.67%	1.39%	1.37%	1.46%
Peri-metro	1.02%	1.13%	1.13%	0.79%	1.12%
Peripheral	0.95%	0.36%	0.61%	-0.31%	0.23%
Canada total	1.11%	1.15%	1.11%	0.77%	1.06%

Employment Growth Rates	1970s	1980s	1991–1996	1996–2001	2001–2006
Metro	3.90%	1.80%	0.65%	2.90%	1.69%
Peri-metro	3.02%	1.78%	0.87%	3.08%	0.73%
Peripheral	3.25%	1.09%	0.69%	0.95%	1.21%
Canada total	3.49%	1.60%	0.71%	2.45%	1.33%

Note: Metro areas are defined as the census metropolitan areas (1991 boundaries) of Toronto, Montreal, Vancouver, Ottawa-Gatineau, Calgary, Edmonton, Winnipeg, and Quebec City.

Peri-metropolitan areas are urban and rural areas within 100 kilometres or so of the metro areas.

Peripheral areas are urban and rural areas beyond 100 kilometres or so from a metro area.

The final trend is out-migration slower than that from Canada's remotest areas: it is well documented that young people from remote areas, faced with poor employment prospects and fewer educational opportunities, move toward Canada's larger cities. Canada's peri-metropolitan areas, in contrast, have benefited from fast employment growth, at least during the 1980s and 1990s, and have access to opportunities in metropolitan areas. This reduces the pressure to migrate, even if opportunities do not materialize for all individuals. As we will see, there are areas of deprivation and poverty in the peri-metropolitan fringe, not only gentrifying pastoral hamlets.

Indeed, the first years of the twenty-first century have apparently not been good for peri-metropolitan regions, at least from an economic perspective: while peripheral regions enjoyed an employment boom (fuelled not only by Alberta's and Newfoundland's oil extraction industries but also by sustained demand for raw materials of all sorts) and while metropolitan regions also enjoyed strong employment growth despite the dot.com crisis, peri-urban employment growth has been slow (Table 5.3). However, population growth has remained strong, although slower than in Canada's major metropolitan areas, which still capture the highest proportion of new population (principally through immigration).

These general trends, although they provide some contextual and historical information, give only a very sketchy idea of the processes underway and do not distinguish between different cities. In order to obtain some insight into underlying processes, and to focus on similarities and differences between individual cities, the industrial structures (see Table 5.4) and socio-demographic profiles found in Canada's peri-metropolitan areas will briefly be described.

Economic Structure

Peri-metropolitan areas enjoy fairly diversified industrial profiles, with a marked presence of manufacturing, warehousing, leisure and entertainment, and a variety of personal services. These areas, as already mentioned, are weekend tourist destinations for the residents of metropolitan areas, residential areas (increasingly for people whose economic activity is focused on the metropolitan area), and sites of industrial production in lower-order manufacturing sectors (which include parts of the agricultural value-added chain such as the manufacture of food and beverages). The agricultural function does not stand out, because even though it is land-intensive it is decreasingly employment-intensive, especially in relation to the service sectors that are expanding in these areas.

The economic profiles of central areas strongly reflect the region where they are located: medium-tech manufacturing (CL14 and 17—see Table 5.4), as well as construction and services (CL20), seem to characterize areas around Toronto. Peri-urban space around Montreal and Ottawa is distinguished by the presence of high-tech manufacturing (CL26). Montreal's peri-metropolitan areas also display a particularly Québécois profile that combines first and second transformation manufacturing, personal services, and finance (CL70). Calgary's surrounding areas, on the other hand, reflect structures that are particular to western Canada—i.e., heavily specialized in the primary sector (CL24 and 32).

Table 5.4 Industrial Profiles Found across Canadian Regions

CL34	Information services, finance, insurance, technical business services, public administration
CL14	Medium-tech manufacturing, high-tech manufacturing, warehousing, all high-order services
CL20	Construction, retail, real estate, professional business services, administrative support, entertainment, hotels and restaurants
CL75	Construction, real estate, professional and technical business services, administrative support, entertainment, hotels and restaurants
CL15	Retail, education, health, other organizations
CL70	First and second transformation, medium-tech manufacturing, retail, finance, health, maintenance and repair.
CL16	First and second transformation, transport, other organizations
CL22	Public administration
CL18	Primary, support to transport, hotels and restaurants
CL19	Support to transport, education, health, public administration, other organizations
CL26	High-tech manufacturing, warehousing, support to transport, maintenance and repair
CL17	First and second transformation, medium-tech manufacturing, warehousing, insurance, maintenance and repair
CL24	Primary, construction, transport, maintenance and repair
CL32	Primary

Note: This classification is based on a 23-sector disaggregation of employment across 413 regions in Canada. Cluster analysis was used to identify industrial profiles. Those most prevalent in peri-metropolitan areas are referred to in the text.

Thus, only cautious generalizations should be made about peri-metropolitan spaces: while it is safe to say that they tend to be diverse and often combine manufacturing, service, and entertainment functions, the nature of this diversity differs depending on the city. Indeed, each metropolitan area is located within a wider regional economy (itself within a wider national and global system): thus, for instance, some of Toronto's outer suburbs encompass functions linked to the North American car manufacturing complex, and, at least in 2006, the strong presence of lower-order manufacturing (CL17) reflects this. Changes in this wider system—e.g., the current massive restructuring of the automotive industry—may very well alter this local profile.

Absolute size and surrounding geographic context also matter: Toronto is part of a wider megalopolis that stretches from Oshawa to Niagara Falls and extends toward Waterloo and London. Hence, within its peri-metropolitan sphere are a large number of small and not-so-small cities that exert their own influence on the location and type of economic activity. There is a high prevalence of service-oriented and urban-type industrial structures within this space, a reflection of the different geography of settlement and development around Toronto when compared to other large but more isolated cities, such as Montreal and Vancouver, that are the sole focus of their respective hinterlands.

These different spatial structures are illustrated by the distribution of population and employment around each urban core (Table 5.5): whereas Montreal, Calgary, and Vancouver are 'contained' within a 60-kilometre radius—jobs and workers are in equilibrium—Ottawa and Toronto have substantially more jobs than people within this radius. For Toronto, this reflects the fact that major peri-metropolitan urban areas (such as Oshawa and Hamilton) are themselves attracting commuters from farther afield. For Ottawa—Canada's capital city and headquarter of many federal agencies—an explanation is less forthcoming but may lie in part with the increasing number of long-distance commutes and flexible working arrangements that civil service jobs allow (and sometimes require).

Table 5.5 Cumulative Percentage of Jobs and Residents across City-Regions

	Kilometres from Core	Toronto	Montreal	Vancouver	Ottawa	Calgary
Residents	15	33.6%	52.3%	41.7%	57.4%	83.4%
(place of residence)	30	55.3%	73.5%	66.9%	66.3%	88.2%
	60	71.0%	90.5%	77.2%	85.1%	93.5%
	100	100.0%	100.0%	100.0%	100.0%	100.0%
Jobs	15	40.2%	66.8%	51.2%	67.8%	87.9%
(place of work)	30	62.6%	79.0%	69.8%	77.6%	89.9%
	60	75.0%	91.2%	77.6%	87.3%	93.7%
	100	100.0%	100.0%	100.0%	100.0%	100.0%
Jobs/residents	15	1.19	1.28	1.23	1.18	1.05
(ratio of percentages)	30	1.13	1.07	1.04	1.17	1.02
	60	1.06	1.01	1.01	1.03	1.00
	100	1.00	1.00	1.00	1.00	1.00

Distance = distance from centroid of municipality to the CBD.
Note: Given the large size of some core municipalities, the 15-kilometre figures are indicative only.

Source: Derived from Statistics Canada commuting data, 2001, ref: 97F0015XCB2001003.

Socio-demographic Structure[8]

Many peri-metropolitan areas share a socio-demographic profile that resembles that of major metropolitan areas: high salaries, a well-qualified population, a younger age profile, and a growing number of children. They have a dynamic profile that many regions would like to have. An even larger number of peri-metropolitan regions enjoy a slightly attenuated version of this profile, a profile also shared by a number of peripheral regions. However, many peri-metropolitan areas also display a less dynamic profile in which, despite population growth and growth in the number of children, incomes, full-time work, and education levels are below average. There is thus considerable contrast in peri-metropolitan regions: concentrations of poverty and low incomes can be found in these areas as well as prosperous communities. With large concentrations of employment in production, particularly in certain areas, the 'old economy' is still going through painful transformations, away from the dynamism at the metropolitan core and from the analytical gaze of most researchers. In many cases, it is traditional men's jobs (in the primary and manufacturing sectors) that are declining, with a rise in lower-paid female jobs in the service sectors. This gives rise to tensions—including questions relating to the identity and function of men in these communities—that are largely unacknowledged but have been documented in a few cases (McDowell 2003).

In different peri-metropolitan locations, a retirement profile is emerging, especially around Toronto and Montreal and also on Vancouver Island close to Victoria (an area that is almost peri-metropolitan according to our definition). In these areas, there are few children and low growth in the 15 to 34 age bracket: however, there is fast growth in the 55 to 75 age category, accompanied by a rapid growth of transfer income (either government transfers or investment income). Furthermore, a closer look at these areas confirms that they correspond to scenic parts of peri-metropolitan space, beyond a comfortable commute distance but within easy reach of a metropolitan area (e.g., areas around Georgian Bay north of Toronto and the lower Laurentians north of Montreal.) The few regions identified are probably not the only ones of their nature: the spatial units used in this analysis are rather coarse, so they cannot pick up the smaller areas and villages where the retirement process is taking hold. However, the fact that some such areas show up despite the spatial units emphasizes this phenomenon's growing importance. There is, of course, evidence of other retirement destinations in Canada (for instance, some regions of the Lower North Shore in Quebec): what characterizes the spaces highlighted in this analysis is that they are within easy reach of metropolitan areas and tend to attract rather well-off retirees.

In short, Canada's peri-metropolitan regions gather some of the fastest-growing and most demographically dynamic areas in Canada. However, not all peri-metropolitan areas are of this nature—a sizeable proportion display the low incomes and growth that also characterize many (but by no means all) peripheral regions and some neighbourhoods in large cities: these regions and neighbourhoods are either housing the remnants of the 'old economy' or have become low-wage service areas for the dynamic metropolitan cores and gentrifying peri-metropolitan spaces.

Of particular importance, despite the low number of regions concerned, is the emergence of peri-urban areas that seem to have become retirement destinations. This reflects a wider trend of residential 'downsizing' (Costello 2007) that is probably occurring across the whole of the peri-urban territory but which—especially for wealthier retirees—focuses on particularly attractive areas just beyond easy commuting distance from metropolitan areas.

Conclusion

This chapter provides a succinct overview of the different scales at which change is taking place in and around metropolitan regions and describes the nature of this change. Although the details of the changes described here are particular to Canadian cities at the turn of the twenty-first century, many of the processes described—competition between new and more traditional uses of space, new economic activity displacing the old, conversion of old buildings to new uses, emerging activities in central locations, the continued importance of centrality and density, the changing uses of space as communication and transport technologies evolve—are variations on well-established patterns of change that have shaped and reshaped urban areas since the Industrial Revolution (Bourne 1971; Pinol 2003).

In Canada, unlike in some North American cities, central areas and CBDs remain the focus of much innovation and activity, and even slow-growth areas are undergoing considerable 'churn'. New activities are emerging from the old, and over the past 10 to 15 years they have become the focus of renewed residential investment. Although this generates problems between competing users and can lead to the displacement of poorer populations in gentrifying areas, it also ensures that the CBDs of Canada's major metropolitan areas are lively, diversified, and, on the whole, safe places in which to work, live, and seek entertainment.

Suburbs, however, are increasingly providing the bulk of new employment and in many metropolitan regions already house the bulk of employment. The internal dynamics of these suburbs cannot be easily summarized but are to some extent described in the individual case studies. On the whole, employment there is organized around key employment poles. Whereas in Toronto these poles are large, are relatively far out, and tend to incorporate high-order service functions as well as manufacturing, retailing, and personal services, in most other Canadian cities these suburban centres are functionally more specialized and somewhat closer to downtown.

Peri-metropolitan spaces, the outer reaches of Canada's city-regions, are also undergoing massive change as their interaction with metropolitan areas intensifies. Their traditional role as sources of materials and food are expanded by recreational activity, residential communities, and, increasingly, retirement destinations. These spaces incorporate both high-income dynamic communities as well as communities of lower socio-economic status: social segregation is being exacerbated by the rural gentrification that is currently occurring under the impetus of demographic (residential choices linked to aging and semi-retirement) and technological (choices linked to distance work, better highways) changes. Poverty and social stress may be less visible in lower-density suburban and peri-metropolitan settings than they are in dense metropolitan neighbourhoods, but they are no less present there.

In short, this chapter illustrates how stability (strong CBDs, persistent suburban centres, average growth in peri-metropolitan spaces) and change ('churning' CBDs, increasing significance of suburbs, changing role of peri-metropolitan spaces) go hand in hand in Canada's largest cities. The driving forces behind both the stability and the change are economic, social, and political: in particular, processes of globalization, social polarization, industrial restructuring, planning, politics, and, of course, conservatism and inertia are shaping the geography of these cities (Desfor et al. 2006; Williams 1999; Catungal, Leslie, and Hii 2009; Shearmur and Alvergne 2003). This chapter has focused on the geographic *outcomes* of these forces and processes: the agency that underpins and (partly) explains these forces and processes, although not addressed in this descriptive chapter, is an important theme to which we return in a variety of ways throughout the case study chapters.

References

Boschma, R. 2005. 'Proximity and innovation: A critical assessment'. *Regional Studies* 39 (1): 61–74.

Bourne, L. 1971. 'Physical adjustment processes and land use succession: A conceptual review and central city example'. *Economic Geography* 47 (1): 1–15.

———, ed. 1982. *Internal Structure of the City: Readings on Urban Form, Growth and Policy.* 2nd edn. New York and Oxford: Oxford University Press.

Bruegmann, R. 2005. *Sprawl: A Compact History.* Chicago: University of Chicago Press.

Bryant, C., and A. Joseph. 2001. 'Canada's rural population: Trends in space and implications in place'. *The Canadian Geographer* 45 (1): 132–7.

Castells, M. 1996. *The Rise of the Network Society.* London: Blackwell.

Catungal, J.-P., D. Leslie, and Y. Hii. 2009. 'Geographies of displacement in the creative city: The case of Liberty Village, Toronto'. *Urban Studies* 46 (5/6): 1095–1114.

Coffey, W., and R. Shearmur. 2001. 'The identification of employment centres in Canadian metropolitan areas: The example of Montreal, 1996'. *Le géographe canadien* 45 (3): 371–86.

Conzen, M. 2006. 'The historicity of Chicago's contemporary metropolitan landscapes at three geographical scales'. In R.P. Greene, M.J. Bouman, and D. Grammenos, *Chicago's Geographies: Metropolis for the 21st century?* Washington, DC: Association of American Geographers.

Costello, L. 2007. 'Going bush: The implications of urban-rural migration'. *Geographical Research* 45 (1): 85–94.

Desfor, G., R. Keil, S. Kipfer, and G. Werkerle. 2006. 'From surf to turf: No limits to growth in Toronto?' *Studies in Political Economy* 77: 131–55.

Edgington, D. 1982. 'Organisational and technological change and the future of the central business district: An Australian example'. *Urban Studies* 19: 281–92.

Evans, G. 2001. *Cultural Planning: An Urban Renaissance?* London: Routledge.

———. 2009. 'Creative cities, creative spaces and urban policy'. *Urban Studies* 46 (5–6): 1003–40.

Florida, R. 2002. *The Creative Class—and How It's Transforming Work, Leisure, Communities and Everyday Life.* New York: Basic Books.

———. 2008. *Who's Your City?* New York: Basic Books.

Gad, G.H.K. 1979. 'Face-to-face linkages and office decentralization potentials: A study of Toronto'. In P.W. Daniels, *Spatial Patterns of Office Growth and Location.* London: Wiley.

Gordon, P., and H.W. Richardson. 1996. 'Beyond polycentricity: The dispersed metropolis, Los Angeles, 1970–1990'. *Journal of the American Planning Association* 62: 289–95.

Gottmann, J. 1974. 'Urban centrality and the interweaving of quaternary activities'. *Ekistics* 29: 322–31.

Grabher, G. 2001. 'Ecologies of creativity: The village, the group, and the heterarchic organisation of the British advertising industry'. *Environment and Planning A* 33: 351–74.

Hall, P. 1998. *Cities and Civilisation.* London: Weidenfeld and Nicolson.

Holmes, J. 2008. 'Impulses towards a multifunctional transition in rural Australia: Interpreting regional dynamics in landscapes, lifestyles and livelihoods'. *Landscape Research* 33 (2): 211–23.

Hutton, T. 2004. 'Post-industrialism, post-modernism and the reproduction of Vancouver's central area: Retheorising the 21st century city'. *Urban Studies* 41 (10): 1953–82.

———. 2008. *The New Economy of the Inner City: Restructuring, Regeneration and Dislocation in the 21st Century Metropolis.* London and New York: Routledge.

———, ed. 2009. 'Trajectories of the new economy: An international comparison of regeneration and dislocation in the inner city'. *Urban Studies* 46 (5/6: special theme issue).

Indergaard, M. 2004. *Silicon Alley: The Rise and Fall of a New Media District.* London and New York: Routledge.

Kaika, M., and K. Thielen. 2005. 'Form follows power: A genealogy of urban shrines'. *City* 10: 59–69.

Keil, R., and D. Young. 2008. 'Transportation: The bottleneck of regional competitiveness in Toronto'. *Environment and Planning C* 26: 728–51.

Ley, D. 1996. *The New Middle Class and the Remaking of the Central City.* Oxford: Oxford University Press.

———. 2003. 'Artists, aestheticization and the field of gentrification'. *Urban Studies* 40: 2527–44.

Lloyd, R. 2006. *Neo-Bohemia: Art and Commerce in the Postindustrial City.* New York and Abingdon, UK: Routledge.

McDowell, L. 2003. *Redundant Masculinities? Employment Change and White Working Class Youth.* London: Blackwell.

Marshall, A. 1927. *Principles of Economics.* London: McMillan.

Naud, D., P. Apparicio, and R. Shearmur. 2009. 'Co-gentrification sociale et économique? La co-localisation de la main-d'œuvre et des emplois de services aux entreprises à Montréal, 1996–2001'. *Cahiers de géographie du Québec* 53 (149): 197–220.

Norcliffe, G., and D. Eberts. 1999. 'The new artisan and metropolitan space: The computer animation industry in Toronto'. In J.-M. Fontan, J.-L. Klein, and D.-G. Tremblay, eds, *Entre la metropolisation et la village global: les scenes territoriales de la conversion*. Québec: Presses de l'Université du Québec.

Olds, K. 2001. *Globalization and Urban Change: Capital, Culture and Pacific Rim Megaprojects*. Oxford: Oxford University Press.

Park, R., E. Burgess, and R. McKenzie, eds. 1925. *The City*. Chicago: Chicago University Press.

Peck, J. 2005. 'Struggling with the creative class'. *International Journal of Urban and Regional Research* 29: 740–70.

Phillips, M. 1993. 'Rural gentrification and the processes of class colonization'. *Journal of Rural Studies* 9 (2): 123–40.

———. 2005. 'Differential productions of rural gentrification: Illustrations from North and South Norfolk'. *Geoforum* 36 (4): 477–94

Polèse, M., and R. Shearmur. 2004a. 'Culture, language and the location of high-order service functions: The case of Montreal and Toronto'. *Economic Geography* 80 (4): 329–50.

———. 2004b. 'Is distance really dead? Comparing industrial location patterns over time in Canada'. *International Regional Science Review* 27 (4): 431–57.

Pinol, J.-L. 2003. *Histoire de l'Europe urbaine II: de l'Ancien Régime à nos jours*. Paris: Seuil.

Pratt, A. 2009. 'Urban regeneration: From the arts "feel good" factor to the cultural economy: A case study of Hoxton, London'. *Urban Studies* 46: (5/6).

Scott, A.J., and M. Storper. 2009. 'Rethinking human capital, creativity and urban growth'. *Journal of Economic Geography* 9 (2): 147–67.

Shearmur, R. 2007a. 'The new knowledge aristocracy: A few thoughts on the creative class, mobility and

urban growth'. *Work, Labour, Organisation and Globalisation* 1 (1): 31–47.

———. 2007b. 'The clustering and spatial distribution of economic activities in eight Canadian cities'. *International Journal of Entrepreneurship and Innovation Management* 7 (2/3/4/5): 111–38.

Shearmur, R., and C. Alvergne. 2003. 'Regional planning policy and the location of employment in the Île-de-France: Does policy matter?' *Urban Affairs Review* 39 (1): 3–31.

Shearmur, R., and M. Charron. 2004. 'From Chicago to L.A. and back again: A Chicago inspired quantitative analysis of income distribution in Montreal'. *Professional Geographer* 56 (1): 109–26.

Shearmur, R., and W. Coffey. 2002. 'A tale of four cities: Intra-metropolitan employment distribution in Toronto, Montreal, Ottawa and Vancouver'. *Environment and Planning A* 34: 575–98.

Shearmur, R., W. Coffey, C. Dubé, and R. Barbonne. 2007. 'Intrametropolitan employment structure: Polycentricity, scatteration, dispersal and chaos in Toronto, Montreal and Vancouver, 1996–2001'. *Urban Studies* 44 (9): 1713–38.

Thrift, N., and K. Olds. 1996. 'Refiguring the economic in economic geography'. *Progress in Human Geography* 20: 311–37.

Torre, A. 2008. 'On the role played by temporary geographical proximity in knowledge transmission'. *Regional Studies* 42 (6): 869–89.

Williams, G. 1999. 'Metropolitan governance and strategic planning: A review of experience in Manchester, Melbourne and Toronto'. *Progress in Planning* 52: 1–100.

Zukin, S. 1989. *Loft Living: Culture and Capital in Urban Change*. New Brunswick, NJ: Rutgers University Press.

Notes

1. Detailed place-of-work data covering 1996, 2001, and 2006 period are available for Montreal, Toronto, and Vancouver. Only the 1996 and 2006 data are available for Ottawa and Calgary.

2. For Calgary and Ottawa, the contiguity criterion is sometimes problematic because of the large area of certain suburban tracts. In these cases, street and land-use maps have been referred to in order to determine whether contiguous tracts in fact constitute single or multiple employment centres.

3. There is a line of thinking that suggests that a more 'objective' way of defining employment centres is to identify statistically significant deviations from a density gradient; however, the threshold of statistical significance is chosen by convention. It is doubtful that any method that

classifies geographic units as 'centres' and 'not centres' can escape from a degree of arbitrariness.

4. This ratio only gives an indication of net commuting, not of actual commuting flows.

5. For a nuanced view of the importance of co-location with regard to these milieu effects, see Boschma 2005 and Torre 2008: these authors argue that co-location may not be a *necessary* prerequisite for intense information and knowledge exchange.

6. Whether a city is major or not depends on the wider context and the purpose of the study: in Canada, an agglomeration of more than 500,000 can be considered 'major' at the national scale, but one of 50,000 could be major in the context of an isolated region.

7. Bryant and Joseph (2001) provide a good over-
    view of the Canadian literature up to 2001.
    Desfor et al. (2006) also discuss these issues in
    the context of Toronto's expansion.

8. This section is based on a factor analysis of a
    series of variables, followed by a cluster analy-
    sis of the factors. The results are available upon
    request.

6

Economic Restructuring and Trajectories of Socio-spatial Polarization in the Twenty-First-Century Canadian City

R. Alan Walks

Introduction

Economic restructuring, globalization, and shifts in welfare state policies and approaches are imprinted in the changing social geography of Canada's cities. The postwar period witnessed massive growth in both the populations and the economies of Canada's five largest cities. However, although Canadian cities remain relatively less socially polarized than their counterparts in the United States,[1] and despite increasing wages and affluence, Canada's five largest cities have also experienced growing social disparities and spatial inequalities at multiple scales of analysis. Regional specialization and outward expansion of many residential and employment uses has been accompanied by gentrification of the inner city, a loss of affordable rental housing, declining relative incomes for recent immigrants, and the deterioration of public spaces and services. Shifts in the distribution of earnings, coupled with the spatial reorganization of Canadian metropolitan areas, have led to growing socio-spatial polarization.

This chapter considers the socio-spatial impact of economic and policy restructuring in Canada's five largest urban regions over the past quarter-century. It begins by situating the processes of globalization, economic restructuring, and the neo-liberal shift in welfare state policies within the context of growing Canadian metropolitan regions. It then empirically analyzes trends in labour market inequalities among occupational and social groups concentrated in metropolitan areas, with an eye to delineating factors producing divergent incomes. Examining the levels of income inequality and poverty in Canada's largest metropolitan areas, the chapter interrogates how inequality has been articulated spatially in the residential fabric and elucidates the effects of economic and policy changes on the geography of advantage and disadvantage in the contemporary metropolis. It concludes by discussing the implications of the shifting social geography of the twenty-first-century city within the context of the contemporary global economic transformation.

Globalization, Neo-liberalism, Economic Restructuring, and the Canadian City

As a country with one of the highest proportions of its GDP arising from trade, as demonstrated by a consistent trade-to-GDP ratio of approximately 82 per cent,[2] as well as one of the globally highest national immigration rates, Canada is intensely linked to the global flows of commodities, investment, and people and in turn highly vulnerable to economic decisions, structures, and shocks originating outside its borders. Foreign direct investment (FDI) into Canada has grown considerably since 1970, reaching $116.7 billion in 2007, more than double the $44.7 billion (Cdn) of just seven years earlier (2000) and more than three times that of the entire decade of the 1970s ($32 billion, in current dollars) (Hurtig 2002; OECD 2009). Yet as the other chapters in this volume attest, globalization of trade and investment flows in the Canadian context has mainly meant *continentalization*—integration with the United States economy and increasing US ownership of Canadian firms and assets. This is not necessarily new. In 1968, roughly 33 to 34 per cent of Canadian assets and revenues were foreign-controlled (mostly by firms in the US), spurring the Trudeau government to enact controls through the Foreign Investment Review Act (FIRA) that brought the level of foreign control down to 21 per cent by 1985. However, by 2000, after the elimination of FIRA and the imposition of the Free Trade Agreement/NAFTA, the level of foreign control rose again to 25.5 per cent of assets and 31.5 per cent of revenues, close to their levels of the mid-1960s and even higher when energy and mining are excluded (Baldwin and Gellatly 2005).[3] Approximately 80 per cent of all Canadian exports flow to the United States in a typical year.[4] Of the approximately $487 billion in foreign investment that flowed into Canada between 1985 and 2002, a full 96.6 per cent was for the foreign takeover of Canadian firms (leaving only 3.4 per cent for investment in new production facilities, firms, or processes), and 64.7 per cent of which went to US companies (Hurtig 2002, 13). In 2001, approximately 69 per cent of all revenue from foreign direct investment in Canada flowed out to firms based in the United States (ibid., 18), and between 1985 and 1995, the percentage of *foreign-controlled* firms with no Canadian shareholders (i.e., that were 100 per cent foreign-owned) increased from approximately half to 85 per cent (ibid., 26). By 2002, the United States controlled approximately 59 per cent of all majority foreign-owned exporting establishments operating in Canada (Byrd 2005).

As the other chapters in this volume attest, globalization and continentalization have been associated with a restructuring of the space-economy. Free(er) trade with the United States after the signing of NAFTA has intensified north–south trade and economic linkages, bringing Canadian metropolitan areas ever more into the fold of a restructuring North American urban system and altering the industrial and occupational structure of many cities (Bourne and Simmons 2003). The decline of labour-intensive manufacturing in eastern Canada and the rise of the 'new west' based on trade in raw materials and commodities has resulted in a regional shift of economic and political power, while the growth of financial and business services has benefited the largest cities. From the early 1980s until the late 2000s, Toronto, then Vancouver, emerged as second-tier 'global cities' with high levels of immigration, international trade connections, and concentrations of headquarters and particular industries (Taylor, Catalano, and Walker 2002). Montreal moved from being the premier national business city to the province of Quebec's 'global city', while Calgary grew quickly to become the business centre for a burgeoning oil and gas industry. Canada's national capital region has specialized in particular high-tech industries and services linked to public administration. In each case, increasing specialization has been accompanied by greater integration into the North American urban system. Canada's dependent position in relation to the

United States, both in its level of foreign ownership and its dependence on trade, make it particularly vulnerable to any economic shocks occurring in the US economy and to protectionist policies enacted by state and federal governments there.

Economic integration has compelled an incremental drift and, in some places, a conscious shift in social policy at all levels of government in favour of the practices and policy approaches characteristic of the northern United States (Courchene and Telmer 1998; Courchene 2001). This has mainly meant the implementation of neo-liberal economic and social policy, although as Brenner and Theodore (2002) note, 'actually existing neoliberalism' is also a messy, inconsistent, incomplete, place-dependent, and politically contested process. 'Neo-liberal times' involve proposals to 'roll back' the welfare state, cut income taxes for corporations and individuals, shift social policy away from universal to targeted social programs, support the 'financial-ization' of economic development and fiscal policy, privatize state assets and services, refashion citizens as 'consumers', and force public service agencies into business management models, among other things (Hackworth 2007; Harvey 1989, 2005; Peck 2004; Peck and Tickell 2002; Wilson 2004). In the eight years between 1988 and 1996 alone, a number of large public energy, mining, transport, and telecommunications firms, worth more than $12.2 billion in total, were privatized at both the federal and provincial levels (McBride 2006, 264). Globalization is linked to the rise of neo-liberal policies in that the latter are often justified with reference to the need to compete internationally in order to maintain high levels of innovation and exports. Of course, globalization and neo-liberalism are conceptually linked, both ideologically and as political projects, through the regulatory and disciplinary functions and activities of international financial organizations such as the International Monetary Fund (IMF), the World Bank (IBRD), the World Trade Organization (WTO) and trade/investment agreements such as the General Agreement on Tariffs and Trade (GATT), the North American Free Trade Agreement (NAFTA), and the not as yet ratified Multilateral Agreement on Investment (MAI) (Panitch 2004).

This is, however, not to say that there has been outright policy convergence among nations or Canadian provinces, nor that such shifts have occurred smoothly and/or by consensus. The policy regimes implemented by provincial governments in Ontario, for instance, first under the NDP led by Bob Rae, then Mike Harris's Progressive Conservatives, and later by Dalton McGuinty's Liberals, have involved both incremental continuities as well as radical shifts in approach. And certain jurisdictions (e.g., Alberta) have been more quick to privatize public assets and companies than others (McBride 2006). Nonetheless, the overall effect has been a creeping neo-liberalism across Canadian jurisdictions, beginning with the federal government in the late 1980s and followed by many provincial regimes over the 1990s. Municipal governments, always 'creatures of the provinces' in the Canadian Constitution, have been compelled either directly by provincial legislation or indirectly by the downloading of responsibility for service delivery and revenue collection to adopt neo-liberal reforms and more entrepreneurial approaches (Boudreau et al. 2007; Clark 2002; Keil 2002; Keil and Kipfer 2003; Kipfer and Keil 2002; Miller 2007). Neo-liberalism in the Canadian context has been characterized by income tax reductions, particularly benefiting the wealthy (Lee 2007), as well as punitive 'roll backs' to welfare state programs and social benefits, which have directly increased the vulnerability of low-income households to economic shocks (Clark 2002; Pulkingham and Ternowetsky 2006; Yalnizyan 1998, 2007). Between 1984 and 1999, the proportion of the federal budget going to 'progressive' social expenditures declined from 19.4 to 12.4 per cent (McBride 2006, 261). Tightening of eligibility criteria has reduced by more than half the proportion of the unemployed receiving federal unemployment benefits, from 74 per cent to only 36 per cent in the nine years leading up to 1997 (ibid., 273). As in the United States, there have been few new programs other than targeted tax credits to make up for the loss of welfare state benefits. Thus, the neo-liberalization of Canadian social, economic, and urban policy, while occasionally taking a different form from that of the United States, has nonetheless been influenced by the globalization and continentalization of public policy. This has produced both similarities and divergences in the policy and political responses evident in different urban regions, as would be expected given the different social and sectoral bases of their economies (Boudreau et al. 2007; Donald 2005).

Although the economies of Canadian urban regions are becoming increasingly integrated with those in the United States, globalization has simultaneously meant strengthening cultural connections between Canadian cities and cities in the post-colonial south. Reforms to Canada's immigration policies in the late 1960s have transformed the population mix of its largest cities as immigrants from East and South Asia, Africa, the West Indies, and Latin America grew from 1.7 per cent of the national population in 1971 to 11.5 per cent in 2006 and made up 59.1 per cent of all immigrants to Canada (Census of Canada, 1971, 2006). The vast majority of immigrants to Canada, more than 70 per cent in 2006, end up in one of Canada's five largest cities (ibid.). Because many immigrants to Canada belong to racialized ethnic groups and self-identify in the census as visible minorities, immigration is said to have changed the 'face' of urban Canada. By 2006, visible minorities made up approximately 42 per cent of the population of Vancouver and Toronto, 22 per cent of Calgary's, and more than 16 per cent of the population in Ottawa-Gatineau and Montreal (Census of Canada, 2006). This stands in contrast with the situation only 25 years before: in 1981 the metropolitan area with the highest proportion of visible minorities (Toronto) still had only 14 per cent of its population identifying as visible minorities (Statistics Canada 2003, 44). Large flows of immigrants come from India, China, Pakistan, the Arab world, and Latin America, and many immigrants now lead transnational lives, maintaining their connections to and positions within the countries of origin (Bernhard, Landolt, and Goldring 2009; Preston, Kobayashi, and Man 2006; Walton-Roberts 2003). Generally speaking, flows of South Asians are highest into the Toronto region (where they comprised 13.5 per cent of the population in 2006), while the concentrations of Chinese and Southeast Asians are strongest in Vancouver (at 19.8 per cent of the population), and immigrants from Arab and West Asian countries are more concentrated in Montreal and Ottawa (where they make up more than 5 per cent of the CMA populations), while blacks from the Caribbean and Africa (at between 5 and 7 per cent) and Latin Americans (2 per cent of the population) are more concentrated in the two largest CMAs, Toronto and Montreal.

Urban Canada in the twenty-first century is very much transformed from the way it was only 30 years ago. Economic restructuring, population growth, immigration, and neo-liberal policy reforms have produced new patterns of social inequality. This chapter examines the impact of such changes on the social geography of the five largest metropolitan areas.

Measuring Social Polarization in the Twenty-First-Century City

This chapter traces shifts in labour market inequalities and poverty over time for the five largest census metropolitan areas in Canada: Toronto, Montreal, Vancouver, Ottawa, and Calgary. Data derive from the cross-sectional 1971, 1981, 1986, 1991, 1996, 2001, and 2006 censuses of Canada. The structural, non-spatial relationships between income and labour market positions are analyzed using data from the public-use micro-sample files (PUMS), which allow for detailed cross-tabulated analyses at the level of households and individuals within entire CMAs, while spatial patterns have been determined from analysis of custom-ordered census data aggregated at the level of census tracts (CTs). Census tracts are spatial units designed by Statistics Canada that house residential populations of a similar demographic, generally ranging in size from 2000 to 8000 people, and that adopt recognizable transportation routes and physical features for boundaries. Although it is clear that census tracts may not necessarily map to the activity spaces or social spaces of functioning neighbourhoods, in the absence of any other spatial units or data aggregated at this level they are accepted as representing the neighbourhood scale in empirical urban research.

Household and per capita incomes are key variables for understanding inequality, since they link one's position within the productive system to the ability to consume. The labour market position of those in different occupational sectors, as well as visible minorities and immigrants, provide particularly important information regarding the ways that economic shifts affect social groups in the city. A series of indices indicating degrees and forms of income inequality and polarization provide a window into how changes occurring within and across sectors of the labour market influence distributive outcomes. Inequality and polarization are related, but nonetheless separate, concepts, and this is reflected in the ways they are measured. Perhaps the best way to visualize the distinction is through the shape of their respective distributions. Inequality typically leads to a pyramid-shaped distribution with a small number of wealthy at the top and a larger number of poor at the bottom, whereas under polarization the population is divided into two large masses of roughly equal size, one wealthy and the other poor, with a declining middle between them, in an hourglass shape. In technical terms, inequality follows the Pigou-Dalton axiom, which states that income inequality measures must increase whenever income is transferred from a poorer to a richer person. Indices of polarization, meanwhile, adhere to both the bipolarity axiom, which states that increasing concentration of the population toward the medians of each of the two 'poles' on either side of the distribution must increase polarization, and the spread axiom, which states that movement away from the centre of the entire distribution must increase polarization (Esteban and Ray 1994; Duclos, Esteban, and Ray 2004). Thus, although indices of income inequality and polarization are often found empirically to shift in tandem, the two concepts are nonetheless distinct and sensitive to different types of movements in the distribution of income.

There are a myriad of different inequality and polarization metrics available, each with its own set of benefits and drawbacks. A triumvirate of three inequality measures, the exponent (EXP), gini concentration ratio (GINICR), and coefficient of variation squared (CV2), is often applied together in analyses of income inequality, since these metrics are sensitive to the lower, middle, and upper ends of the income distribution, respectively (Heisz 2007; Wolfson 1997). Measures of income polarization have appeared in the literature more recently and tend to be more difficult to calculate using traditional data sources. In particular, the Wolfson and Esteban and Ray (ER) polarization indices cannot be calculated using spatially aggregated census data divided into income bands or ranges, which is the form in which custom-ordered datasets at the level of census tracts are typically produced (see Wolfson 1997; Esteban and Ray 1994). Indices that can be calculated using spatially aggregated data may have peculiarities and drawbacks. For instance, the Wang-Tsui (WT) polarization index is more sensitive to shifts in the upper end of the income distribution (Wang and Tsui 2000), while another common statistic often employed to gauge the level of polarization—the ratio of the income of the top to bottom deciles—does not actually adhere to either the Pigou-Dalton axiom (of inequality) or the bipolarity or spread axioms (of polarization). To deal with these issues, this author has designed the Walks polarization index to be equally sensitive to movements at both the lower and upper ends of the income distribution while at the same time adhering to the bipolarity and spread axioms, although it has the drawback of not being able to handle incomes of zero.[5] Nonetheless, this index can be calculated using custom data grouped into income bands and aggregated spatially, and its resulting values correlate strongly with the Wolfson and ER indices, thus lending itself well to spatial analyses of income polarization (see Walks, forthcoming).

Growing Unequal: From Uneven Development to Social Polarization

Canada has been tagged as one of the five best countries in which to live, at least as measured by the United Nations Human Development Index since the mid-1980s (UNDP 2008). Canada's historically strong and diversified economy, high wages, investment in health and education, and urban quality of life are clearly related. Canadian cities have often enjoyed praise for their low levels of crime, high quality of services and environmental amenities, strong support for multiculturalism and immigration, and indeed relatively low levels of social disparity. However, economic restructuring and shifts in public policy over the 1990s and 2000s have conspired to increase the level of inequality across Canada, and there is evidence of mounting social tension, deteriorating infrastructure, increasingly insecure and precarious work, and declining living standards for visible minorities and new immigrants (Galabuzi 2006; Vojnovic 2006; Yalnizyan 2007; FCM 2008). Heisz (2007) calculates that for Canada as a whole, the distribution of family income trended toward greater equality in most years between 1977 and 1989, with the level of inequality reaching its lowest point (a gini coefficient less than 0.28) at the very end of the 1980s. However, this pattern reversed in the 1990s, and from 1990 to 2004 the level of inequality grew persistently (to a gini just under 0.32), as did the Wolfson index of polarization (from 0.235 to 0.265). The entirety of the growth in inequality and polarization in Canadian income is due to changes in the income distribution among those of working age, since levels of inequality actually declined during this time among seniors (ibid.). While similar trends are evident in other countries, the growth of inequality has been more dramatic in Canada than elsewhere. The Organisation for Economic Co-operation and Development (OECD 2008) demonstrates that income inequality as measured by the gini coefficient has grown faster in Canada since the mid-1990s than in any other OECD member country, except (by a slight amount) Finland, and the latter country still ended up with a level far below that for Canada by the end of the period (the gini was only 0.265 in Finland in 2004).

Growing social inequality has been driven not only by (continued) divergence of fortunes between traditionally slower- and faster-growing cities and regions but increasingly by widening disparities among social groups and classes within Canada's metropolitan areas. This is demonstrated through analysis of the household income data provided by Statistics Canada for each decennial census. In each of the five largest metropolitan areas, a similar trajectory of varying, but nonetheless upward trending, income inequality and polarization among households is discernable between 1980 and 2000 (Figures 6.1a and 6.1b, respectively). Index values of zero indicate perfect equality, while values approaching 1.00 indicate maximum inequality/polarization. While the recessions of the early/mid-1980s and early/mid-1990s led to swelling levels of inequality in most metropolitan labour markets during the middle of each decade, only to fall back subsequently, a pattern is evident whereby economic growth following each recession rarely makes up for the rise of inequality in the preceding period. Every CMA showed higher levels of both inequality and polarization in 2000 than was evident in either 1990 or 1980. These indices, furthermore, underestimate the true extent of inequality and polarization growth because of limitations in the PUMS data.[6]

The extent of poverty and deprivation is one indicator of the health of an urban labour market. In large and growing Canadian metropolitan areas, indicators of poverty have tended to rise and fall with the business cycle, peaking during the recession of the early/mid-1990s but falling subsequently. While not a true measure of poverty, the Low Income Cutoff (LICO) is often used as a proxy for a poverty line. According to analysis of the LICO, only Toronto and Vancouver consistently retained relatively heightened levels of poverty following the 1990s recession (although still falling from their 1996 peaks), while after peaking in 1996, Montreal, Ottawa-Gatineau, and Calgary all saw low income rates return to late 1980s levels when poverty was historically low (Figure 6.2a). Examination of the proportion of households with incomes below $20,000 (in constant 2000 dollars) provides another window into the shifting state of low income and the differences between CMAs (Figure 6.2b). This metric points to divergence between the two main immigration cities and the other CMAs (Figure 6.2b). For instance, Ottawa-Gatineau has seen the number of such households decline fairly steadily as a proportion of the total since 1980, while Calgary and Montreal saw their proportions of such low-income households crest in 1990 and 2000 respectively. In Toronto and Vancouver, however, the proportion of such low-income households has risen over the period, reaching its highest level in the most recent (2006) census. What such statistics do not show, however, is the depth of poverty for the households affected. According to Yalnizyan (1998; 2007), the income gap for households with incomes below the LICO across Canada has widened persistently since the early 1990s. Thus, while only Toronto and Vancouver

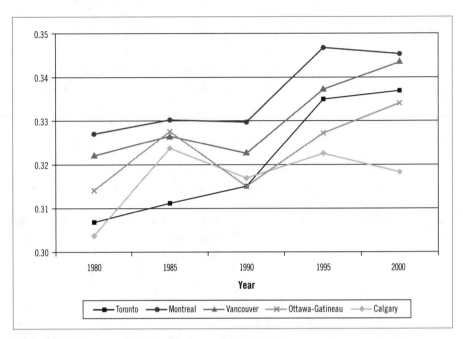

Figure 6.1a Gini concentration ratio (GINI CR) (inequality).

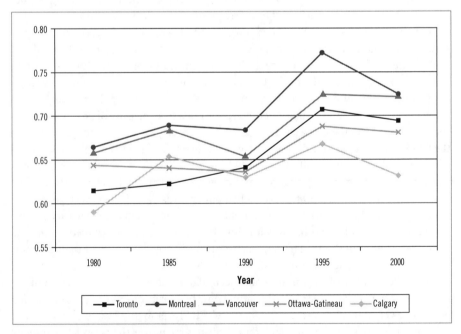

Figure 6.1b Walks polarization index (polarization).
Figure 6.1 Measures of income inequality and polarization (non-spatial) among all households, 1980–2000.

Source: Calculated from Census of Canada Public Use Microsample Files (PUMS), 1981, 1991, 2001.

Note: The PUMS file for the 2006 census (2005 data) was not yet available at the time of writing.

may have witnessed significant growth in poverty to 2006, the depth of poverty has in many cases worsened for those who find themselves poor.

Growing affluence and the increase in high-income households constitute the other components of a polarizing labour market. This has been of particular importance to the rapid growth of income inequality and polarization in Canada's cities since the late 1990s when the federal and many provincial governments reduced the tax burden disproportionately on high-income earners (Heisz 2007; Lee 2007). The proportion of households with high incomes (more than $100,000 in 2000 constant dollars) has grown significantly since 1980 (Figure 6.3a). Of the five largest CMAs, all but Montreal ended the period with proportions of such high-income households well above the Canadian metropolitan average of 16 per cent. And not only are there more rich households, but in recent years the rich have become significantly richer. For most of the postwar period, between 1945 and 1990, the share of income going to the top 5 per cent of Canadian income earners (not households) remained at roughly 24 per cent, but this rapidly ballooned to 29 per cent after 1990 (Saez and Veall 2005). Even among the rich, the greatest spoils have gone to the richest, with the wealthiest 0.1 per cent doubling their share of total income between 1990 and 2000 (ibid.). In the face of flat or slow growth in the proportions of low-income earners, the rapid increase in high-income earners has led to significant declines in the proportions of households with middling incomes (defined as incomes between $20,000 and $99,999 in constant 2000 dollars) and thus to overall income polarization (Figure 6.3b).[7]

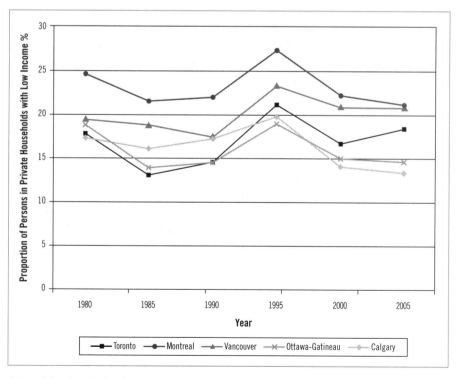

Figure 6.2a Population in private households with low income (% below the LICO).

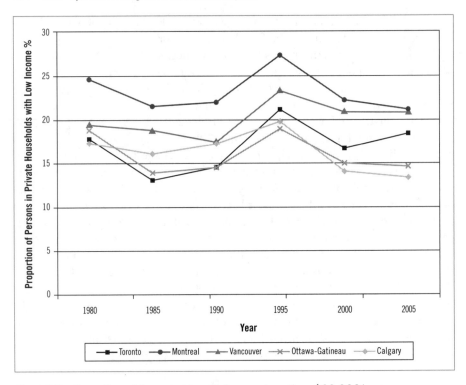

Figure 6.2b Proportion of households with income less than $20,000/year (constant 2000 dollars).

Figure 6.2 Proportion low income, 1980–2005.

Source: Calculated from Census of Canada, 1981–2006, Custom Tabulations E1171, E982; per cent below the LICO from Census of Canada, various years.

Neo-liberalism, which seeks both to reduce the barriers to accumulation (through reduced taxation and regulation) and to promote greater rewards to (the right kind of business-related) 'talent', is directly implicated in the gross polarization of earnings. It has justified both the tax reductions as well as massive bonuses and salaries for top-level management, while median earnings have mostly remained flat or have grown only slowly. Between 2006 and 2007, the incomes of the CEOs of Canada's top 100 firms, virtually all of whom reside in one of the five largest CMAs, grew from an average of $8.6 million to $10.4 million per year, with roughly 55 per cent of this deriving from the cashing of stock options, which are taxed at half the regular income tax rate (Mackenzie 2007; 2009). Between 1998 and 2007, the top 100 CEOs saw their average income rise from approximately 104 times to about 259 times the income of the average Canadian worker (Mackenzie 2009), an increase that outpaced inflation by 70 per cent. In 2007 alone, the top 100 CEOs received a 22 per cent pay increase (ibid.). Because those employed in managerial occupations (whom Richard Florida includes in the 'creative class'; see Florida 2005) can use their control over the levers of decision-making (via company boards of directors, the information they supply to shareholders, direct and indirect influence through school ties, social clubs, and so on), they are able to acquire

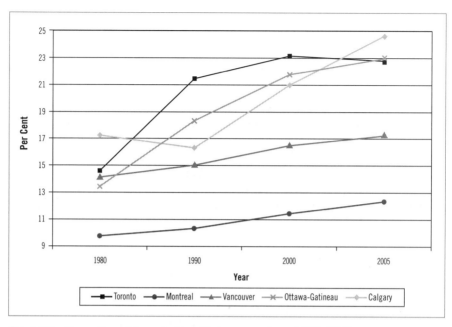

Figure 6.3a Proportion of households with incomes above $100,000/year (constant 2000 dollars).

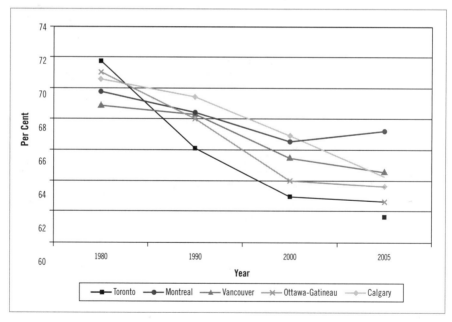

Figure 6.3b Proportion of households with incomes between $20,000 and $99,999 (constant 2000 dollars).

Figure 6.3 Polarization in incomes: From declining middle to ballooning top.

Source: Calculated by the author from Census of Canada, 1981–2006, Custom Tabulations E1171, E982.

more of the pie for themselves, in turn leaving less for those employed in the remaining occupations. Census data corroborates this phenomenon (Table 6.1). The ratio of average per capita employment incomes (compared with the CMA averages) for most of the major urban occupational groups either declined over time (for construction and

Table 6.1 Ratios of Average Employment Income, by Occupation (Compared to Total CMA Average Employment Income for the Total Labour Force in Each Census Year)

Census Metropolitan Area	1980	1990	2000	2005	1980	1990	2000	2005
	Managerial				Professionals			
Montreal	1.684	1.531	1.806	1.868	1.316	1.247	1.300	1.287
Ottawa-Hull/ Gatineau	1.761	1.457	1.701	1.749	1.294	1.228	1.250	1.278
Toronto	1.757	1.485	1.898	2.067	1.308	1.270	1.265	1.304
Calgary	1.724	1.606	1.872	2.198	1.259	1.336	1.323	1.331
Vancouver	1.583	1.605	1.702	1.834	1.266	1.254	1.294	1.295
	Manufacturing and Utilities				Construction and Transport			
Montreal	0.854	0.903	0.777	0.872	1.061	1.051	0.958	0.916
Ottawa-Hull/ Gatineau	0.900	0.885	0.772	0.870	0.949	0.926	0.824	0.744
Toronto	0.919	0.889	0.723	0.761	1.076	1.001	0.902	0.814
Calgary	0.961	0.925	0.728	0.731	1.030	0.961	0.879	0.760
Vancouver	1.053	0.994	0.786	0.795	1.119	1.067	0.981	0.921
	Arts, Literary, and Sport				Clerical			
Montreal	0.956	0.951	0.834	0.755	0.727	0.722	0.774	0.792
Ottawa-Hull/ Gatineau	1.037	1.019	0.834	0.806	0.698	0.744	0.747	0.747
Toronto	0.973	0.931	0.823	0.668	0.704	0.713	0.701	0.716
Calgary	0.804	0.793	0.712	0.567	0.649	0.715	0.695	0.653
Vancouver	0.901	0.869	0.825	0.736	0.717	0.733	0.763	0.784
	Sales		*Sales/Service		Services		*Sales/Service	
Montreal	0.946	0.893	0.630	0.590	0.683	0.646	0.630	0.590
Ottawa-Hull/ Gatineau	0.776	0.727	0.540	0.516	0.687	0.679	0.540	0.516
Toronto	0.983	0.951	0.588	0.541	0.617	0.618	0.588	0.541
Calgary	1.008	0.918	0.549	0.480	0.612	0.554	0.549	0.480
Vancouver	1.029	0.926	0.628	0.592	0.596	0.612	0.628	0.592

Source: Calculated from Census of Canada: Public Use Microsample Files, 1981, 1991; Special Interest Tabulations, 2001, 2006.

Note: Ratios are expressed in relation to the CMA average household income (CMA = 1.000).
* Sales and services are aggregated together in the Special Interest Tabulations but are separated in the Microsample data.

transport, artistic and literary occupations, sales and services, and, except in the 2000–5 period, manufacturing and processing occupations) or remained essentially flat (clerical jobs and professional occupations in medicine and health, teaching, natural and applied sciences, and social sciences and law). Yet the managerial class was able to turn around the relative decline in their employment incomes that occurred between 1980 and 1990 to their significant advantage after 1990 and through to 2005. In all CMAs except Ottawa, managerial employment incomes shot up significantly, particularly in Calgary (and note that this measure does *not* include the significant investment incomes, stock options, and so on promoted by neo-liberal policies and open to many working in this occupational segment). Thus, at least for the Canadian metropolis, Hamnett's (1994; Hamnett and Cross 1998) hypothesis linking globalization and urban wage inequality to 'professionalization' is demonstrably incorrect. Instead of professionalization leading to income and occupational polarization, it is the disproportionate share paid to the managerial sector that best explains income divergence. The professional middle class has maintained remarkably stable relative incomes over time. The result is a clear polarization of employment earnings between the managerial class and other workers after 1990 (visualized for all five CMAs simultaneously in Figure 6.4).

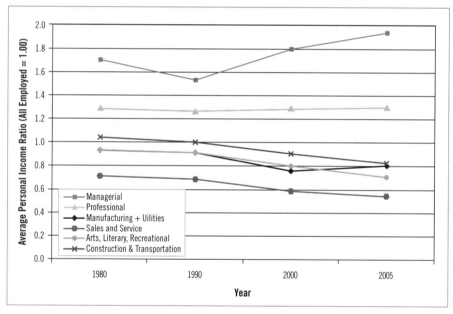

Figure 6.4 Average personal income ratios (compared to CMA-wide average) for selected occupational groups, average for all five CMAS.

Source: Calculated from Census of Canada, Public Use Microsample Files, 1981, 1991; Special Interest Tabulations, 2001, 2006.

Notes: Ratios are expressed in relation to the CMA average household income (CMA = 1.00).

Sales and services are aggregated together in the Special Interest Tabulations but are separated in the Microsample data. Aggregated sales and services data for 1981 and 1991 are calculated as a weighted estimate using the separated data for those years.

The other major factor driving social polarization on the ground in large Canadian cities is the increasing divergence in incomes between racialized workers and the white labour force. Despite Canada's reputation as one of the most welcoming and tolerant nations in the world toward immigrants, with a well-developed policy of multiculturalism (Adams 2007), since the 1980s successive cohorts of immigrants have seen their incomes fall relative to those of the native-born (Mok 2009). Visible minorities have likewise seen their levels of income relative to the white populations decline overall (Galabuzi 2006). Even with significant visible minority populations (more than 41 per cent of the populations of both Toronto and Vancouver), elite channels and political power structures are highly disproportionately dominated by native-born whites (Ray and Peake 2001), a situation characterized by Galabuzi (2006) as 'economic apartheid'. Since most visible minorities residing in Canadian cities are also immigrants to Canada (or their children), it is understandable that the income trajectories of recent immigrants and visible minorities would be very similar (Figures 6.5a and 6.5b). As expected by scholars of globalization processes occurring in cities (Sassen 2001), there have been declines in the purchasing power of visible minorities and recent immigrants in comparison with that of the non-visible minority and native-born populations. The relative employment incomes of visible minorities fell by an average of 23.5 per cent across the five CMAs between 1980 and 2000 and those of recent immigrants (compared to native-born) by 26.5 per cent.[8] Declining incomes would appear related to metropolitan size and level of visible minority concentration, with Ottawa and Calgary, two metros with smaller flows of international migrants, posting lesser declines (and even slight reversals in the 1990s), while the larger cities reveal a steeper and more consistent deterioration of minority incomes.

Economic restructuring and predominant immigration policies are also linked to growing occupational bifurcation and income polarization both within and between minority groups, not just between the white and non-white populations. Visible minorities are concentrated not only in occupational sectors with low and/or declining incomes that are vulnerable to economic shocks (manufacturing, sales and services) but also in well-paid professional occupations that place their workers firmly in the middle class (natural and applied sciences, health and medicine). On the other hand, minorities are notably under-represented among the occupational groups cited as driving gentrification (e.g., arts and literary, social sciences and law) (Table 6.2). Perhaps most important is the low proportions of minorities working in management occupations. The managerial elite is still disproportionately white, and this partially explains the increasing income gap between whites and minorities.

Thus, while Canadian cities maintain positive reputations for their relative levels of interracial harmony, tolerance, and equity, labour market dynamics within them reveal the potential for polarization, marked by increasing class and racial divisions in the twenty-first century. This phenomenon is felt particularly strongly in Canada's most 'global' cities, Toronto and Vancouver.

Spatially Dividing Cities: From Social to Spatial Polarization

Increasing social divisions are imprinted in the landscape of Canada's metropolitan areas. As the managerial class employs their 'talent' to claim higher shares of the total

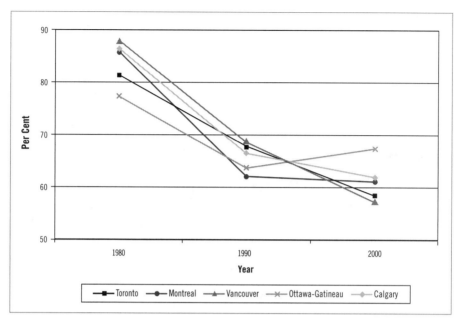

Figure 6.5a Average per capita incomes of recent immigrants as a % of native-born incomes.

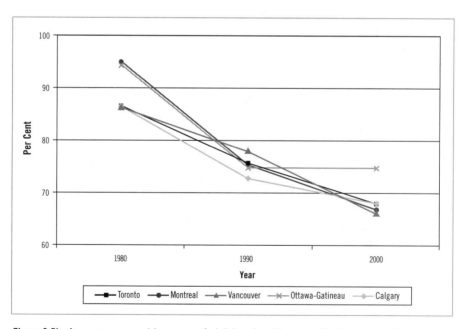

Figure 6.5b Average personal incomes of visible minorities as a % of non-minority incomes.
Figure 6.5 Declining incomes of recent immigrants and visible minorities.

Source: Calculated from Census of Canada, Public Use Microsample Files, 1981, 1991, 2001.

Note: Recent immigrants are those who immigrated to Canada in the decade preceding the census. Minority status is a self-identified variable in the census. The PUMS file for the 2006 census (2005 data) was not yet available at the time of writing.

Table 6.2 Visible Minority Concentration within Occupational Sectors, Five Largest CMAs, 2006

WHITES (%)	Vancouver	Calgary	Toronto	Montreal	Ottawa-Gatineau
Total Labour Force	59.6	77.2	58.3	83.8	83.3
Management	65.8	83.6	70.1	88.4	88.4
Business, Finance, Admin.	61.5	80.5	58.3	86.7	86.2
Natural and Applied Sciences	59.9	75.3	52.3	83.0	77.7
Health Occupations	57.1	75.7	51.9	81.2	80.5
Social Science, Law, Religion	70.1	83.8	70.7	87.6	85.2
Art, Culture, Recreation	74.1	87.4	75.4	90.1	90.1
Sales and Services	52.2	70.4	55.5	81.1	79.4
Trades, Transport	65.5	80.6	61.7	87.8	87.6
Manufacturing and Processing	36.9	48.9	31.7	69.9	71.8

VISIBLE MINORITIES (%)	Vancouver	Calgary	Toronto	Montreal	Ottawa-Gatineau
Total Labour Force	40.4	22.8	41.7	16.2	16.7
Management	34.2	16.4	29.9	11.6	11.6
Business, Finance, Administrative	38.5	19.5	41.7	13.3	13.8
Natural and Applied Sciences	40.1	24.7	47.7	17.0	22.3
Health Occupations	42.9	24.3	48.1	18.8	19.5
Social Science, Law, Religion	29.9	16.2	29.3	12.4	14.8
Art, Culture, Recreation	25.9	12.6	24.6	9.9	9.9
Sales and Services	47.8	29.6	44.5	18.9	20.6
Trades, Transport	34.5	19.4	38.3	12.2	12.4
Manufacturing and Processing	63.1	51.1	68.3	30.1	28.2

VISIBLE MINORITIES (LQ)	Vancouver	Calgary	Toronto	Montreal	Ottawa-Gatineau
Management	0.85	0.72	0.72	0.72	0.69
Business, Finance, Admin.	0.95	0.86	1.00	0.82	0.83
Natural and Applied Sciences	0.99	1.08	1.14	1.04	1.34
Health Occupations	1.06	1.07	1.15	1.16	1.17
Social Science, Law, Religion	0.74	0.71	0.70	0.76	0.89
Art, Culture, Recreation	0.64	0.55	0.59	0.61	0.59
Sales and Services	1.18	1.30	1.07	1.16	1.23
Trades, Transport	0.85	0.85	0.92	0.75	0.74
Manufacturing and Processing	1.56	2.24	1.64	1.86	1.69

Source: Special Interest Profile Tables of the Census of Canada, 2006.

Note: The location quotient (LQ) indicates the ratio of the percentage of visible minorities in an occupational sector compared to the visible minority percentage in the total labour force.

income pie, they have been able to outbid others in key consumption markets like housing at the same time that neo-liberal reforms undercut the production of rental housing for those whose incomes have stagnated or declined, leading to eroding affordability and growing homelessness (Walks 2006). Simultaneously, declining relative incomes for successive cohorts of recent immigrants have led to diminishing levels of homeownership (Haan 2005; Mok 2009), increasingly trapping poorer immigrant segments in decaying high-density rental districts (Walks and Bourne 2006). Polarization in incomes becomes articulated spatially as increasing polarization between owners and renters (Figure 6.6) and thus ultimately between the neighbourhoods and municipalities where tenants are concentrated on the one hand, and the mostly newer (post-1970s) suburban areas where homeownership dominates as well as gentrifying inner-city districts where rental housing has been de-converted to homeownership tenure on the other hand (see Walks and Maaranen 2008a for an examination of gentrification processes in Toronto, Montreal, and Vancouver). As a result of gentrification and housing market dynamics, immigrant reception neighbourhoods are increasingly shifting out of the inner cities and into postwar suburban neighbourhoods marked by aging concentrations of low-cost rental housing (Ray, Halseth, and Johnson 1997; Smith and Ley 2008; Walks and Maaranen 2008b).

Changes in the spatial concentration of immigrants and minorities, however, do not always point toward increasing racial or ethnic segregation. Canadian CMAs have

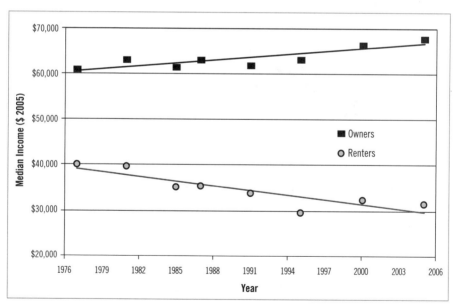

Figure 6.6 Homeowners getting richer, renters getting poorer: Median income by tenure, 1977–2005, all of Canada (with trend lines added).

Source: Calculated by the author from CMHC 2001; Census of Canada, 1996, 2001, 2006.

Note: Median household incomes are expressed in 2005 constant dollars.

demonstrated distinct trajectories with regard to the level and direction of visible minority segregation since 1986 (when visible minorities were first tracked in the census) (Figure 6.7). The Toronto region has exhibited a clear movement toward increasing segregation and unevenness in every census period (as measured by the index of segregation[9]), while Vancouver also shows creeping increases in segregation levels over the study period. In the other CMAs, however, visible minorities tended toward growing segregation only during the 1990s. By 2006, there was perfect correlation between degrees of segregation and place within the urban hierarchy, with the largest region (Toronto) revealing the highest level of segregation, while segregation declines with each reduction in metropolitan size. Furthermore, although there has been a trend toward the concentration of racialized groups in high-density apartment districts, this has not meant the development of ghettos in any Canadian city (see Walks and Bourne 2006). Indeed, the neighbourhoods containing the strongest concentrations of the largest visible minority groups (namely, Chinese and South Asians) often have average and in some cases above average incomes (ibid.). This is particularly true in the large 'global' cities, where higher-income ethnic suburban communities have formed ('ethnoburbs'), suggesting that high levels of immigration, per se, have not been a serious problem (Walks and Bourne 2006). As noted above, visible minorities are concentrated not only

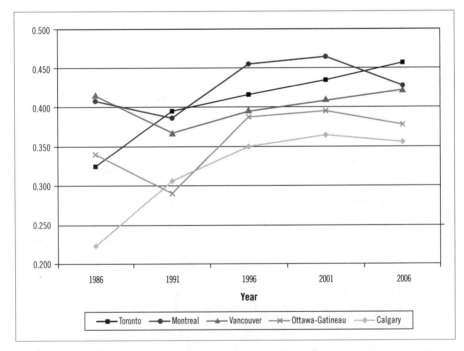

Figure 6.7 Neighbourhood (census tract) segregation of visible minorities, 1986 to 2006.

Source: Calculated from Census of Canada, 1986, 1991, 1996, 2001, 2006.

Note: Index of segregation values vary between 0 (no segregation) and 1.00 (complete segregation).

in occupations with declining incomes, including manufacturing and sales and services, but as well in respected professional middle-class occupations, particularly in applied sciences and health. This class polarization among immigrants contributes to income polarization across neighbourhoods. Whereas those employed in professional occupations are able to congregate in ethnoburbs, the poorest segments of each minority group are becoming concentrated in the lowest-cost and thus least desirable neighbourhoods.

Neo-liberal policy reforms can be implicated in the increasing segregation of race and class within the growing metropolis. Regressive tax reductions and reforms, coupled with the roll back of the welfare state, have reduced the health and well-being of lower-income households by reducing the quality and assortment of public services available to them, often forcing them into low-paying and precarious jobs (which particularly affects new immigrants). The limiting and targeting of welfare state benefits reduces access to income among many low-income earners, while relaxation of regulation in the realms of foreign investment, financial services, and inner-city land-use conversion encourages capital accumulation through speculation, gentrification, displacement, and dispossession (Harvey 2005). The downloading of social service costs, including those for social housing, onto municipalities (as has occurred in Ontario) only exacerbates the tendency toward spatial class polarization, as intra-urban migration is further fuelled by greater unevenness in property tax levels, the quality of public services and spaces, and concerns about the protection of property values. Growing class inequality among and within immigrant cohorts fuels the racialization of poverty. Income divergence and access to quality services become capitalized into the housing market, exacerbating the price and status differences between neighbourhoods, in turn encouraging affluent households to remove themselves from low- and middle-income areas, particularly areas with elevated levels of racialized poverty (see Atkinson and Blandy 2007).

A picture of how the geography of advantage and disadvantage is articulated within the large and growing Canadian metropolis can be seen through a comparison of high-income neighbourhoods (those with average household incomes more than 150 per cent of their CMA average) with the level of concentration of low-income household by neighbourhood (Figures 6.8 to 6.12). Within Toronto, a strong sectoral pattern of low income, often dubbed the 'U' of poverty, has remained within the (amalgamated) City of Toronto, although it now resembles more of a donut shape. However, in Toronto's case the 'hole' in the centre of the donut contains many of the highest-income neighbourhoods in the region and indeed in all of Canada (Figure 6.8), while the circular bread of the donut is where the poor are found, inverting the typical geography of poverty found in many cities in the US. Many census tracts in the 'inner' suburban areas of Scarborough, North York, and Etobicoke, now amalgamated into the City of Toronto, have seen their incomes decline persistently over time, and they make up the largest block of very low-income neighbourhoods in the region, although the lowest-income tracts also tend to be those with very high proportions of social rental housing. Neighbourhoods in municipalities bordering the central city, particularly those in Markham, Mississauga, and Brampton (which have the highest proportions of foreign-born residents) also exhibit declining incomes and relatively high concentrations of low income. Overall, 38 per cent of all census tracts have rates of low income above 20 per cent, while areas with a high concentration of poverty (typically 40 per cent or greater, according to the literature on concentrated poverty) make up 3.6 per cent of tracts. In addition to the historically high-income neighbourhoods in the centre of the City of Toronto (and in central Etobicoke), new high-income neighbourhoods have emerged in a number of gentrifying areas, including the Beaches and Bloor West Village, while very wealthy neighbourhoods are also found in southwest Mississauga and east Oakville, in and around Aurora, and increasingly in the exurban reaches near Caledon, King City, and Milton. The Toronto region is thus one in which a high level of neighbourhood diversity characterizes each suburban zone.

Montreal is the region (out of the five examined here) that has displayed the highest levels of poverty historically. Low-income neighbourhoods are found across the region

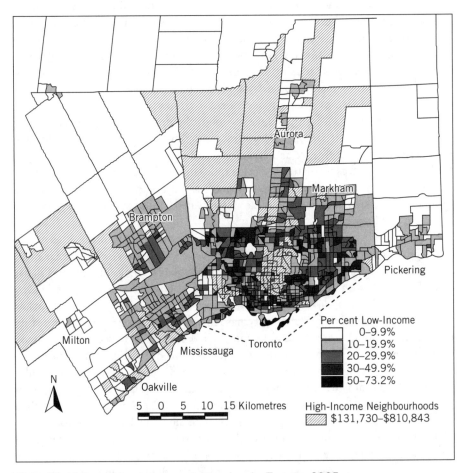

Figure 6.8 High- and low-income neighbourhoods, Toronto, 2005.

Source: Created by the author from Census of Canada, 2006.

Note: High-income neighbourhoods are defined here as census tracts with average household incomes more than 150 per cent of the CMA average household income. The per cent low-income refers to the percentage of the population in private households with incomes below Statistics Canada's Low Income Cutoff.

but particularly in the most central areas to the east, north, and west of the central business district, as well as in the old municipality of Montréal Nord to the northeast of the CBD (Figure 6.9). These are also the areas where recent immigrants and visible minorities tend to be most concentrated. There are also poor neighbourhoods spread throughout both the 'inner' suburbs (on Montreal island) and the 'outer' suburbs in old areas of Laval on the north shore, in and around Longueil on the south shore, and in the many smaller older towns that are located within commuting distance of the city. A full 52 per cent of all census tracts display levels of low income above 20 per cent, and a very large proportion (13.8 per cent) reveal high levels of concentrated poverty (with low income rates of 40 per cent or greater). The central communities/municipalities of Westmount, Outremont, and Mount Royal tend to contain the highest-income

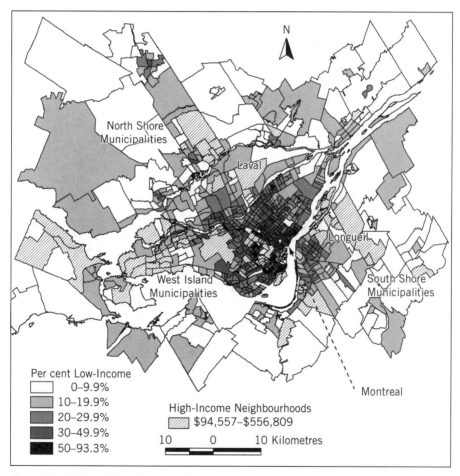

Figure 6.9 High- and low-income neighbourhoods, Montreal, 2005.

Source: Created by the author from Census of Canada, 2006

Note: High-income neighbourhoods are defined here as census tracts with average household incomes more than 150 per cent of the CMA average household income. The per cent low-income refers to the percentage of the population in private households with incomes below Statistics Canada's Low Income Cutoff.

neighbourhoods, as do many (now 'de-merged') municipalities on the West Island. However, as in Toronto, a number of exurban and outer-suburban areas exhibit very high incomes, including newly emerging areas of wealth on Laval island. Thus, as in Toronto, there is a clear pattern of rich and poor neighbourhoods but nonetheless characterized by much diversity within different zones of the region, as well as within municipalities and sub-municipal districts. Only within Toronto and Montreal do we find a few census tracts that can be doubly classified as containing both high-income and disproportionately low-income communities.

Vancouver reveals a particular geography of advantage. The spread of low income across Vancouver's landscape is notable (Figure 6.10), and a large proportion of tracts (50.1 per cent, more than half) had rates of low income above 20 per cent in 2006,

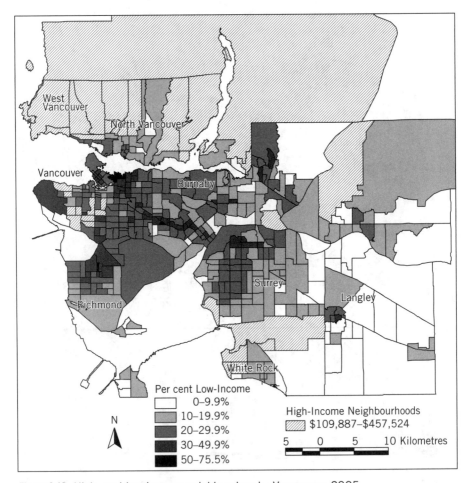

Figure 6.10 High- and low-income neighbourhoods, Vancouver, 2005.

Source: Created by the author from Census of Canada, 2006

Note: High-income neighbourhoods are defined here as census tracts with average household incomes more than 150 per cent of the CMA average household income. The per cent low-income refers to the percentage of the population in private households with incomes below Statistics Canada's Low Income Cutoff.

although only 4 per cent had low-income rates of 40 per cent or more (i.e., about the same as Toronto). Areas of low income extend from the CBD outward all the way to Richmond and through Burnaby, New Westminster, and Surrey to Langley. However, the highest concentrations of low income (greater than 50 per cent) are located in Gastown and the Downtown Eastside, the best known area of concentrated poverty in the city. High-income neighbourhoods are located in the traditionally wealthy areas of Shaughnessy Heights, Kerrisdale, and Point Grey to the south and west of the CBD and in suburban and exurban outer areas along the north shore, including West Vancouver, Anmore, and North Vancouver, and to a lesser extent to the south in White Rock. In Vancouver, the richest neighbourhoods are mainly located in well-known centrally located areas, or alternatively at fringes of the region.

The smaller metros of Ottawa-Gatineau and Calgary share similarities as well as differences in their residential geographies (Figures 6.11 and 6.12). Typically, the western suburbs and exurban areas of both regions contain the highest-income neighbourhoods (Nepean in Ottawa, the southwest suburbs and Rocky View in Calgary), while the eastern and northern zones of each region contain distinctly working-class populations and low-income neighbourhoods (northeast Calgary and Vanier and Hull in the Ottawa-Gatineau area). Meanwhile, each also reveals distinct geographies of affluence and poverty. Ottawa contains some very high-income neighbourhoods close to the CBD, both across the Ottawa River near the old CBD of Hull and in Rockcliffe, and these neighbourhoods often abut areas of concentrated poverty such as Vanier in Ottawa and the core of old Hull. In Ottawa, 30.4 per cent of census tracts have low-income rates above 20 per cent, while a relatively high proportion (6.4 per cent) are neighbourhoods with low-income rates above 40 per cent. In Calgary, the CBD itself contains high concentrations of poverty but otherwise reveals relatively lower levels of, and fewer areas associated with, low income, even while low-income households are becoming more segregated over time (Figure 6.10). Calgary has the smallest proportion of tracts (25.6 per cent) with low-income rates above 20 per cent in our sample, and a very small proportion (0.5 per cent) of tracts reveal a high concentration of poverty. However, although Calgary has traditionally exhibited a more equitable distribution of income, it has witnessed rapid increases in levels of social and spatial inequality.

Changes in indices of neighbourhood income inequality and polarization reveal the extent to which the five largest metropolitan areas have become more residentially segregated by class over time (Table 6.3). The statistics most sensitive to the upper end of the income distribution (the CV2 index of inequality and the WT index of polarization) reveal a particularly stark picture of growing segregation. The gini concentration ratio (more sensitive to shifts in the middle of the income distribution) is the only one of the indices to contain any ambiguity, suggesting that it is persistent changes occurring in the richer and poorer census tracts (i.e., at the extremes), more than in middle-class neighbourhoods, that are having a disproportionate influence on the spatial transformation of Canadian metropolitan housing markets. Changes in neighbourhood residential income segregation in each CMA are compared to their relative starting positions in 1970 in Figures 6.13a and 6.13b. Regardless of the measure employed, it is the Toronto region that has witnessed the most rapid increases in spatial income segregation, as measured through indices of spatial inequality and

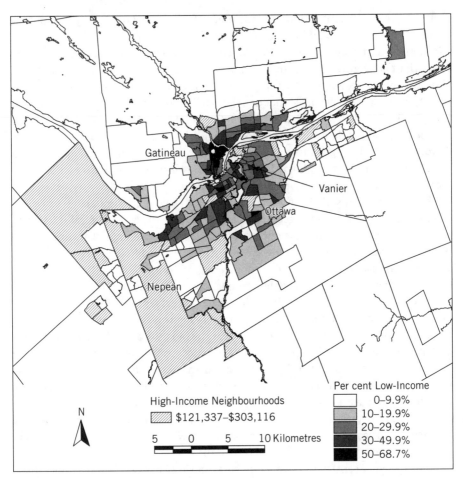

Figure 6.11 High- and low-income neighbourhoods, Ottawa-Gatineau, 2005.

Source: Created by the author from Census of Canada, 2006.

Note: High-income neighbourhoods are defined here as census tracts with average household incomes more than 150 per cent of the CMA average household income. The per cent low-income refers to the percentage of the population in private households with incomes below Statistics Canada's Low Income Cutoff.

polarization. However, the stories told by such indices vary over time and space and by the measure examined. Neighbourhood-level *inequality* only begins to grow rapidly in the 1980s and 1990s (and in the western metropolises particularly after 2000), whereas it was during the 1970s that the initial rapid growth of spatial *polarization* for most CMAs took place. Changes over time indicated by the Walks polarization index (which is equally sensitive to both high and low incomes) suggest that whereas the two Ontario metropolises—the Toronto and Ottawa regions—continued to polarize spatially over the entire period through 2005, the level of neighbourhood divergence had begun to level off in the Montreal region by 1990, while the CMAs in western Canada saw very slow or even declining levels of spatial polarization through 2005.

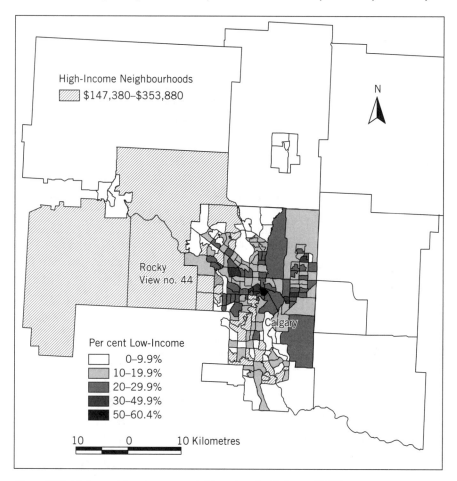

Figure 6.12 High- and low-income neighbourhoods, Calgary, 2005.

Source: Created by the author from Census of Canada, 2006

Note: High-income neighbourhoods are defined here as census tracts with average household incomes more than 150 per cent of the CMA average household income. The per cent low-income refers to the percentage of the population in private households with incomes below Statistics Canada's Low Income Cutoff.

Although Canadian metropolitan areas are becoming more spatially segregated by class, the same processes and mechanisms are not necessarily operating in identical fashion in each place. A window on this is provided by Figure 6.14, which simultaneously plots the segregation trajectories of rich (as indicated by the index of dissimilarity for households with inflation-adjusted incomes of more than $100,000 per year) and poor (as indicated by the index of dissimilarity for those with incomes below the LICO) in each city.[10] While all CMAs exhibit a general trend toward increasing spatial polarization, the individual trajectories diverge considerably, particularly after 1990. The two easternmost metros witnessed increases in the segregation of poverty first, followed later by growing concentration of the wealthy in the 2000s.

Table 6.3 Indices of Census Tract/Neighbourhood (Spatial) Income Inequality and Polarization, Five Largest CMAs, 1970–2005

CMA	1970	1980	1990	2000	2005
EXP (sensitive to lower end)					
Toronto	0.380	0.383	0.386	0.395	0.401
Montreal	0.384	0.385	0.388	0.391	0.394
Vancouver	0.382	0.384	0.385	0.387	0.391
Ottawa-Gatineau/Hull	0.380	0.382	0.384	0.388	0.388
Calgary	0.383	0.383	0.386	0.387	0.399
GINI CR (sensitive to middle)					
Toronto	0.211	0.345	0.349	0.345	0.344
Montreal	0.184	0.290	0.262	0.288	0.307
Vancouver	0.209	0.295	0.335	0.381	0.381
Ottawa-Gatineau/Hull	0.232	0.339	0.315	0.269	0.298
Calgary	0.153	0.307	0.264	0.328	0.276
cv2 (sensitive to upper end)					
Toronto	0.285	0.309	0.336	0.452	0.526
Montreal	0.334	0.326	0.360	0.402	0.444
Vancouver	0.303	0.313	0.320	0.344	0.427
Ottawa-Gatineau/Hull	0.278	0.280	0.296	0.341	0.340
Calgary	0.297	0.294	0.329	0.360	0.472
WT Polarization Index (sensitive to upper end)					
Toronto	0.416	0.459	0.491	0.547	0.586
Montreal	0.440	0.484	0.513	0.545	0.562
Vancouver	0.418	0.476	0.502	0.529	0.568
Ottawa-Gatineau/Hull	0.447	0.477	0.490	0.538	0.543
Calgary	0.431	0.458	0.492	0.508	0.608
Walks Polarization Index (equally sensitive to both ends)					
Toronto	0.175	0.234	0.253	0.277	0.285
Montreal	0.209	0.265	0.297	0.300	0.298
Vancouver	0.206	0.250	0.260	0.245	0.242
Ottawa-Gatineau/Hull	0.216	0.261	0.285	0.300	0.308
Calgary	0.235	0.255	0.284	0.260	0.263

Source: Calculated from Census of Canada, Custom Tabulations Tables E1171, E982.

Note: Values range from zero (complete spatial equality) to 1.00 (maximum inequality/polarization). Income data in each census year is for the prior tax year (i.e., the 2006 census contains income data for 2005).

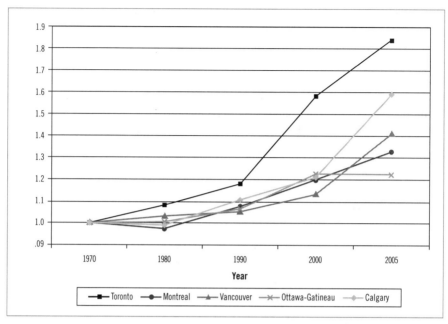

Figure 6.13a Coefficient of variation squared (CV2) (more sensitive to the upper end).

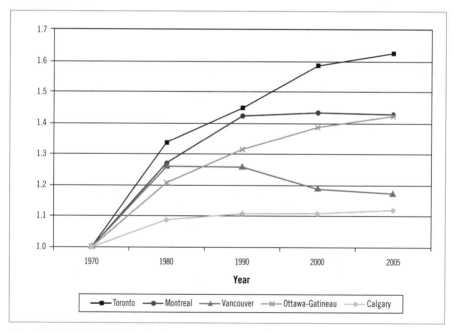

Figure 6.13b Walks Polarization Index (sensitive to both ends of the distribution).
Figure 6.13 Indices of neighbourhood income inequality and polarization (segregation) among census tracts.

Source: Calculated from Statistics Canada, Custom Tabulations from the Census, 1971–2006, Tables E1171, E982.

Toronto and Vancouver, the two regions with the highest proportions of immigrants and visible minorities, held stable or reversed their levels of spatial segregation among the poor through the (late) 1990s but afterward saw renewed concentration of the affluent. These dynamics may be partly explained by gentrification processes in these cities whereby professionals moved into initially still below-average-income inner-city neighbourhoods, subsequently displacing many existing low-income tenants and becoming ever more concentrated in the areas already gentrified through the turn of the millennium (Walks and Maaranen 2008a). Revealing a trend distinct from the other CMAs, Calgary demonstrates persistent increases in the segregation of poverty,

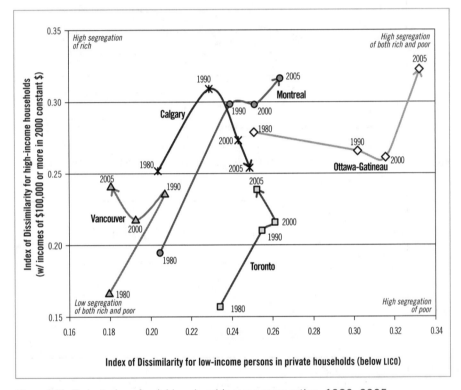

Figure 6.14 Trajectories of neighbourhood income segregation, 1980–2005.

Source: Calculated by the author from the Census of Canada, 1981, 1991, 2001, 2006 and Custom Tabulations from the Census, Tables E1171, E982.

Note: This chart shows changes in the Index of Dissimilarity for high-income households (those with income of $100,000/year or more, in 2000 constant dollars) and low-income persons in private households (proportion below the Low Income Cutoff, LICO). The latter is created by Statistics Canada as a way to gauge the level of low income and is often accepted as a proxy for the rate of poverty. This is much preferable to simply examining the proportion of households with incomes below a particular threshold, since the latter does not take into account household size, metropolitan size, or household circumstances. The Index of Dissimilarity ranges between 0 and 1.00, with 0 indicating a perfectly even distribution whereas 1.00 indicates complete segregation (all those in the relevant group are concentrated in one census tract). Thus, movement toward the top and right sides of the chart indicate increasing segregation of high-income and low-income households, respectively. Note that this does not take into account the size of each of these groups (high-income or low-income). Thus, in Ottawa-Gatineau these groups are highly segregated, but because they make up a smaller proportion of the total household sector, the overall inequality level in this CMA reported in Figure 6.13 is lower than that of other cities.

but after 20 years of increasing spatial concentrations of wealth, a reversal or desegregation of the rich after 1990.

Large, growing Canadian metropolises have witnessed a transformation in their economies, polities, and social composition, and this is imprinted in the urban landscape. The evidence demonstrates that economic and social transitions have been accompanied by growing labour market inequality and neighbourhood income polarization. Part of the story involves the gentrification of the inner cities, particularly of the largest urban regions. As large Canadian cities witnessed growth in finance, various business services, and other white-collar occupations, it spurred demand for residential spaces near the CBD. However, the resulting gentrification of the inner cities has removed valuable rental housing from the market and hurt the immigrant reception function of the inner city. In Toronto and Vancouver, and to a lesser extent Montreal, gentrification has been shown to lead to declining levels of income mix, ethnic diversity, and immigrant concentrations in affected neighbourhoods (Walks and Maaranen 2008b). The decentralization of manufacturing has likewise influenced the transformation of older working-class districts, either refashioned as gentrified or redeveloped neighbourhoods in the inner city or as the new immigrant reception neighbourhoods in the older postwar suburbs. Processes of under-maintenance, conversion to tenancy, and neglect now characterize many inner-suburban neighbourhoods, a number of which contain large and mixed minority populations. It is these populations that tend to suffer the most during times of economic volatility. Between October 2008 and October 2009, for instance, while the Canadian-born saw their average incomes drop by an estimated 2.2 per cent (mostly because of growing unemployment), immigrants who arrived in Canada during the previous five years saw their incomes drop by 12.9 per cent (Statistics Canada 2009b). Middle- and high-income visible minorities and whites, on the other hand, continue to concentrate in newer, outer suburban areas, although not necessarily in the same ones. Gated communities of various stripes are proliferating in the outer suburbs of Toronto, Calgary, and, particularly, Vancouver, some catering to the ever-better compensated managerial class that can command disproportionate attention within the housing market. Levels of neighbourhood divergence may have reached the point at which one's place of residence begins to independently feed back, through its influence on residents' life chances, into the systems producing and reproducing both social and spatial inequality, although the latter is an empirical question outside the scope of this chapter.[11]

Conclusion

Canadian cities have prospered significantly in the postwar period. Canada's metropolitan areas were able to develop robust and diverse economies, an enviable quality of life, and on balance a tolerant and welcoming social milieu. The expansion of the welfare state and public health care through the early postwar period, and its maturation through the 1980s, provided a significant level of ontological security to many urban residents.

However, the restructuring of urban economies, coupled with the neo-liberal weakening or dismantling of many welfare state protections, has produced greater variability and vulnerability within urban labour markets. The ability of the managerial class to acquire an ever-greater share of total income, along with the growing racialization of poverty, has led to new forms of class and racial inequality in the Canadian city. Overall, levels of income inequality and polarization in the largest Canadian regions were lowest before the recession of the early 1980s and rose thereafter, even though the level of inequality in the rest of the country continued to decline until 1989. The segregation of income reached its lowest and most equal levels earlier during the early 1970s. Since that time, labour market inequality grew among urban residents, and this drove the self-segregation of higher-income groups as well as the involuntary segregation of lower-income households, a portion of which became further concentrated in the aging rental stock in older postwar neighbourhoods. The racialization of poverty and the increasing income gap between whites and visible minorities, driven partly by declining opportunities for recent cohorts of immigrants, has only exacerbated spatial polarization processes. Regressive changes to income taxation, the neo-liberal restructuring of social benefits, the abandonment of a commitment to building affordable housing on the part of the federal and many provincial governments, and the downloading of responsibility for service delivery together work to produce and re-produce geographies of advantage and disadvantage within Canadian cities. While it is difficult to discern the distinct contributions of occupational restructuring, an aging society, demographic changes, and neo-liberal policies to growing income inequality, we can say that the interaction of these forces is responsible for the growing class polarization and neighbourhood divergence unearthed in this analysis.

Thus have Canadian cities entered the twenty-first century. It is with this in mind that we ponder the continued restructuring of urban economies in the 2000s and beyond in the context of globalization, continentalization, and neo-liberalism. Globalization in Canada has largely meant continentalization. However, this makes Canada's economy particularly dependent upon, and thus vulnerable to, demand for manufactured products and natural resources in the United States and on a favourable exchange rate between the Canadian and US dollar. Declining demand and US dollar volatility thus threaten the viability of Canada's main productive sectors. Between November 2002 and October 2009, Canada experienced a net loss of more than 573,000 jobs in manufacturing as production processes became internationally uncompetitive and were closed or transferred overseas or south of the border (Statistics Canada 2009b). Many important industrial and commercial assets have been liquidated and sold to foreign multinational firms, from canneries, to automotive parts suppliers, to (Nortel's) communications technology. While many sectors remain viable, the thinning of urban productive bases has negative consequences with regard to both the level of inequality and the geographic articulation of unemployment and poverty. This places downward pressure on wages in the manufacturing and service sectors and, by extension, in much of the arts and other occupational sectors, leading to continued deterioration in the lower ends of the income spectrum. This has the effect of further differentiating neighbourhoods containing working-class populations and industrial districts, many of which also house large visible minority populations, in turn fuelling their continued downward filtering. If at the same time the managerial class is able to continue extracting large shares of the total income pie for itself, it will continue to improve its position in relation to labour and housing markets. In this event, polarization could begin to tear at the residential fabric as elites seek to insulate themselves and their property investments from growing urban economic and political vulnerabilities, either in traditionally wealthy or newly gentrified neighbourhoods or in exurban and gated estates.

The future of Canada's cities thus remains cloudy. Economic imbalances came to a head in mid-2008, heralding the beginning of a transformative global 'recession' and exposing the weaknesses of both the global financial system and the productive economy of the United States, Canada's main trading partner and customer. The federal and many provincial governments responded with attempts to stem the bleeding by reducing income taxes and socializing private sector losses and debts, leading to an explosion of public sector deficits. As the ramifications of the state taking on the losses of the capitalist class play out within a neo-liberalized political environment, we could very well see further attacks on the welfare state, as well as income declines for both public sector professionals and workers in struggling industries, perhaps combined with the continued deification of the 'creative' entrepreneurial and managerial class. If the factors currently driving inequality are allowed to continue unabated, labour markets in urban Canada will further polarize between an increasingly vulnerable, marginalized, racialized, and residualized low-income labour force and the increasingly wealthy financial and managerial elite whose claims on wealth (bonuses, stock options, real estate assets, and so on) have been first and foremost protected by state responses. Burgeoning inequality could become cemented into the urban landscape, resulting in growing neighbourhood stratification, shelter poverty (see Stone 1993),[12] fiscal crises in municipal services, and deteriorating public spaces. If the capacity for integrating immigrants within the fabric of the city becomes overwhelmed, there is also the potential for heightened racial tension and restrictive immigration policies. Canadian cities thus begin the twenty-first century in a potentially precarious state. Without significant investment in income redistribution, economically and environmentally sustainable public infrastructure, and new innovative productive sectors, among other things, urban Canada is likely to suffer declining productivity, reduced quality of life, growing disadvantage, and most of all, greater socio-spatial inequality.

References

Adams, M. 2007. *Unlikely Utopia: The Surprising Triumph of Canadian Pluralism*. Toronto: Viking.

Atkinson, R., and S. Blandy. 2007. 'Panic rooms: The rise of defensive homeownership'. *Housing Studies* 22 (4): 443–58.

Baldwin, J.R., and G. Gellatly. 2005. *Global Links: Long Term Trends in Foreign Investment and Foreign Control in Canada 1960 to 2000*. Catalogue no. 11-622-MIE, no. 008. Ottawa: Statistics Canada.

Bernhard, J.K., P. Landolt, and I. Goldring. 2009. 'Transnationalizing families: Canadian immigration policy and the spatial fragmentation of care-giving among Latin American newcomers'. *International Migration* 47 (2): 3–31.

Boudreau, J., P. Hamel, B. Jouve, and R. Keil. 2007. 'New state spaces in Canada: Metropolitanization in Montreal and Toronto compared'. *Urban Geography* 28 (1): 30–53.

Bourne, L.S., and J. Simmons. 2003. 'New fault lines? Recent trends in the Canadian urban system and their implications for planning and public policy'. *Canadian Journal of Urban Research* 12 (1): 22–47.

Brenner, N., and N. Theodore. 2002. 'Cities and the geographies of "actually existing neoliberalism"'. *Antipode* 34 (4): 349–79.

Byrd, C. 2005. *Foreign Control of Canada's Merchandise Exports, 2002*. Catalogue no. 65-507-MIE, no. 004. Ottawa: Statistics Canada.

CIA (Central Intelligence Agency). 2009. 'The world factbook'. Washington, DC: CIA. https://www.cia.gov/library/publications/the-world-factbook/fields/2195.html?countryName=Canada&countryCode=ca&regionCode=na&#ca.

CMHC (Canada Mortgage and Housing Corporation). 2001. *Residualization of Rental Tenure: Attitudes of Private Landlords toward Low-Income Households.* CMHC Research Highlights, Socio-economic Series issue 93. Ottawa: CMHC.

Clark, D. 2002. 'Neoliberalism and public service reform: Canada in comparative perspective'. *Canadian Journal of Political Science* 35 (4): 771–93.

Courchene, T.J. 2001. 'Ontario as a North-American region-state, Toronto as a global city-region: Responding to the NAFTA challenge'. In A.J. Scott, ed., *Global City-Regions*, 158–92. Oxford: Oxford University Press.

Courchene, T.J., and C. Telmer. 1998. *From Heartland to North American Region State: The Social, Fiscal and Federal Evolution of Ontario.* Toronto: Centre for Public Management, University of Toronto.

Donald, B. 2005. 'The politics of local economic development in Canada's city-regions: New dependencies, new deals, and a new politics of scale'. *Space and Polity* 9 (3): 261–81.

Duclos, J., J.M. Esteban, and D. Ray. 2004. 'Polarization: Concepts, measurement, estimation'. *Econometrica* 72 (6): 1737–72.

Esteban, J.M., and D. Ray. 1994. 'On the measurement of polarization'. *Econometrica* 62 (4): 819–51.

FCM (Federal of Canadian Municipalities). 2008. 'The infrastructure challenge'. Online statement. http://www.fcm.ca/english/View.asp?mp=467&x=707.

Florida, R. 2005. *Cities and the Creative Class.* New York: Routledge.

Galabuzi, G.-E. 2006. *Canada's Economic Apartheid: The Social Exclusion of Racialized Groups in the New Century.* Toronto: Canadian Scholars Press.

Haan, M. 2005. 'The decline of the immigrant home-ownership advantage: Life-cycle, declining fortunes and changing housing careers in Montreal, Toronto and Vancouver, 1981–2001'. *Urban Studies* 42 (12): 2191–2212.

Hackworth, J. 2007. *The Neoliberal City: Governance, Ideology, and Development in American Cities.* Ithaca, NY: Cornell University Press.

Hamnett, C. 1994. 'Social polarization in global cities—theory and evidence'. *Urban Studies* 31 (3): 401–24.

Hamnett, C., and D. Cross. 1998. 'Social polarisation and inequality in London: The earnings evidence, 1979–95'. *Environment and Planning C: Government and Policy* 16 (6): 659–80.

Harvey, D. 1989. 'From managerialism to entrepreneurialism: The transformation of urban governance in late capitalism'. *Geografiska Annaler* 71 (B): 3–17.

———. 2005. *A Brief History of Neoliberalism.* Oxford: Oxford University Press.

Heisz, A. 2007. *Income Inequality and Redistribution in Canada: 1976 to 2004.* Catalogue no. 11F0019MIE, Research Paper no. 298. Ottawa: Statistics Canada.

Hurtig, M. 2002. *The Vanishing Country.* Toronto: McClelland and Stewart.

Keil, R. 2002. '"Common-sense" neoliberalism: Progressive Conservative urbanism in Toronto, Canada. *Antipode* 34 (4): 578–601.

Keil, R., and S. Kipfer. 2003. 'The urban experience and globalization'. In W. Clement and L. Vosko, eds, *Changing Canada: Political Economy as Transformation*, 335–62. Montreal: McGill-Queen's University Press.

Kipfer, S., and R. Keil. 2002. 'Toronto Inc? Planning the competitive city in the new Toronto'. *Antipode* 32 (2): 227–64.

Lee, M. 2007. *Eroding Tax Fairness: Tax Incidence in Canada, 1990 to 2005.* Toronto: Canadian Centre for Policy Alternatives.

McBride, S. 2006. 'Domestic neo-liberalism'. In V. Shalla, ed., *Working in a Global Era: Canadian Perspectives*, 257–77. Toronto: Canadian Scholars Press.

Mackenzie, H. 2007. *Timing Is Everything: Comparing the Earnings of Canada's Highest-Paid CEO's and the Rest of Us.* Toronto: Canadian Centre for Policy Alternatives.

———. 2009. *Banner Year for Canada's CEOs: Record High Pay Increase.* Toronto: Canadian Centre for Policy Alternatives.

Marshall, K. 2005. 'How Canada compares in the G8'. *Perspectives* June 2005. Catalogue no. 75-001-XIE. Ottawa: Statistics Canada.

Miller, B. 2007. 'Modes of governance, modes of resistance: Contesting neoliberalism in Calgary'. In H. Leitner, J. Peck, and E.S. Sheppard, eds, *Contesting Neoliberalism*, 223–49. London: Guilford Press.

Mok, D. 2009. 'Cohort effects, incomes, and home-ownership status among four cohorts of Canadian immigrants'. *The Professional Geographer* 61 (4): 527–46.

OECD (Organisation for Economic Co-operation and Development). 2008. *Growing Unequal? Income Distribution and Poverty in OECD Countries.* Report 8108051E. Zurich: OECD.

———. 'Stat extracts' (Tables for 'Foreign direct investment flows by industry' and 'Macro trade indicators: Trade-to-GDP ratio'). Zurich: OECD. http://stats.oecd.org/index.aspx.

Oreopoulos, P. 2008. 'Neighbourhood effects in Canada: A critique'. *Canadian Public Policy* 34 (2): 237–58.

Panitch, L. 2004. 'Globalization and the state'. In L. Panitch, C. Leys, A. Zuege, and M. Konings, eds, *The Globalization Decade.* London: Merlin Press.

Peck, J. 2004. 'Geography and public policy: Constructions of neoliberalism'. *Progress in Human Geography* 28 (3): 392–405.

Peck, J., and A. Tickell. 2002. 'Neoliberalizing space'. *Antipode* 34 (3): 380–404.

Preston, V., A. Kobayashi, and G. Man. 2006. 'Transnationalism, gender, and civic participation: Canadian case studies of Hong Kong immigrants'. *Environment and Planning A* 38 (9): 1633–51.

Pulkingham, J., and G. Ternowetsky. 2006. 'Neoliberalism and retrenchment: Employment, universality, safety-net provisions, and a collapsing Canadian welfare state'. In V. Shalla, ed., *Working in a Global Era: Canadian Perspectives*, 278–96. Toronto: Canadian Scholars Press.

Ray, B., and L. Peake. 2001. 'Racialising the Canadian landscape: Whiteness, uneven geographies, and social justice'. *The Canadian Geographer* 45 (1): 180–6.

Ray, B., G. Halseth, and B. Johnson. 1997. 'The changing "face" of the suburbs: Issues of ethnicity and residential change in suburban Vancouver'. *International Journal of Urban and Regional Research* 21 (1): 75–99.

Saez, E., and M.R. Veall. 2005. 'The evolution of high incomes in Northern America: Lessons from Canadian evidence'. *The American Economic Review* 95 (3): 831–49.

Sassen, S. 2001. *Global City: New York, London, Tokyo.* 2nd edn. Princeton, NJ: Princeton University Press.

Smith, H.A., and D. Ley. 2008. 'Even in Canada? The multiscalar construction and experience of concentrated immigrant poverty in gateway cities'. *Annals of the Association of American Geographers* 98 (3): 686–713.

Statistics Canada. 2003. *Canada's Ethnocultural Portrait: The Changing Mosaic.* Census Analysis Series. Catalogue no. 96F0030XIE2001008. Ottawa: Statistics Canada.

———. 2009a. 'Data on imports, exports and trade balance of goods on a balance-of-payments basis, by country or country grouping'. CANSIM table 228-003. http://www40.statcan.gc.ca/l01/cst01/gblec02a-eng.htm.

———. 2009b. 'Canada's employment downturn'. *Statistics Canada Canadian Economic Observer.* 22 (11). Catalogue no. 11-010-X. http://www.statcan.gc.ca/daily-quotidien/091112/dq091112a-eng.htm.

Stone, M.E. 1993. *Shelter Poverty: New Ideas on Housing Affordability.* Philadelphia: Temple University Press.

Taylor, P.J., G. Catalano, and D.R.F. Walker. 2002. 'Measurement of the world city network'. *Urban Studies* 39 (13): 2367–76.

UNDP (United Nations Development Programme). 2008. *Human Development Report, 2007–2008.* New York: UNDP.

Vojnovic, I. 2006. 'Urban infrastructures'. In T. Bunting and P. Filion, eds, *Canadian Cities in Transition: Local Through Global Perspectives,* 123–37. Toronto: Oxford University Press.

Walks, R.A. 2006. 'Homelessness, the new poverty, and housing affordability'. In T.E. Bunting and P. Filion, eds, *Canadian Cities in Transition: Global through Local Perspectives,* 419–37. Toronto: Oxford University Press.

———. Forthcoming. *Income Inequality and Polarization in Canada's Cities: An Examination and New Form of Measurement.* Research Paper. Toronto: University of Toronto Cities Centre.

Walks, R.A., and L.S. Bourne. 2006. 'Ghettos in Canada's cities? Racial segregation, ethnic enclaves, and poverty concentration in Canadian urban areas'. *The Canadian Geographer* 50 (3): 273–97.

Walks, R.A., and R. Maaranen. 2008a. *The Timing, Patterning, and Forms of Gentrification and Neighbourhood Upgrading in Montreal, Toronto, and Vancouver, 1961–2001.* Research Paper 211. Toronto: University of Toronto Centre for Urban and Community Studies/Cities Centre.

———. 2008b. 'Gentrification, social mix, and social polarization: Testing the linkages in large Canadian cities'. *Urban Geography* 29 (4): 293–326.

Walton-Roberts, M. 2003. 'Transnational geographies: Indian immigration to Canada. *The Canadian Geographer* 47 (3): 235–50.

Wang, Y., and K. Tsui. 2000. 'Polarization orderings and new classes of polarization indices'. *Journal of Public Economic Theory* 2 (3): 349–63.

Wilson, D. 2004. 'Toward a contingent urban neoliberalism'. *Urban Geography* 25 (8): 771–83.

Wolfson, M.C. 1997. 'Divergent inequalities: Theory and empirical results'. *Review of Income and Wealth* 43 (4): 401–21.

Yalnizyan, A. 1998. *The Growing Gap: A Report on the Growing Inequality between Rich and Poor in Canada.* Toronto: Centre for Social Justice.

———. 2007. *The Rich and the Rest of Us: The Changing Face of Canada's Growing Gap.* Toronto: Canadian Centre for Policy Alternatives.

Notes

1. Note that this is not saying much. Cities in the United States reveal the highest levels of inequality and spatial segregation in the developed world. The national gini coefficient for income (0.408 in 2000) in the United States is much higher than that of any other developed country in the

global north and has been so on a continuous basis. The gini for Canada is much lower, at 0.326 in 2000 and 2004 (see United Nations Human Development Report, various years; Heisz 2007).

2. The trade-to-GDP ratio sums the total of exports and imports, then divides by the GDP. It measures the degree of a country's 'openness' to or 'integration' with the world economy (OECD 2009). Canada's ratio (82.2) in 2007 was higher than that of many other trade-dependent countries such as Japan (26.5) but smaller than Germany's (93.1) and those of a few other much smaller countries (e.g., Ireland, Czech Republic, Netherlands). Canada is particularly dependent upon exports of commodities, and this dependence has increased. Between 2004 and 2008, the proportion of Canada's GDP that came from merchandise exports rose from 32 to 35 per cent. This is below the level for Germany (40 per cent) and similar to China's (32.5 per cent) but much higher than the level of dependence on trade found in other well-known trade-dependent countries like Japan (15.2 per cent) and Russia (28.1 per cent) (Marshall 2005; CIA 2009; OECD 2009).

3. In the 1960s, more than 60 per cent of Canadian mining and 70 per cent of energy firms were foreign-controlled. After the creation of FIRA, the level of foreign-control in these two sectors declined significantly and continued declining through the 1980s and 1990s. Baldwin and Gellatly (2005) provide evidence showing that for other sectors (excluding energy), the level of foreign control rebounded to levels roughly seen in the 1960s. Importantly, foreign control grew significantly between 1988 and 1998 in the following sectors: services, food and transportation services, consumer goods and services, wood and paper, machinery and equipment, transportation equipment, and chemicals and textiles. Almost 70 per cent of all Canadian assets in the latter sector are foreign-controlled (Baldwin and Gellatly 2005, 25). Only in the energy and communication sectors did the proportion of foreign control decline through to 2000. Note that the energy and communication sectors were most heavily targeted for privatization over the 1990s and 2000s (McBride 2006).

4. The share of Canada's exports flowing to the United States reached more than 82.5 per cent in the late 1990s and again in 2003. It dropped back to 80 per cent by 2006 (Statistics Canada 2009a), and then the economic crisis that came into focus in 2008 saw the proportion of exports flowing to the US drop significantly to 75.5 per cent (OECD 2009). This was a result of declining US demand rather than growing demand outside of Canada.

5. The Walks polarization index is calculated by taking log of the absolute deviation between each income data value (i.e., census tract, or household, income value) and the median income of the population base. The use of the log prevents the inclusion of income values of zero (in similar fashion to the requirements of entropy indices of inequality). However, in most developed nations, households with incomes of zero in the census are the result of the ability to deduct self-employment expenses from base income, and government subsidies, pensions, and welfare payments typically raise the real incomes of even the lowest-income households above zero. Furthermore, this issue tends not to arise with the use of aggregated household-level data. When aggregated into income bands within spatial units such as census tracts, for instance, the lowest income values are never zero unless the data is suppressed. Thus, the fact that zero incomes must be excluded from the calculation does not present as much of a problem as this might sound at first. See Walks, forthcoming, for more information.

6. The main limitation is that incomes in the PUMS data are capped at the top end at $200,000. Thus, the increasing number of millionaires in Canada are misrepresented in the PUMS data as having incomes of only $200,000, which obviously dampens any measurements of inequality and polarization derived from this dataset. The other main limitation is that the PUMS files accessible to universities involve a 3 per cent sample of the entire census.

7. While the average size of households has declined over time, leading to a greater diversity of the household sector and contributing to household-level inequality, the fact is that almost identical trends in inequality are evident among individuals (in fact, the latter are slightly more dramatic), suggesting that age and household size are only minor factors fuelling inequality over time.

8. The straight per capita incomes from all sources (employment, investments, and government sources) declined in real terms adjusted for inflation (in 2000 constant dollars) by 18.9 per cent for visible minorities and by 21.8 per cent for recent immigrants between 1981 and 2001 (calculated by the author from the PUMS data from the 1981, 1991, and 2001 censuses; the 2006 microdata was not yet available at the time of writing).

9. The index of segregation is similar to the index of dissimilarity, except that it compares concentration of the target group with the concentration of the remaining population. The base calculation is the same (sum of the absolute differences between the shares of the target and non-target group in each tract, divided by two).

10. The two metrics used to gauge the trajectories of segregation of the rich and poor by necessity have slightly different bases. Importantly, Statistics Canada does not have a strict definition of low income for households as they do for families and individuals. However, they do measure the proportion of persons in private households that live with low income, and so I have used this here. Meanwhile, Statistics Canada has no definition of high income for any unit and provides no guidance on how to define the affluent. So I have arbitrarily defined high-income households as those with incomes of $100,000 and over (in 2000 constant dollars).

11. There is significant literature on neighbourhood effects in the United States and the United Kingdom. Past evidence suggests that neighbourhoods have little independent effect on residents' life chances in Canadian cities (Oreopoulos 2008). However, those empirical studies were completed during times of relatively greater social and spatial equity. The current situation calls for new research on this topic.

12. Shelter poverty exists when high housing costs, rather than unemployment or disproportionately low wages, spurs diminishing purchasing power and limited consumption of basic needs on the part of the poor. See Stone 1993.

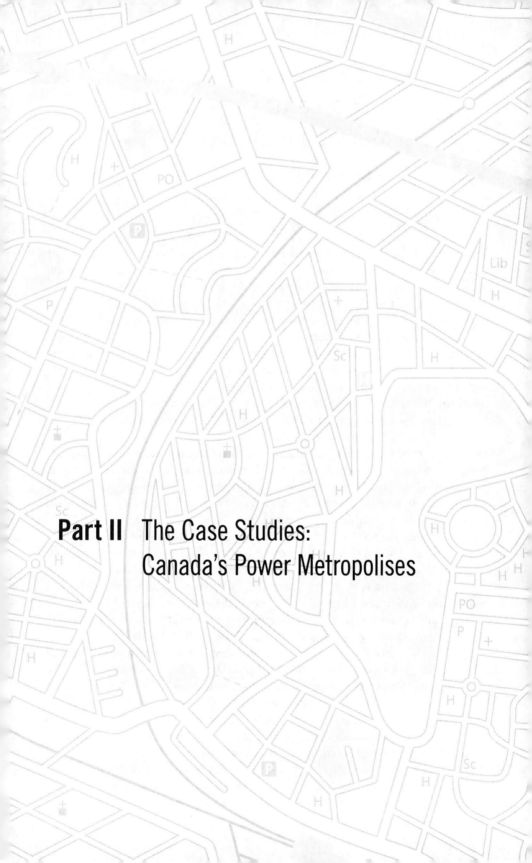

Part II The Case Studies: Canada's Power Metropolises

7 Case Studies Overview: A Profile of Canada's Major Metropolitan Areas

Jim Simmons and Larry S. Bourne

This chapter links the urban system and thematic analyses provided in Chapters 1 through 6 with examples of the processes of urban growth and change operating within individual Canadian cities. Following the multi-scalar approach outlined in Chapter 1, the purpose of this chapter is to set a comparative framework for the individual case studies to follow. The editors have selected Canada's five largest metropolitan economies as case studies for more intensive analysis: Toronto, Montreal, Vancouver, Ottawa-Gatineau, and Calgary. These cities, or more accurately city-regions, are Canada's largest urban concentrations, with relatively high rates of recent growth, and they represent distinctive examples of the principal variations in the experience of urbanization and metropolitan development in evidence across the country.

They also serve as the principal gateways to the global economy and act to organize the economic landscape in Canada. Collectively, they also demonstrate most of the important urban economic transformations that are taking place within Canada, as described in the first chapters, and serve as diverse laboratories for assessing the local consequences of those transformations. Jointly, they account for 43 per cent of the country's population, 45 per cent of total employment, and 62 per cent of the recent population growth—within only 0.3 per cent of Canada's land area.

Table 7.1 summarizes the main characteristics of the five case study cities. For each city, and for ease of comparison, we provide simple indicators relating to population size, growth rate, employment, employment to population ratios, land-area densities, immigrant populations, linguistic character, average household income levels, and total market size (income in $billion). These indicators offer standard benchmarks that not only are useful for comparative purposes but direct particular emphasis to differences in city size, geographical setting, culture, and market size.

Note that two versions of Toronto are included here. The first, which we call Toronto 1, is the census metropolitan area (CMA) as defined by Statistics Canada, based on the same procedures that are used for the other four urban areas and for all metropolitan areas in the country (see the discussion in Chapter 3). Toronto 2 is the principal definition that is used in the Toronto case study in Chapter 10. It combines the Toronto CMA with two adjacent and closely integrated CMAs, Hamilton and Oshawa, in order to capture the full extent of the current built-up area in the region and the geographical spread of regional labour markets. These three CMAs also form the urbanized core of a much larger region, the Greater Golden Horseshoe (GGH), sometimes referred to as Toronto 3, which is now frequently employed as the planning region for decisions

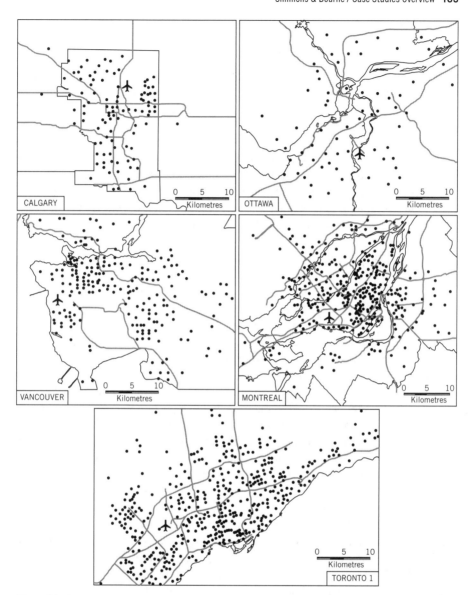

Figure 7.1 Population distribution, case study cities, 2006. (One dot equals 10,000 persons.)

Source: Statistics Canada. *Census of Canada, 2006.*

on infrastructure and transportation investments and growth management by the provincial government in Ontario. Either one of the first two versions of Toronto is considerably larger than the other case study cities, with 40 per cent more population than the Montreal region and more than twice that of metropolitan Vancouver.

In general, the larger the CMA, the higher the overall density of population in the city, but Figure 7.1 also suggests that the actual distributions of population for these cities are much more diverse, both among the cities and within them. Calgary's population is largely confined within the municipal boundary of the central city. Ottawa-Gatineau, in contrast, is bi-nodal in form, with two distinct central cores, but since both the Ontario and Quebec parts of the CMA were recently amalgamated into single-tier municipalities by their respective provincial governments, the municipal boundaries are now essentially the same as the CMA. It is also densely settled near the two city centres but with widely dispersed populations farther away from the core. The three largest cities each display rather different population distributions and patterns of development. The high density of Vancouver's central area is offset by extensive low-density suburbs to the south and east. Montreal's development has dispersed from the initial settlements on the island in an irregular fashion, shaped in part by the availability of transportation facilities and in part by physical restrictions. Toronto has maintained relatively high densities within the built-up areas, but new low-density suburbs are expanding rapidly to the north and the west.

Why These Cities?

While these five cities are an obvious choice as the five largest cities in the country, they also have considerably more economic, social, and cultural significance for Canadians than smaller places. Selecting these five, however, does not in any way devalue the contributions of other cities, large and small, to the Canadian economy or to the country's self-image and its economic and social vitality. For precisely this reason, the first six chapters have focused on analyses of the entire Canadian urban system—the 33 metropolitan areas and the rest of the 144 urban places recognized by Statistics Canada. These places are individually important in their own right, and when viewed as part of a larger national urban system, they provide critical context for the case studies to follow. Changes in the relative position of cities within the national urban hierarchy provide a particularly valuable measure of differential urban change.

In their essay on Montreal (Chapter 8), Richard Shearmur and Norma Rantisi explain how the shift in Montreal's relative position within the national hierarchy came about and how it has affected Montreal's economy, politics, and sense of itself. For much of Canada's history, the role of dominant metropolis belonged to Montreal, and this significant historical fact, combined with the city's successful resurgence as the francophone metropolis, warrants a special place in this volume. At the same time, a myriad of francophone corporations (the so-called Quebec Inc.) have developed specialized roles in such fields as transportation (Bombardier), resource technology (SNC-Lavalin), and other sectors associated with the new economy.

The significance of Ottawa-Gatineau in the national urban hierarchy has grown with the increased importance of government within the Canadian economy, as demonstrated by Caroline Andrew, Brian Ray, and Guy Chiasson in Chapter 9. The composite metropolitan area, encompassing both sides of the Ottawa River, now ranks fourth among metropolitan areas in terms of population, and the links with federal government research agencies have spawned major research facilities for high technology as government intersects with the private sector.

Toronto emerged as the dominant Canadian metropolis during the past half-century: Canada's candidate global city. Toronto is distinguished within the national urban system not only by its size but also by its unequalled concentrations of corporate headquarters, banking and finance, and business services, as well as its exceptional underboundedness within the Greater Golden Horseshoe, creating a host of problems concerning taxation, infrastructure provision, and planning. Over the past decade or so, Toronto has also emerged as a leading centre of creative industries and cultural production. In spatial terms, the Toronto city-region is also marked by the size and extent of its suburban (and ex-urban) municipalities and communities, a phenomenon that has no counterpart anywhere else in Canada. At the same time, Toronto's central area and inner city have experienced a redefining social reconstruction process, including new economy workers, gentrifiers, and empty-nesters seeking the exceptional amenities of the core, housed in the city's burgeoning condominium districts and loft conversions. This story of exceptionalism and global city formation was prepared by Larry Bourne, John Britton, and Deborah Leslie (Chapter 10).

Canada has several other large cities that could have been selected—notably Edmonton, Quebec City, and Winnipeg, as well as Halifax (which although much smaller is symbolically and substantively important as the regional centre of the Atlantic region). These cities, and others of substantial size (e.g., London, Kitchener-Waterloo, Saskatoon, Windsor, Victoria), are clearly distinctive and offer their own trajectories of growth and change. As such, they warrant a volume of their own. But space here is limited, and most of the functions provided by these cities are also evident in the case study cities. Calgary and Vancouver, for example, exemplify the roles of strong regional capitals and service centres: they dominate their regional economies while providing the links between those economies and both national and global markets. Calgary's population is only 50,000 or so larger than Edmonton's and is equally and deeply embedded in the energy sector, although specializing in different industries, and Winnipeg has a long history as the gateway to the West. Nonetheless, it is Calgary that has attracted the corporate headquarters, the financial connections, the explosive growth, and all the publicity that goes with big money in a flashy and high-growth economic sector. This fascinating growth experience is documented and dissected in Chapter 11, prepared by Byron Miller and Alan Smart.

Vancouver has a much longer tradition as the unchallenged metropolis of British Columbia and as Canada's gateway to the Pacific, while it continues to host public utilities based on natural resources (notably hydro power and gas) and a number of the mining and forestry corporations and shipping companies that drive the provincial economy (a roster of head offices significantly diminished from Vancouver's apogee as corporate control centre for the resource sector in the 1970s and 1980s)—even as the city's emerging new roles in technology, new media, higher education, and cultural industries become more prominent. One feature of Vancouver that will clearly affect its future development is the extraordinary influx of Asian immigrants who have transformed the city in recent years, incorporating infusions of capital, entrepreneurial skills, and connectivity. The impacts of this inflow are given particular emphasis in Chapter 12, co-written by Trevor Barnes, Tom Hutton, David Ley, and Markus Moos.

Consider the wide differences even among the case study cities as revealed in Table 7.1. Canadian cities are remarkably different places, illustrating the rich diversity of urban experience that follows from their locations, histories, economies, and social character. It is for this reason that detailed case studies are essential in a volume such as this; national urban trends and averages are obviously important in themselves, but they often conceal too much. The case studies help us to understand why cities in Canada differ and on what particular attributes, and they illustrate how similar processes of urban economic development can produce quite different results and patterns on the ground.

The first and most striking factor is the wide variation in population size even among these five places. The Toronto region is five times larger than Ottawa-Gatineau or Calgary, and this differential carries over to the economy (in terms of total employment and market/income size). Second, there are also significant differences in growth rates—not just over the past five years but over 20 or more years. Toronto and Calgary have grown strongly for some time, while Montreal and Ottawa have lagged the others but with more consistent rates of growth. Vancouver's growth, in contrast, tends to be more variable, characterized by both boom and bust periods.

Third, each metropolitan area has a distinctly different migration regime. While immigration is a major factor in recent population growth in all cases, especially for Toronto and Vancouver and to a lesser extent Calgary and Montreal, the latter two metropolitan areas rely more on domestic (internal) migration. The cumulative impacts of immigration over time are substantial to the extent that the populations of Toronto and Vancouver are now more than 45 per cent foreign-born, while those of the other three cities approach or just exceed 20 per cent. High growth rates—in large part driven by high levels of immigration—also distinguish the case study cities from most other Canadian cities, as Chapter 3 demonstrated.

Table 7.1 Comparing the Case Study Cities

Measure	Toronto 1	Toronto 2	Montreal	Vancouver	Ottawa	Calgary
Population, 2006	5113.1	6136.6	3635.6	2116.6	1130.8	1079.3
Growth Rate, 2001–6	9.2%	8.8%	5.3%	6.5%	5.9%	13.4%
Employment	2627.4	3144.9	1835.8	1104.8	601.5	632.0
Employment/Population	0.514	0.512	0.505	0.522	0.532	0.586
Land Area (kms²)	5904	33,203	4259	2877	5716	5107
Population Density (persqkm)	866.1	245.9	853.6	735.6	197.8	211.3
Immigrants, 2001–6	411.3	437.9	164.4	165.7	37.8	62.8
% Francophone	1.2	1.2	66.4	1.2	32.9	1.5
% Foreign-Born	45.7	41.8	20.6	39.6	18.1	23.6
Total Income ($b)	158.2	188.4	96.2	59.9	36.3	40.8
Average Household Income $C	87,800	86,200	63,000	73,300	80,800	98,300

Source: Statistics Canada. *Census of Canada*, 2006.

Notes: Urban areas are as defined in the 2006 census, with Toronto 2 including Hamilton and Oshawa CMAs. Population, employment measured in 1000s.

In 2006, 1$C = 0.882$US.

Finally, consider the variations in household and total income in Table 7.1. Rapidly growing cities tend to have younger populations because they attract more in-migrants and are able to retain existing residents, and they typically show higher proportions of their populations in the labour force and lower unemployment levels. These attributes in turn translate into higher incomes overall. On the other hand, they also have large immigrant populations that are not always easy to employ, at least in the short term and especially in occupations appropriate to their skill levels. This tends to deflate average wages. Calgary has the highest average income levels in the group by far, since it has a youthful and engaged population, while Montreal and Vancouver lag behind, but for quite different reasons. The Toronto region ranks second in average incomes, reflecting its combined role as a financial and management centre, which leads to higher average incomes, and as a major gateway for new immigrants whose integration into the labour market may be a slow process, especially during economic downturns. The latter also tends to depress wages, at least in certain sectors of the economy and for certain social groups, as demonstrated by Alan Walks in the previous chapter (Chapter 6) and in the subsequent Toronto case study (Chapter 10).

Defining the Cities

Each one of Canada's urban areas, especially when viewed as a city-region, is a variable combination of several regional governments and local municipalities—i.e., political jurisdictions. The only consistent functional definitions available that allow for direct comparisons are the census metropolitan areas defined by Statistics Canada as part of the population census. They are based on the identification of commuting regions (or sheds) using place-of-work data from the previous census—as discussed in Chapter 3. For the most part, as is evident in Figure 7.1, census definitions tend to over-bound the physically built-up area while underestimating the geographical extent of current commuter sheds (using place-of-work data) and the trade areas of the urban core.

In some instances, these census units are not only adjacent but tightly integrated. Simmons and Bourne (2007) have labelled these larger urban areas with adjacent urban places as *megacities*: clusters of interdependent urban centres that share workers and facilities, such as airports, universities, sports teams, shopping malls, and rail and road infrastructure. For example, Montreal is next to Saint-Jean-sur-Richelieu, Salaberry de Valleyfield, and Joliette, while Vancouver is closely linked to Squamish, Abbotsford, and Chilliwack. Should these areas be treated as separate urban centres even when their functional interconnections are strong and their economic futures mutually interdependent?

The ambiguity of the current Statistics Canada definition has created an awkward situation in southern Ontario, and specifically in the Toronto region. An unwillingness to integrate the nearby CMAs of Hamilton and Oshawa into the Toronto definition forces students and policy analysts to expand their study regions in a variety of ways. This paper simply combines the three CMAs into one unit, while many other Toronto researchers use a five-county region: the Greater Toronto Area (GTA), consisting of the City of Toronto and the four regional government units (the former counties of Halton, Peel, York, and Durham) that surround it, and more recently, research has focused on the much larger Greater Golden Horseshoe extending from Peterborough in the east to Niagara Falls in the southwest. Obviously, using different definitions generates different empirical results, and these differences further complicate any comparisons of urban development and economic growth.

Employment and Spatial Structure

Since the focus of the case studies is the changing structure and attributes of the metropolitan economy, Figure 7.2 compares the distribution of employment in the five cities. While each city has evolved a distinctive map of employment—which becomes even more variable if the various economic sectors are disaggregated—they do share several common patterns (Shearmur et al. 2007). First, the downtown core remains as the major employment node—typically anchored by financial and corporate offices in Toronto, Montreal, Vancouver, and Calgary and by government offices in Ottawa. This concentration permits the elaboration of traditional core-oriented transit systems and commuter routes that encourage higher residential densities in the central city. In this sense, the case studies follow conventional paths of development but with a distinctively Canadian character.

At the same time, each city has developed significant secondary employment concentrations in the suburbs, with at least one node near the airport and others serviced by suburban expressways. The differences in city size are also evident in the varying levels of complexity of the employment distributions. Calgary and Ottawa-Gatineau still show relatively simple and centralized employment geographies, with higher proportions of jobs in central areas, while the complexity of Toronto and Montreal defies easy generalization, with multiple employment nodes strung along a network of expressways. In the Toronto 2 region, new employment is located in an arc around the northern fringe of the urbanized area about 20 kilometres from downtown. Montreal has developed two major peripheral nodes, one about 15 kilometres west of downtown near the airport in Dorval and St. Laurent and the other (substantially smaller) to the northwest in Laval. Vancouver, in contrast, has generated smaller suburban employment nodes that are widely dispersed and oriented to institutional and commercial sites rather than industrial zones.

The Themes of the Case Studies

As suggested in Chapter 3, the larger Canadian cities share several common patterns of development over the postwar period. These cities have all witnessed relatively rapid population growth, initially by way of high birth rates and subsequently through domestic and international migration. This, in turn, has created pressure to generate employment of all kinds and has encouraged investment by both the public and private sectors—investment that might be described as rapid rather than thoughtful. The landscapes of these cities increasingly represent the needs and values of the past 30 years. At the same time, as argued in the earlier chapters, the cities have been further transformed by the transition from primary/secondary employment to the services sector, a transformation that has both affected the roles of large cities within their national economies and modified the economic landscapes within the cities themselves. In recent years, these cities have been further transformed by waves of

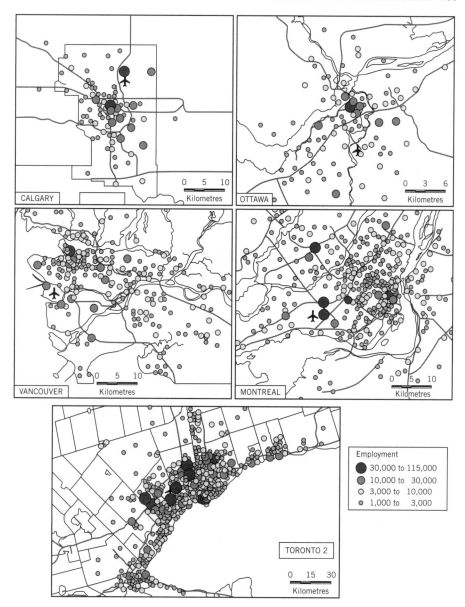

Figure 7.2 Employment distribution, case study cities, 2006.

Source: Statistics Canada. *Census of Canada, 2006*.

immigration that have created new cultural contexts and identities for them, altering workplaces, neighbourhoods, and lifestyles for all their residents.

Each one of the case study cities has followed a somewhat different path through these massive transformations, as will be seen in the chapters to follow. Montreal has become the francophone commercial capital, with specializations in resource and transportation technology and strong economic ties to European capitals. Toronto is now the financial and corporate centre of the country, having accomplished a massive transformation of the regional economy from manufacturing to financial and business services. With its Canadian hegemony firmly established, the city is trying to define an expanded international role in the face of intense global competition for resources, investment, and talent. For Ottawa-Gatineau, the growth of the federal role in Canada's culture and economy has created a national political capital with associated high-tech specializations. Still, the Ottawa-Gatineau region grows when the federal government grows and stalls when the federal government ceases to grow. Calgary is now the dominant regional capital for the Prairie region and the headquarters for Canada's energy sector. Its future will depend on international markets as they are shaped by resource prices, carbon controls, and peak oil. Vancouver, as Canada's Pacific gateway with significant corporate and family links to Asian countries, should continue to thrive as these countries play a larger role in the international economy in general and Canada's economy in particular. Thus, Toronto and Ottawa tend to grow in step with the national economy, while Montreal, Calgary, and Vancouver are more closely linked to regional growth rates in Quebec, the Prairies, and BC, respectively, and to the expansion of specific economic sectors and industries.

Population Projections

The importance to Canada of the country's largest cities, and particularly the case study cities, has been outlined above and is widely cited in the broader public policy literature (Brender and Lefebvre 2006; Courchene 2007; Wolfe 2009). The degree of importance, however, has been confirmed by recent population projections issued by Statistics Canada for 2031 (Statistics Canada 2010). The projections for the case study cities and Edmonton—the million+ population cities—are summarized in Table 7.2. Only the modest (i.e., low) and moderate (reference) projections are provided here. The aggregate population growth of the six cities over the 25-year period from 2006 to 2031 is 5.8 million for the low projection and 7.7 million for the reference projection. These are very substantial growth estimates; indeed, they represent more than 80 per cent of the country's anticipated population growth. If we use the definition of Toronto 2 (as defined above) rather than the CMA, the aggregate growth totals 6.1 and 8.0 million, respectively. The Greater Toronto region (the Toronto, Hamilton, and Oshawa CMAs) alone is expected to add between 3 and 4 million people over the period. Some of the serious challenges involved in accommodating this magnitude of growth are discussed in Chapter 10. The anticipated growth in other cities is also substantial: Montreal and Vancouver will likely add roughly 1 million new residents, Calgary more than 600,000, and Ottawa-Gatineau and Edmonton between 300,000 and 400,000.

In relative terms, the six megacities represented 47 per cent of the total Canadian population in 2006 (using Statistics Canada's own base population); this is expected to increase to nearly 55 per cent in the low projection and more than 56 per cent in the high projection in 2031. In other words, more than half of the Canadian population will then live in the country's six largest city-regions. Increasingly, Canada's future is tied to the growth of the country's largest urban places, which we examine in detail in the following chapters.

Table 7.2 Population Projections for Canada's Largest City-Regions: The Case Study Cities, Including Toronto 1 and 2, Plus Edmonton, 2006–31

City-Region/CMA	Population Projections 2031		
	Base (2006) 000s	Low 000s	Reference 000s
Toronto 1	5320	8016	8864
Montreal	3680	4541	4900
Vancouver	2181	3195	3483
Calgary	1118	1731	1864
Ottawa-Gatineau	1167	1474	1574
Edmonton	1069	1434	1529
Toronto 2*	6,382	9,306	10,244
Canada	32,522	39,521	42,078
% of nation:			
with Toronto 1	44.7%	51.6%	52.8%
with Toronto 2*	47.9%	54.8%	56.1%

* Includes the Toronto, Hamilton, and Oshawa CMAs.

Source: Statistics Canada. Catalogue no. 91-551-X, 2010.

Such projections are based on numerous assumptions and expectations and thus should be treated with some caution. The assumptions, typically, include continued economic growth, low but stable fertility levels, high levels of immigration, and moderate to low domestic migration rates (as the rural and small-town populations decline proportionally), among others. Most future growth, in both population and the labour force, will come through immigration. All of these assumptions are of course open to debate, and all are likely to fluctuate in importance over the forecast period. The estimates themselves may also turn out to be too high, for various and largely unpredictable reasons. Nevertheless, the projections in Table 7.2 do illustrate one plausible scenario of the country's future, a future that is almost certain to be dominated by the largest metropolitan regions. Moreover, as the overwhelmingly dominant destinations of new immigrants, these cities are also where the increasing ethno-cultural diversity of the Canadian population will be most pronounced and most visible. They are also the settings where the social, cultural, and economic adaptations required by this increasing diversity will be resolved.

Template

It is important to note that there is no single template for comparative research in general or the case studies to follow. The authors of each chapter were given carte blanche to develop the empirical analyses and overall narrative that they felt would best capture the changing character of their city and the larger city-region and that would at the same time provide the essential ingredients to understanding the diverse trajectories of urban growth in Canada in the twenty-first century. Each author, however, was asked to highlight a number of common indicators: namely, population growth, employment structure and occupational change, governance issues, and recent public policy initiatives, emphasizing the policies that have influenced economic and physical development in the past and will likely continue to do so in the future. The resulting chapters are as varied and stimulating as the individual places they describe and analyze.

References

Brender, N., and M. Lefebvre. 2006. *Canada's Hub Cities: A Driving Force of the National Economy.* Ottawa: Conference Board of Canada.

Courchene, T. 2007. 'Global futures for Canada's global cities'. IRPP Policy Papers 8, 2. Montreal: Institute for Research on Public Policy.

OECD. 2009. *Territorial Review: Toronto.* Paris: OECD.

Shearmur, R., et al. 2007. 'Intrametropolitan employment structure: Polycentricity, scatteration, and dispersal and chaos in Toronto, Montreal and Vancouver, 1996–2001'. *Urban Studies* 44: 1713–38.

Simmons, J.W., and L.S. Bourne. 2007. 'Living with population growth and decline'. *Plan Canada* 47 (2): 13–21.

Sloan, J., ed. 2007. *Urban Enigmas: Montreal, Toronto and the Problem of Comparing Cities.* Montreal: McGill-Queen's University Press.

Statistics Canada. 2010. *Projections of the Diversity of the Canadian Population, 2006–2031.* Catalogue no. 91-551-X. March. Ottawa: Ministry of Industry.

Taylor, Z., and M. Burchfield. 2010. 'Growing cities: Comparing urban growth patterns and regional growth policies in Calgary, Toronto and Vancouver'. Report 2.15.1. Toronto: Neptis Foundation.

Wolfe, D. 2009. *The Geography of Innovation: Twenty-First Century Cities.* Toronto: Conference Board of Canada.

8 Montreal: Rising Again from the Same Ashes

Richard Shearmur and Norma Rantisi

Introduction: Language, Industry, and Primacy

Situated at the confluence of the St Lawrence and Ottawa rivers, Montreal owes its origin to its position as the furthest inland port (before the construction of the Lachine Canal and, later, the St Lawrence Seaway) that trans-Atlantic ships could reach because of the rapids that surround Montreal island. Consequently, Montreal first developed in the seventeenth century as a trading post where resources—principally fur—gathered in Canada's interior could be shipped out to European markets. Indeed, fur, leather, and, more generally, clothing have remained important sectors in Montreal's economy since its foundation.

Of course, all cities are unique in terms of their hinterland and location. Montreal is distinguished by three further factors that, taken together, have guided the path of its economic development and continue to influence it into the early twenty-first century. These factors are not amenable to short-term policy intervention and have little to do with current discourses on creativity, globalization, and the knowledge economy: what has shaped and is shaping Montreal's economic trajectory is the way in which its unique geographic endowments, combined with a number of other long-term (or structural) factors—the key ones are described below—are interacting with and adapting to the wider social and economic changes that are affecting cities throughout Canada and, indeed, throughout the world. In other words, and stated more generally, a city's economy evolves in a path-dependent way as it adapts to (but also, in an incremental way, fashions) the world around it: this adaptive process depends upon its past history, its current industrial structure, and the decisions (constrained, but not determined, by structural and geographic factors) made by leaders, business people, and individuals.

The first of three factors that distinguish Montreal is its bicultural identity. Whereas most multicultural cities clearly possess one dominant, or at least founding, culture, Montreal has both an English and a French heritage.[1] Founded in 1642 at a location identified by Champlain in 1611 (Marsan 1990), it was still a small and somewhat rudimentary colonial outpost when the British conquered New France in the 1760s. Tension between these competing colonialists has shaped Montreal's destiny over the ensuing 250 years, bringing to the fore tensions between the rural and the urban, between Catholicism and Protestantism, between agriculture and industry, and between French and English cultural backgrounds and views of the world. This complex and multi-faceted issue is often reduced, in politics and popular perception, to the language question (should Montreal be anglophone or francophone?), with language perceived as the vector of culture and cultural dominance. Some of the ways that this has played out over the past 30 years will be explained in the following sections.

A second key element in Montreal's history that continues to shape its current destiny—in particular, the way in which its industries are adapting to the challenges of globalization and the knowledge economy—is the fact that Canada's industrial revolution occurred here. In 1820, Montreal was still a fairly minor colonial outpost dealing mainly in the export of staple products to Europe. By 1860, it had become a major industrial metropolis, paralleling the development of other cities on North America's eastern seaboard (Lewis 2000). No other major city in Canada industrialized so early or so extensively: Toronto, which at the time was a central place for southern Ontario's growing agricultural economy and was becoming a transport, trade, and banking hub, had virtually no export-oriented manufacturing until the early twentieth century. Montreal's dominance in what would now be considered traditional manufacturing sectors but were in fact the high-tech industries of the day was boosted after Confederation (1867) by high tariffs on imported manufactured goods (particularly from the US) and by the completion of the trans-Canada railway, which widened its internal market to cover the entire Canadian territory (Pomfret 1981). From the 1860s to the 1970s, Montreal was Canada's economic and financial capital. The legacy of this industrial heritage continues to shape Montreal's current economic possibilities and trajectory.

The final main structuring element that underpins Montreal's economic development is in fact a consequence of the first two. Already by the 1950s, Toronto was catching up with Montreal as a financial centre (Germain and Rose 2000, 29 et seq.), and by the early 1970s its population had nearly overtaken Montreal's (Coffey and Polèse 1993; Polèse and Shearmur 2004). Notwithstanding these signs, Montreal still perceived itself as having an edge, since it dominated Canada's railways, its financial system, and, of course, its manufacturing sector. As Canada's largest city in 1971, this perception was still grounded in contemporary facts if not in its trajectory.

However, during the 1960s the economic and cultural dominance of anglophone communities in Quebec was being challenged by the francophone majority: after throwing off the yoke of the Catholic church, the Quiet Revolution turned to what was perceived as the cultural—and even colonial—yoke of 'the English' (Levine 1990). The rapid, and peaceful,[2] rise of a francophone elite, which culminated in the election of René Lévesque's separatist government in 1976 and the first referendum on Quebec independence in 1980, encaptured the trends that were already apparent. Financial control of Canada's economy, together with a large number of (usually well qualified and connected) anglophones, migrated to Toronto during the 1970s. By the early 1980s, Montreal was hard hit by recession in its manufacturing industries, had definitively lost is primacy in the Canadian urban system, and was on the brink of seceding from Canada: this abrupt alteration of its position in the urban hierarchy—together with continued uncertainty about its position in Canada—is of fundamental importance in interpreting the evolution of its economy over the past 30 years (Polèse and Shearmur 2004).

These three factors, which are unique to Montreal, are of course not the only important structural trends that have limited and shaped the evolution of its economy. For instance,

Montreal, like other Canadian cities, attracts large numbers of immigrants from across the world (McNicoll 1993; Germain and Poirier 2007), and these immigrants form a cultural mosaic that interconnects, penetrates, and complicates the fundamental division between English and French heritage. As with other cities, the transition from a manufacturing economy dominated by large firms to a service-oriented economy in which manufacturing, if it occurs, tends to be performed by networks of smaller firms—i.e., the transition from 'Fordism' to 'post-Fordism'—brought about major changes in the city's industrial structure (Coffey 1994). And the move toward increased international trade and globalization, exemplified by the Free Trade Agreement with the US (signed in 1989), NAFTA (signed in 1991), and the transformation of the expanding GATT agreements into the World Trade Organization (founded in 1999), has affected the composition and destination of the city's production and exports. As the knowledge economy takes hold and as the complexification of value chains begins to affect manufacturing and service sectors alike, Montreal's economy continues to adapt.

These trends can only serve as a backdrop to the description of the way in which Montreal's economy has evolved since the early 1980s. Each merits a chapter of its own, and Germain and Rose's (2000) excellent book provides a more global picture of Montreal's recent evolution as a city. The focus here is on Montreal's economy, and the chapter is structured as follows. In the first section, a general description is given of Montreal's evolving industrial structure, population, and employment levels. Montreal's position relative to Toronto is examined in some detail. The second section focuses on the way in which these changes have played out over space: in particular, the continued resilience—and indeed growth—of Montreal's downtown is described, and certain explanations are evoked. The third section focuses on a sector that epitomizes some of the history, challenges, and possibilities inherent in Montreal's economy: the textile and clothing sector. The way in which one of Montreal's most traditional industries is suffering from, but also adapting to, the early twenty-first century context of globalization and innovation-based competition is illustrative of the more general way in which Montreal can (and, in our opinion, should) build upon its existing economic and cultural heritage to forge a unique way forward. To conclude, we draw some lessons from Montreal and attempt to show how these lessons may be of wider relevance to the way the economies of metropolitan areas are currently being theorized.

A Brief Overview of Recent Trends

In 2006, Montreal's[3] population was 3.45 million, having risen slowly from 2.76 million in 1971 (Figure 8.1). Despite this rate of growth, which is quite rapid relative to many European city regions, it is rather slow by North American standards. Toronto, which in 1971 boasted only 2.70 million inhabitants, has expanded to more than 5 million. Furthermore, Montreal's population, which was younger than Toronto's in the early 1970s, is aging more rapidly: the working-age population in Montreal is growing more slowly than the population as a whole, and this is probably attributable to Toronto's greater ability to attract both international and internal migrants. Although the economic and cultural processes are intertwined, it is undeniable that Quebec's language laws limit its attraction for external migrants (who in general already speak some English and often see more advantages, at the North American scale, in improving this language than in learning French).[4] Furthermore, the 1970s, and to a lesser extent the 1980s and 1990s, saw a sizeable out-migration, principally of anglophones worried by the possibility of separation. Indeed, the net loss of anglophones from Quebec between 1976 and 1986 was about 150,000, and even in the late 1990s, about 30,000 net (i.e., about 7.5 per cent of all remaining anglophones) departed (Jedwab 2004; Bourhis 2008). This loss of mainly working-age people, often to the benefit of Ontario and Toronto, may have played a role— albeit a partial one—in Montreal's more rapid aging. It almost certainly accounts for the relative drop in formal education levels among Montrealers (Figure 8.2): indeed, in the early 1970s Montreal's population had a very similar proportion of graduates to Toronto's. By the early 1980s, Montreal had dropped behind (despite the founding of the University of Quebec at Montreal in the late 1960s): since then, Montreal has produced or attracted as many graduates as Toronto, but the gap that emerged in the 1970s has never closed.

Another worrying trend—but one that may well be undergoing a reversal in the wake of the recent financial crisis—is the gap in wage levels between Montreal and Toronto (Figure 8.2). From very similar levels in the 1970s, Toronto's wages—often assumed to be indicative of productivity levels—increased beyond Montreal's in the 1980s and 1990s. However, this trend is not necessarily particular to Montreal: indeed, Vancouver's wage levels are also included in Figure 8.2, and it can be seen that Canada's other major city has also fallen behind Toronto. Thus, the trend in work income may be attributable to Toronto's accession to (quasi) world-city status, whereas Montreal and Vancouver remain important regional centres where wages tend to be set locally rather than in relation to global trends.

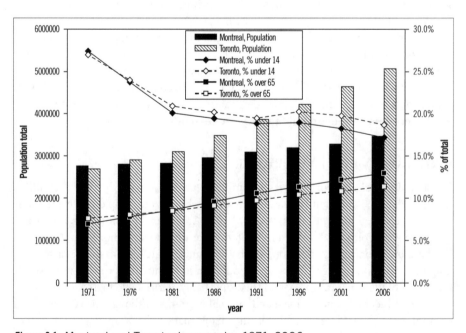

Figure 8.1 Montreal and Toronto demography, 1971–2006.

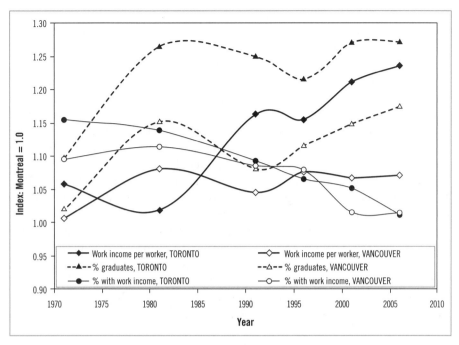

Figure 8.2 Toronto and Vancouver in relation to Montreal, various indicators, 1971–2006.

This illustrates one of Montreal's psychological failings: as already mentioned, the loss of primacy to Toronto is a key component of Montreal's economic trajectory that has been accepted with difficulty. Montreal often compares itself to Toronto, and this chapter reflects this common discourse (Coffey and Polèse 1999). However, as Montreal accepts its regional status—as it is increasingly doing—and compares itself with similar cities, then its relatively low wage levels (for instance), while not a particularly desirable attribute, can be interpreted in a more favourable light.

Indeed, the history of the past 30 years has in many ways been one of Montreal adapting (in terms of size and function) to the reality of this regional role, and despite the gloom that often prevails when recollections of past glory cloud the issue, in fact Montreal has performed remarkably well since the mid-1990s. This is reflected primarily in its employment and graduate figures. Its workforce is acquiring human capital at the same rate as Canada's other major agglomerations (Figure 8.2). Furthermore, since the early 1990s its employment rate (the percentage of 15- to 65-years-olds with work income) has increased, and by 2006 the long-standing gap with Toronto and Vancouver had closed. Indeed, its relatively smaller labour force (Figure 8.1), combined with its diversified and thriving economy, have ensured that increasing numbers of people are now economically active; this may in turn put some upward pressure on wages and productivity, since it would seem that Montreal's transition to regional status is just about complete and it is now ready to move forward on a new trajectory.

Montreal also benefits from the province of Quebec's indirect intervention in the economy: in particular, the universal daycare system (all babies and toddlers have access to $7-a-day daycare) ensures that women and men are not compelled to interrupt their careers when starting a family. While the system has important welfare implications for families and children—the early involvement of most children in constructive play, socialization, and educational activities will probably have long-term beneficial effects on Quebec society (Campbell et al. 2002)—in the shorter term and in a narrowly economic sense, it ensures that there is a more efficient use of existing male and female human capital and of experienced workers. The current rise in Quebec's birth rates and increase in participation, although undoubtedly caused by a variety of factors, are occurring concurrently with the implementation of these universal welfare programs, and there is some evidence of a causal link (Milligan 2005).

To sum up, the city of Montreal is facing broad challenges similar to those faced by Toronto and other cities: an aging population, the quest for a skilled workforce, the widening and deepening of globalization, tertiarization, and changing modes of production. However, it is facing these challenges with its own historical baggage, which comprises both opportunities and constraints. In particular, its bicultural heritage, which may be an advantage in terms of creativity and the generation of new ideas, has also been a problem in terms of the political uncertainty that it has generated. This in turn has had a strong influence on its position within the Canadian urban system and on its demographic and economic evolution. It is to the city's economy that we now turn.

The Evolution of Montreal's Industrial Structure

As emphasized by Coffey and Polèse (1999), it can be misleading to analyze a city's industrial or occupational structure out of context. Trends and structures that appear unique may in fact be less unique than they first appear. Indeed, despite Montreal's clear specialization in traditional and high-tech manufacturing and Toronto's increasing dominance of the financial and professional services sectors, the differences between the two cities are smaller than the similarities. In both 1971 and 2006, the two cities' industrial structures were the most strongly correlated among the five case study cities, and the broad specializations that characterized Montreal in 1971 still did in 2006 (Table 8.1). Indeed, in 2006 both cities's industrial structure was very highly correlated with its 1971 structure, and the matrix of 1971 to 2006 correlations is almost symmetric, indicating very slow relative structural changes both within and between cities (Table 8.1). Thus, any changes discussed in the next paragraphs should be understood as marginal, although this does not mean that they are not indicative of long-term shifts in economic structure. Rather, it highlights the (almost) glacial pace of these changes and puts into perspective any short-term policy attempts to modify economic structures (by, for instance, attempting to attract knowledge workers in the hope of kickstarting the new economy) of large cities. These slow and marginal changes are nevertheless highly significant precisely because their direction is unlikely to change in the short to medium term. It should also be emphasized that an analysis of location quotients partials out any general changes affecting the economy of the whole country (Canada is used as the reference for these location quotients) and also abstracts from changes in absolute size of each city (which were discussed in the previous section).

With these caveats in mind, how has Montreal's industrial structure evolved? Although it is still the city with the highest concentration of traditional industries

(Table 8.1), this specialization is declining when considered relative to Canada: these industries are generally no longer located in cities and, in the case of Quebec, have tended to concentrate along the US border east of Montreal and south of Quebec City. At the same time, Montreal has increased its relative specialization in high-tech manufacturing—principally aeronautics, pharmaceutics, and, to some extent, communication equipment: it now quite clearly dominates Canadian cities in this type of production.

The relative decline of traditional manufacturing in Montreal (Figure 8.3) should not mask its crucial importance to the city's economy: in 2006, these sectors (which include textiles, clothing, oil refining, food, beverages, and other first and second transformation industries) contained 108,000 jobs, about 6.5 per cent of all jobs in the city. Likewise, the city's high-tech sector, while a propulsive industry in terms of image and value added, is sometimes over-emphasized: it comprised only 41,000 jobs in 2006. However, not only do these jobs require high qualifications (32 per cent of them are in scientific, social scientific, or artistic [SSA] occupations), but it seems that this occupational structure may have also spilled over into the medium-technology sectors (e.g., non-pharmaceutical chemicals, transport equipment, machinery, plastics): indeed, these sectors have seen the proportion of SSA occupations increase from 8.2 per cent in 1990 (equivalent with Toronto) to 15.2 per cent in 2006 (against 12.1 per cent in Toronto) (Figure 8.4).

Almost all types of manufacturing suffered from employment decline during the 2001 to 2006 period: after enjoying historically low exchange rates with the US in the late 1990s, the Canadian dollar shot up to par in the early 2000s. However, job loss has been moderate in the medium- and high-tech sectors, where employment rose in the late 1990s to return to its long-term level in 2006 (Figure 8.3). The lower-tech sectors, textiles and clothing in particular, suffered very severe job losses: more than 20,000 jobs disappeared between 2001 and 2006, particularly in the clothing (12,000 jobs out of an initial 40,000) and textile (3000 out of 10,500) sectors. The end of the transitory tariff exemption agreement on textiles in 2005 exposed Montreal's clothing and textile sectors to the brunt of worldwide competition. Just as the occupational profile in medium-tech sectors is changing to reflect the increased importance of knowledge and innovation, so the textile and clothing sectors are adapting to this new imperative: in the context of major job loss just described, the proportion of SSA employees in Montreal's textile industry has risen from 3.9 per cent to 5.8 per cent (between 2001 and 2006) and in clothing from 4.8 per cent to 7.4 per cent. In the following section, this sector is taken as a case study to illustrate in more detail the ways in which it—and, by extension, a large portion of Montreal's traditional economy—is adapting.

Whereas manufacturing employment has remained stable or has declined over the past 35 (and particularly over the past five) years, service employment has increased rapidly (Figure 8.3). This is evident in Montreal just as it is in Toronto, reflecting general trends across the whole economy. However, the differences in trajectory are instructive. Whereas the financial sector (FIRE) in Montreal basically ceased to grow in the early 1980s, this sector has continued to grow in Toronto: it is not only the relative concentration of financial employment but its absolute level that is fast evolving in Toronto. High-tech producer services, on the other hand, have continued to grow rapidly in both cities: only in the 1970s did this sector grow faster in Toronto. Since

Table 8.1 Industrial Structure of Canada's Five Largest Agglomerations, 1971–2006

	1971					2006				
	Montreal	Toronto	Vancouver	Ottawa	Calgary	Montreal	Toronto	Vancouver	Ottawa	Calgary
Primary sectors	0.09	0.18	0.36	0.24	0.99	0.28	0.21	0.38	0.21	1.56
Traditional manufacturing	1.27	0.97	0.93	0.37	0.57	1.03	0.86	0.76	0.25	0.61
Medium-tech manufacturing	0.97	1.54	0.62	0.13	0.43	0.99	1.37	0.68	0.37	0.66
High-tech manufacturing	2.31	2.38	0.46	1.01	0.35	2.70	1.44	0.61	1.44	0.55
Construction	0.76	1.02	1.10	0.98	1.37	0.73	0.87	1.03	0.78	1.33
Transport	1.34	0.87	1.46	0.51	1.08	1.03	1.02	1.15	0.70	1.13
Warehousing and wholesale	1.19	1.34	1.52	0.63	1.32	1.25	1.39	1.23	0.55	1.13
Retail	0.97	1.03	1.11	0.88	1.05	1.06	0.93	0.97	0.91	0.93
Personal and other services	1.11	1.23	1.31	1.13	1.27	0.95	1.09	1.08	1.04	0.96
Communication services, printing	1.33	1.40	1.23	1.42	1.08	1.46	1.48	1.45	1.21	1.07
Hotels and restaurants	0.97	0.85	1.19	0.85	1.08	0.89	0.86	1.21	0.91	0.99
Leisure (cinemas, performance . . .)	1.07	1.07	1.07	0.95	1.06	1.06	0.98	1.18	1.06	1.09
Finance, insurance, real estate	1.39	1.59	1.43	0.99	1.31	1.09	1.59	1.25	0.80	1.02
Professional producer services	1.44	1.69	1.52	0.84	1.37	1.21	1.53	1.36	1.03	1.20
High-tech producer services	1.42	1.49	1.99	1.48	2.74	1.26	1.39	1.43	1.71	2.15
Education and related services	0.94	0.89	0.91	1.11	1.08	1.04	0.91	1.05	1.01	0.84
Health and social services	1.08	0.85	1.05	0.86	1.00	1.06	0.79	0.91	0.96	0.83
Public administration	0.71	0.72	0.64	3.93	0.94	0.77	0.59	0.66	3.70	0.51

a) Correlation of 1971 LQs with 1971 LQs						a) Correlation of 2006 LQs with 2006 LQs				
	Montreal	Toronto	Vancouver	Ottawa	Calgary	Montreal	Toronto	Vancouver	Ottawa	Calgary
Montreal	1.00					1.00				
Toronto	0.88	1.00				0.64	1.00			
Vancouver	0.29	0.21	1.00			0.11	0.60	1.00		
Ottawa	-0.05	-0.06	-0.03	1.00		0.13	-0.07	0.00	1.00	
Calgary	-0.04	-0.01	0.79	0.17	1.00	-0.26	0.05	0.43	-0.16	1.00

b) Correlation of 1971 LQs with 2006 LQs					
	Montreal	Toronto	Vancouver	Ottawa	Calgary
Montreal 2006	0.92	0.76	0.32	0.08	-0.18
Toronto 2006	0.85	0.89	0.28	0.05	-0.12
Vancouver 2006	0.01	0.52	0.89	-0.04	0.58
Ottawa 2006	0.08	-0.13	0.02	0.97	-0.17
Calgary 2006	-0.19	0.20	0.64	0.17	0.86

Notes: The top part of this table presents location quotients (LQs) for each sector. This is the ratio of two percentages: the percentage of the city's employment in sector i divided by the percentage of Canada's employment in sector i.

The bottom part of the table presents correlations of: a) 1971 with 1971 LQs and 2006 with 2006 LQs and b) 1971 with 2006 LQs.

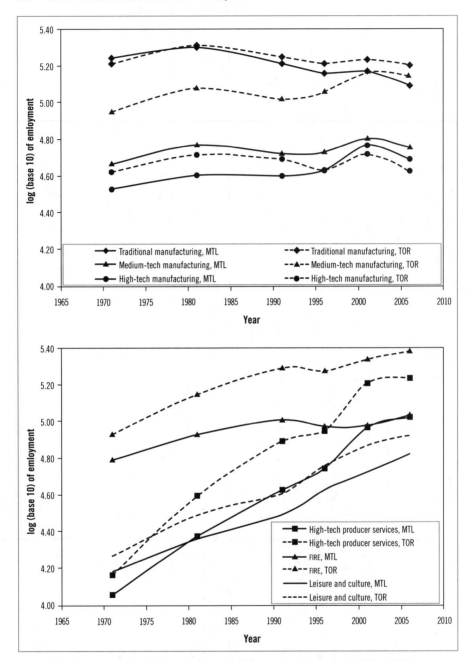

Figure 8.3 Employment in selected sectors, Montreal and Toronto, 1971–2006.

then, Montreal has held its own, reflecting not only connections between engineering consultants and high-tech manufacturing but also the growth of successful international consulting firms such as SNC Lavallin and CGI. These companies—and other, smaller but just as successful ones—enable young Montrealers to begin high-flying

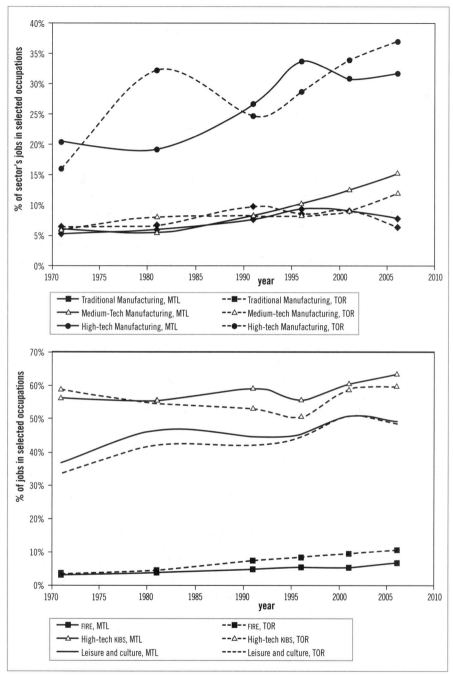

Figure 8.4 Social science, scientific, and artistic occupations in selected sectors, 1971–2006.

international careers in the city in an environment where English and French are both used: this is an important consideration, given the reluctance of many (but far from all) Montrealers to 'go into exile' in Toronto or elsewhere. The rise of 'Quebec Inc.' (Fraser 1987; Legault 2009)—the name often given to a somewhat corporatist constellation of leading francophone business people and politicians, and (often public sector) unions who wrested control of the economy from the anglophone economic elites in the 1970s and 1980s—has real effects for qualified Quebecers who can now work from a home base using their own language (at least within the office) and participate actively in cutting-edge economic activities. Whether or not this effect is quantifiable in any way is perhaps less important than the impact the rise of these major consulting firms (and the same can be said of major high-tech manufacturers such as Bombardier) has had on Quebec's collective psyche. Although these firms have been growing—often with state help—over the past 20 to 30 years, they have come into their own since the mid-1990s. The combination of a (very close) defeat for the separatist agenda in the second referendum (held in 1995) and strong employment growth since then (Figure 8.2) has enabled 'Quebec Inc.' to engage in global economic activity with a confidence that the recession of the early 1980s and Montreal's slow acceptance of its loss of Canadian primacy had undermined. It should also be noted that this corporatism has a negative side to it: collusion between Quebec's close-knit economic and political elites has led to pervasive corruption and influence peddling, which is slowly coming to light as Quebec's economy globalizes (*The Economist* 2009).

The leisure and cultural sector, which comprises employment in the arts, theatre, cinema, professional sport, and so on, is overrepresented in Montreal (Table 8.1). Furthermore, notwithstanding Vancouver's rise between 1971 and 2006, it is interesting to note that Montreal has maintained its specialization in this sector, whereas Toronto has witnessed a relative decline. Polèse and Shearmur (2004) argue that in an increasingly globalized knowledge economy, services that can be easily traded will tend to rise to the top of the urban hierarchy. However, one of the barriers to trade in services is culture: certain services, in particular in the cultural sector where language is an important factor but also in certain consulting services for which interpersonal understanding is required, cannot be easily traded across different cultures. This would explain why Montreal has retained its cultural activities, since it is the apex of the francophone urban hierarchy in North America. Toronto is the apex for financial transactions denominated in Canadian dollars (hence its strength in the financial sector). Its relative weakness as a cultural centre may reflect the fact that many English cultural products are imported from New York and Los Angeles. Calgary and Vancouver's strength in this sector in 2006 probably reflects a consumption effect, which is also reflected in their relative specialization in the hotel and restaurant sector.

Apart from the role played by language, provincial government policy has also been a critical factor in shaping the buoyancy of the cultural sector of Montreal. Since the 1970s, with the rise of Quebec nationalism, culture has come to be viewed as a means of affirming French-Canadian identity and consequently as a domain for public involvement. A particularly important initative in this regard is the Cultural Policy of Quebec, which was published in 1992. This policy called for increased support for— and promotion of—Quebec cultural products, prompting the establishment of two key organizations: SODEC, which provides support to cultural entreprises, and Conseil des arts du Québec, which provides support to individual artists.

Moreover, within the policy Montreal's position as a cultural metropolis and 'special place within Quebec' is highlighted (Ministère des affaires culturelles 1992, 132). Indeed, Montreal has been a major beneficiary in terms of funding, and the policy has paved the way for the development of parallel initiatives and programs at the municipal level (Brault 2002; Leslie and Rantisi 2006).

The case of cultural policy, like that of daycare services, points to Quebec's more interventionist government. Taxation rates in Quebec are substantially higher for medium- to high-earning individuals than they are in the rest of Canada, and this leads to a stronger presence of government in many walks of life. The benefits of this intervention are strongly debated, and it is not our intention to enter the fray in this chapter. However, the stronger presence of public service jobs (education, public administration, health and social services) in Montreal than in most other large cities except for Ottawa (Table 8.1) probably reflects this different balance between the public and private sectors in Quebec.

To conclude this section, a word must be said about Montreal's (as opposed to the province of Quebec's) government. In general, most economic geographers would not consider a metropolitan area's particular form of government as relevant to its overall economic performance. The debate between neo-regionalists (e.g., Savitch and Vogel [2004], who argue for metropolitan-level governance on grounds of equity, efficient regional-level functions, and shared fiscal burden) and those who favour a more fragmented governance (e.g., Sancton [2000], who argues that fragmentation enables individuals and firms to select the combination of taxes and services that best suits them) is by and large not connected to whether or not a metropolitan economy functions properly. Although both sides of the debate often resort to the argument that metropolitan economies would benefit from their preferred regime, there is little evidence to support either contention (Carr and Feiock 1999; Morgan and Mareschal 1999). In fact, empirical evidence tends to suggest that provided local institutions within a metropolitan area *cooperate*—whether formally or informally—then the economy will perform better than if there is antagonism (Nelson and Foster 1999; Paytas 2001). It is therefore unfortunate that the politically motivated fusion of municipalities on Montreal island (see Figure 8.5), which was forced on unwilling suburbs in 2001, has led to almost a decade of acrimony and partial backtracking. Although anglophone and francophone suburbs alike were forced into the island-wide municipality and unhappiness with the undemocratic procedures was spread across the board (but was by no means universal), the debate was framed by some commentators and politicians as an anglophone versus francophone one. This served to stoke the historic antagonism between these two founding cultural groups and was useful in isolating opponents to the fusions who were branded as 'the privileged English' (see Prémont 2000; Marsan 2000).

The particular form of local government imposed upon Montreal was probably not ideal, since the new mega-municipality, which has since been partially dismantled, only covered the island of Montreal (28 municipalities) and left about 70 off-island metropolitan municipalities outside of the arrangement.[5] However, the more fundamental problem is that it has caused a decade of confusion over exactly how Montreal should be run and over who is responsible for what within the city. Thus, citizens and businesses are unhappy with basic services from snow-clearing to pothole-filling and urban transport, and the business community—despite a decade of relatively good performance in Montreal—feels

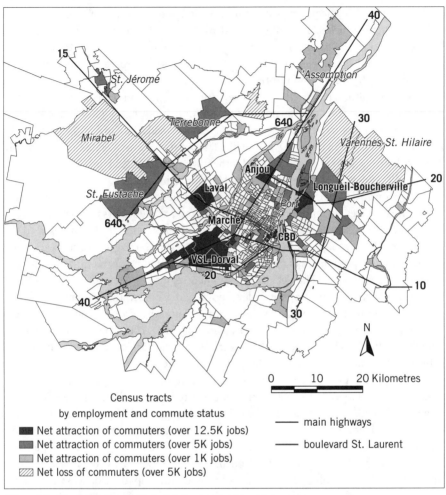

Figure 8.5 Montreal: Employment poles and major highways.

that the city lacks leadership and vision for the future (Côté and Séguin 2009). The expansion of 'Quebec Inc.' into global markets and the sense of opportunity that has propelled it is therefore tempered by mismanagement of Montreal's municipal affairs.

Whatever the form of government, Montreal's economy occupies space irrespective of municipal boundaries (Shearmur and Motte 2009), and it is to this that we now turn.

Where Does the Economic Activity Take Place?

In this section, we examine where economic activity in Montreal takes place, where it is growing, and where it is declining. Recent work led by William Coffey and Richard Shearmur (Coffey and Shearmur 2001a, 2001b; Shearmur and Coffey 2002a, 2002b; Terral and Shearmur 2008; Shearmur et al. 2007) has extensively covered this topic for the 1981 to 2001 period, and their work is summarized and updated in this section.

The basic geography of Montreal's economy is illustrated in Figure 8.5. Two structures are superimposed. The first is a polycentric structure organized around Montreal's six largest employment poles (see Table 8.2). These six poles gather 33 per cent of all jobs. The second structure relates to major transport axes: smaller employment centres, and most of the remaining jobs in the metropolitan area, are located within a mile or so of the major axes portrayed in Figure 8.5 (Shearmur et al. 2007; Terral and Shearmur 2008). Overall, this geography has remained quite stable since the early 1980s (Shearmur and Coffey 2002a; 2002b).

It can readily be seen from Table 8.2 that despite the overall stability of Montreal's urban form, changes occurred between 1996 and 2006.[6] Not unexpectedly, there is evidence of very rapid suburban growth: indeed, the 'residential' job centres (i.e., census tracts that gather more than 5000 jobs but are not net attractors of commuters) have nearly doubled in size and are specializing not only in retail but increasingly in manufacturing activities. Furthermore, it is evident that there is substantial dispersed job growth in the outer suburbs (Table 8.3)—as one moves further away from the CBD, growth rates increase in all sectors except for leisure and communication and producer services.

However, this suburban growth must be qualified. Indeed, despite evidence of rapid growth in smaller employment centres, Montreal's CBD has expanded substantially, adding a further 40,000 jobs to its initial 200,000. The other major employment centres have also expanded, and overall these centres have increased their share of jobs from 32.8 per cent to 33.2 per cent. Similarly, the rest of the CMA, the portion that is not identified as an employment pole in Figure 8.5, has shrunk in relative weight from 60.0 per cent to 58.2 per cent.

All this points to the fact that employment in Montreal is *both* sprawling (in its outer suburbs) *and* polarizing (in and around its city centre and key employment poles). The CBD dominates in terms of high-order services, finance, and leisure—to which can be added health services and higher education (numbers not shown). Manufacturing employment is rapidly declining in central parts of the agglomeration, more rapidly outside of the employment poles, and growing fast in the medium to outer suburbs, but in a mixed polarized and non-polarized fashion (Table 8.3). Throughout the agglomeration, retail is evolving in a polarized fashion as large retail centres increasingly take over the market shares of smaller shops, whereas leisure and cultural activities are growing in a more dispersed fashion. Overall, if employment totals are considered, it is clear that employment poles are consolidating within a 15-kilometre radius of the CBD but are tending to grow at a similar or slower rate to that of surrounding areas in the extended suburbs. These patterns cannot be fully explored in this chapter, but they are further evidence of the complex way in which metropolitan form is evolving (Shearmur et al. 2007): a simple story of polycentrism, sprawl, or dispersal is not sufficient to capture the dynamics of metropolitan change.

The centre's strength is partly attributable to Montreal's geography (it is an island) and to its dense late-Victorian urban fabric. It is also attributable to some major government-led development projects. Despite some controversy over their cost and over their effectiveness at encouraging small start-ups and entrepreneurs (notwithstanding promotional claims, the projects are mainly geared toward large corporate clients and well-established firms), the Cité du Multimédia (just south of the core CBD), the Cité International and

Table 8.2 Employment Poles in the Montreal CMA: Employment and Selected Structure, 1996–2006

	Employment		% of CMA		Selected location quotients, 1996							Selected location quotients, 2006						
	1996	2006	1996	2006	A	B	C	D	E	F	G	A	B	C	D	E	F	G
CBD	203,740	243,675	14.1	14.0	0.20	0.53	0.08	2.97	2.12	1.76	0.40	0.20	0.21	0.08	2.78	2.24	1.68	0.38
VSL-Dorval	154,370	174,615	10.6	10.1	1.32	1.63	4.44	0.36	0.97	0.97	0.77	1.36	2.08	4.13	0.46	0.87	0.61	0.83
Carrefour-Laval	39,005	54,005	2.7	3.1	1.23	1.92	0.80	0.99	0.93	0.62	1.55	1.37	1.72	1.06	0.72	0.82	0.50	1.73
Anjou	29,080	41,310	2.0	2.4	3.07	2.27	0.30	0.45	0.62	0.38	0.95	2.39	2.55	0.67	0.61	0.47	0.30	1.26
Longueuil-Boucherville	16,140	24,150	1.1	1.4	0.78	1.70	0.55	0.53	0.66	0.35	1.62	1.01	1.56	0.81	0.51	0.56	0.39	2.09
Marché-Central–Chabanel	33,270	37,440	2.3	2.2	3.73	0.93	0.60	0.59	0.75	0.27	0.87	3.16	1.23	1.48	0.96	0.69	0.49	1.34
sub-total primary poles	475,605	575,195	32.8	33.2	1.09	1.18	1.62	1.56	1.41	1.17	0.72	1.05	1.21	1.57	1.51	1.39	1.01	0.84
Côte-des-Neiges	17,950	23,870	1.2	1.4	0.04	0.05	0.02	0.30	0.22	0.12	0.29	0.06	0.05	0.00	0.22	0.30	0.23	0.23
East-Port	12,220	22,045	0.8	1.3	1.80	0.68	0.05	0.75	0.42	0.98	0.16	1.70	0.51	0.02	1.19	1.21	1.31	0.26
Saint-Eustache	10,050	15,275	0.7	0.9	1.25	5.16	0.72	0.25	0.38	0.34	0.65	2.07	2.24	0.91	0.37	0.40	0.23	1.34
Saint-Jerome	11,415	13,250	0.8	0.8	1.15	1.66	0.14	0.71	0.55	0.56	1.36	1.26	0.85	0.00	0.62	0.38	0.38	1.54
sub-total secondary poles	51,635	74,440	3.6	4.3	0.94	1.55	0.19	0.49	0.37	0.46	0.57	1.17	0.78	0.19	0.61	0.60	0.57	0.70
Saint-Michel	5905	6705	0.4	0.4	5.43	0.87	2.05	0.25	0.37	0.19	0.68	7.19	0.84	1.13	0.49	0.22	0.17	0.59
Terrebonne	1510	6660	0.1	0.4	3.28	1.67	0.76	0.10	0.75	0.41	0.28	3.70	2.56	0.27	0.28	0.43	0.61	0.96
Angrignon	6480	6605	0.4	0.4	1.58	2.02	0.20	0.52	0.24	0.46	2.53	2.14	1.67	0.00	0.84	0.21	0.63	2.96
Longueuil-centre	5370	6290	0.4	0.4	0.09	0.36	0.06	1.43	1.10	0.48	1.77	0.11	0.14	0.06	1.46	0.94	0.77	1.49
Plateau	3810	5320	0.3	0.3	0.00	0.10	0.00	0.81	0.83	0.68	0.43	0.00	0.07	0.00	1.00	0.61	1.09	0.40

	1996	2006	% 1996	% 2006	A	B	C	D	E	F	G	A	B	C	D	E	F	G
Boul-Tachereau	3320	5265	0.2	0.3	0.11	0.00	0.00	0.53	0.23	0.12	1.26	0.00	0.06	0.00	0.23	0.30	0.55	1.14
Saint-Hubert	4010	5255	0.3	0.3	1.27	1.36	0.69	0.77	0.53	1.47	1.85	2.93	1.03	1.97	0.91	0.48	2.13	1.22
Montreal-Nord	4490	5145	0.3	0.3	0.02	0.04	0.00	0.27	0.32	0.18	0.09	0.03	0.04	0.00	0.16	0.19	0.21	0.12
sub-total isolated poles	34,895	47,245	2.4	2.7	1.53	0.82	0.51	0.62	0.52	0.49	1.26	2.18	0.87	0.42	0.68	0.42	0.74	1.15
Mirabel-Terrebonne	11,170	16,495	0.8	1.0	0.41	0.37	8.01	0.28	0.53	1.19	0.37	1.00	0.92	10.00	0.42	0.60	0.79	0.59
Mascouche	2970	5955	0.2	0.3	0.86	0.42	0.88	0.59	0.60	0.42	1.17	0.93	0.96	0.13	0.51	0.30	0.27	2.69
Varennes-Saint-Hilaire	3215	5060	0.2	0.3	0.29	2.72	0.61	0.26	0.91	0.23	0.85	1.12	2.88	0.16	0.32	0.96	0.38	1.04
sub-total residential poles	17,355	27,510	1.2	1.6	0.47	0.81	5.42	0.33	0.61	0.88	0.60	1.01	1.29	6.05	0.42	0.60	0.60	1.13
Rest CMA	870,300	1,010,030	60.0	58.2	0.94	0.88	0.64	0.75	0.84	0.96	1.17	1.17	0.90	0.90	0.73	0.90	0.84	0.90
TOTAL	**1,449,790**	**1,734,420**	**100.0**	**100.0**														

Notes: A = traditional manufacturing; B = medium-tech manufacturing; C = High-tech manufacturing; D = FIRE; E = producer services; F = leisure, arts, and communications; G = retail. Location quotients greater than 1.05 are in bold.

Employment poles are defined using 2006 data: this introduces a bias toward identifying growth within employment poles. This bias is stronger the smaller the spatial unit used (Shearmur et al. 2007) and the smaller the overall employment growth: census tracts are usually large enough to minimize this problem, and the considerable growth over the 1996–2006 period minimizes the chances that major 1996 poles are no longer identified as such in 2006. However, a complete analysis would examine the same data from the perspective of poles defined in 1996.

Table 8.3 Job Distribution and Growth by Distance from the CBD and by Employment Pole, Montreal, 1996–2006

	Manufacturing			High-order services		Leisure and retail		Total	Total increase
	A	B	C	D	E	F	G	H	
Jobs within employment poles, 1996									
0–5 km	23,315	13,655	1340	49,490	55,370	34,035	32,915	430,505	
5–10 km	47,705	15,775	12,600	18,910	26,315	13,125	51,340	356,570	
10–15 km	37,035	20,060	14,830	9580	15,940	10,725	37,260	270,705	
15–20 km	15,005	10,710	8635	6630	10,495	4595	24,545	153,945	
20–25 km	6140	4570	2350	3700	5330	2870	16,595	85,490	
25 km+	12,675	9905	4560	6145	8495	4555	24,765	152,575	
Jobs outside employment poles, 1996									
0–5 km	19,265	8045	830	9475	18,385	16,525	21,460	205,005	
5–10 km	24,345	11,365	10,110	14,300	19,145	10,130	39,130	259,475	
10–15 km	17,395	8160	2580	7370	8855	5575	25,215	154,445	
15–20 km	4720	2645	1260	3335	3940	2025	11,875	66,210	
20–25 km	5610	3480	810	3130	4190	2645	13,570	69,705	
25 km+	8980	5855	1440	5125	6900	3350	20,870	115,460	

% Growth within employment poles, 1996–2006	A	B	C	D	E	F	G	Total	n
0–5 km	−39.3	−45.9	−36.9	18.8	26.0	61.7	0.1	14.6	62,975
5–10 km	−27.8	3.7	−16.2	33.0	11.4	34.1	9.1	10.4	36,990
10–15 km	−17.8	47.9	−24.9	33.9	19.2	−9.8	10.2	16.5	44,685
15–20 km	−4.2	46.6	39.2	25.0	11.9	21.1	35.7	32.6	50,250
20–25 km	−10.7	40.6	5.5	38.9	45.0	40.1	13.3	29.5	25,215
25 km+	25.3	39.5	45.0	48.5	44.0	44.3	43.6	42.3	64,515

% Growth outside employment poles, 1996–2006	A	B	C	D	E	F	G	Total	n
0–5 km	−44.1	−41.4	−56.6	21.2	18.4	88.3	−4.5	7.6	15,610
5–10 km	−35.1	−5.7	−27.8	21.5	4.1	19.1	0.8	6.0	15,480
10–15 km	−15.6	39.5	−1.7	16.0	16.7	−2.2	−9.4	10.9	16,910
15–20 km	2.9	105.1	55.6	26.7	37.3	30.6	16.8	35.1	23,210
20–25 km	−15.3	38.5	113.0	39.6	48.1	40.3	24.3	35.2	24,525
25 km +	6.8	62.6	42.0	43.6	45.2	50.1	26.5	38.1	43,995

Note: The greater of each pair of growth figures (they are compared by sector and distance from the CBD) is highlighted in bold.

A = traditional manufacturing; B = medium-tech manufacturing; C = high-tech manufacturing; D = FIRE; E = producer services; F = leisure, arts, and communications; G = retail.

Conference Centre (built above a now-covered urban highway), and the Cité du Commerce Electronique (a partly completed office complex that now houses one of Quebec Inc.'s key players) have all contributed to maintaining the CBD's viability as a business location. Furthermore, current projects (such as the development of a downtown cultural district) and ongoing festivals (such as the Jazz Festival and Just for Laughs) ensure that downtown remains a leisure destination for Montrealers and tourists alike. However, just as the large business projects have not been able to provide cheap and flexible accommodation for dynamic start-ups, there is a danger that the over-formalization of Montreal's cultural scene may push out the more exploratory and experimental artistic productions.

The detailed geography of particular sectors will not be further addressed in this chapter. Tables 8.2 and 8.3 provide an idea of the geography of some of them. It is worth emphasizing that the city's key development corridors, boulevard Saint-Laurent and the Lachine Canal (which runs along Highway 20 as it exits the CBD to the southwest), which have structured Montreal's spatial economy since the 1860s (Lewis 2000), continue to be important axes of development. The Marché Central/Chabanel district, at the northern end of boulevard Saint-Laurent, is a key clothing and textile district: it is slowly moving away from traditional industries and into medium- and high-tech manufacturing and retail and has preserved its employment levels despite jobs losses in its core sectors. The Angrignon employment pole—the visible portion of a mixed residential and industrial district that stretches along Lachine Canal between the CBD and Ville Saint-Laurent–Dorval—has maintained its traditional manufacturing function but is consolidating its position as a retail district and seems also to be strengthening its financial services. Despite the stability of Montreal's internal geography, changes are occurring as neighbourhoods adapt to the evolving economy.

A Case Study: Clothing in Montreal[7]

At the beginning of this chapter, we noted that clothing has been an important sector for the Montreal economy since the city's foundation. The late nineteenth century, in particular, marks a defining period when the adoption of new technologies and the influx of immigrants (primarily Italian and German Jewish immigrants with skills in the trade) contributed to the city's transformation from a local to a national apparel centre (Lewis 2000). However, since the 1980s the city's status as a clothing manufacturing capital has been under threat. Free trade agreements and the removal of tarifs and import duties (mentioned above) have led to a signficant rise in imports and a contraction of manufacturing jobs. The local producers that have survived face intense competition from low-wage overseas producers, the volatility of global markets, and a shrinking base of suppliers and support services.

The severity of the crisis can not be overstated; as noted above, more than 10,000 jobs were lost between 2003 and 2005 alone (see also King 2008). However, the crisis is also triggering change. A new industrial model—or more precisely, set of models—is emerging out of the ashes. Established, mass-market producers are altering their strategies and, in a 'post-industrial' vein, shifting core activities from manufacturing to service-oriented functions (e.g., distribution and marketing). At the same time, we are witnessing the emergence of a dynamic, post-Fordist cluster of small-scale, independent designers who are competing on the basis of style, quality, and exclusivity. These trends are also tied to broader changes in the industry's intra-metropolitan geography and the development of public and non-profit support institutions. In what follows, an overview of these trends is provided, as well as a consideration of how Montreal's distinct features—its industrial heritage, biculturalism, and status as a regional city—are shaping prospects for renewal.

Responses to the Crisis by the Mass-Market Segment

Among the more established clothing manufacturing firms, the primary response to current challenges has been to reduce manufacturing activities by outsourcing to contractors overseas. While the geography of overseas production is complex and can rely on a number of factors (e.g., the availability of fabrics or the presence of workers experienced in certain types of trimming), producers are increasingly looking to Asia because of the low cost of labour (Friede 2005; Millstein & Co. Consulting Inc. 2007; personal interviews).

In most cases, these manufacturers still oversee the production process; design/patternmaking occurs in-house, and many of the preassembly functions (e.g., the marking and cutting of the fabric and the production of prototypes) are performed in-house or locally. In the case of higher-end manufacturers, they may even keep some of the assembly (sewing) activities local (or regional) for many product segments, particularly for their time-sensitive items or for items for which they want to ensure quality production. However, for the most part, the assembly process is now performed in other countries, and clothing firms are devoting more time and resources to the management of a global supply chain and, increasingly, to the marketing and branding of their products[8] (Milstein & Co. Consulting Inc. 2007). Accordingly, as the data on SSA above suggests, a demand for seamstresses has been giving way to a demand for production managers, quality control officers, logistics managers, and fashion marketers.

These service-oriented activities are viewed as new sites of value added for manufacturers who want to expand their presence in the US in order to compensate for the loss of their domestic sales to imports. Physical and cultural proximity to the US market (particularly to New York City) and a long-standing reputation as reliable suppliers with quality goods confers benefits on Canadian manufacturers relative to other overseas producers (Milstein & Co. Consulting Inc. 2007; author's buyer survey). Moreover, manufacturers that continue to produce at least some of their products locally are well positioned to accommodate the seasonal nature of the US industry (four to six seasons, as compared to two to four in Canada) and to meet demands on the part of US retailers for the quick replenishment of popular lines.

One Montreal firm that is an example of this new paradigm is Peerless Clothing, the largest North American manufacturer of men's suits, with more than $50 million in annual sales. The company, which exports nearly all its products to the US, produces its own line, Peerless, but also manufactures as a licensee for some major US brands (Anon. 2005). Nearly half of its production takes place in its Montreal facility, and half is done offshore (Cloutier 2007). Peerless competes on the basis of technological innovation (its suits are fused rather than sewn) and quality and, increasingly, on the basis of service and distribution. With a distribution centre in Vermont equipped with the latest electronic data interchange software, the company can service customers within a 48-hour period. Peerless has also upgraded the logistics technology in its Montreal plant to allow for a quick replenishment system, and it is building its brand through the distribution of catalogues and attendance at major trade shows (Anon. 2005). To quote Peerless vice-president Elliot Lifson: 'Our industry is not just a factory, there's a new paradigm—we have to stress innovation, creativity, design and distribution being done here' (King 2008).

Changes in the orientation of mass-market clothing producers are also inscribed in the metropolitan landscape. Chabanel, a district in the northern part of the city, has been the centre of garment production for Montreal—and for Canada more generally—for much of the postwar period. It constitutes the greater part of the Marché-Central–Chabanel employment pole (see Figure 8.5) and still has a significant concentration of manufacturing jobs (see Table 8.2). Within the past 10 years, however, the district has been undergoing a gradual transformation from an industrial to a commercial centre (Friede 2005). In the premier fashion building, 555 Chabanel West, for example, more than 90 per cent of the space is now dedicated to showrooms—i.e., spaces where firms market their clothing to local and international buyers. In addition, the district is attracting other users, such as banking and insurance: with the municipal government investing $17 million to revitalize the district and make it appealing for a diverse set of uses, it is likely to see further change (Maughan 2008).

Consequently, new production districts in metropolitan Montreal are emerging as alternatives to Chabanel. In particular, Ville Saint-Laurent in the northwest (see the VSL-Dorval employment pole in Figure 8.5) and the Saint-Michel district in the northeast (just west of Anjou; see Figure 8.5 and Table 8.2) are becoming the preferred sites for larger-scale clothing manufacturers because of the availability of vacant land and government subsidies (personal interviews with clothing manufacturers). In contrast to Chabanel, which is already built up with multi-storey buildings, these locations offer firms the possibility of constructing single-storey factories that can accommodate new technologies (e.g., automated handling systems) and allow for future expansion.

The Rise of A New Generation of Actors

Along with a process of adaptation on the part of existing firms, the Montreal clothing industry is characterized today by the rise of a new set of actors who have become the icons of Montreal clothing to the outside world—the fashion designers. Although there have always been fashion designers in the industry, it is the growing number of designers who operate as independent entrepreneurs that is unprecedented. In the past, budding designers had to work for an established clothing producer in Chabanel for a number of years, often in the assembly of garments, before working their way up to the rank of designer or starting their own business. Today, a number of designers are starting their own businesses immediately after they graduate from one of the local design schools (personal interviews with designers). These designers are generally responsible for the conceptualization and production of their products and in the early phases may have a few people helping them. As they grow and establish a larger clientele, they work with local contractors to meet production demands, since their in-house production capabilities are limited. The independent designers are oriented primarily to the local market and have a strong command of local tastes because they work closely with local independent boutiques and/or sell in their own boutiques (Rantisi, forthcoming).

A confluence of factors have contributed to the rise of this fashion design scene. The first factor is the availability of affordable workspace in the form of older, loft buildings—a consequence of the process of deindustrialization that Montreal has been experiencing. The second factor relates to the presence of an industrial infrastructure, such as contractors, button and trim suppliers, fabric suppliers, and machine/repair shops, on which these small-scale producers can draw. This is particularly significant, since many of the designers have short production runs and require quick and easy access to supplies. The local design schools, which are increasingly looking to balance their focus on the aesthetic side of the industry with greater training in the business and technical sides, represent a third factor. And finally, as we discuss below, a set of public and non-profit institutions have emerged within the past decade to highlight the significance of design for the industry and provide designers with financial and material support as well as greater visibility.

Interestingly, this community of independent designers has a parallel existence with (but for the most part separate from) the mass-market producers described above. While mass-market producers compete on their ability to supply large volumes of goods with quality service at a competitive price, independent designers produce limited runs of items and compete on the basis of their styles and quality fabrics and production. These differences are reinforced by the distinct markets they serve, with mass-market producers selling to department and discount stores and independent designers selling to boutiques and specialty stores.[9]

Differences between these two segments are also evident in their respective geographies. While the majority of mass-market producers have their headquarters and showrooms in Chabanel, a significant number of independent designers have gathered in the formerly industrial, inner-city neighbourhoods of the Plateau and Mile End (both located to the southeast of Chabanel). Designers are drawn to these areas because of the studio and work/live spaces, the presence of other cultural workers, and the concentration of boutiques, galleries, cafés, restaurants, and music venues (Rantisi and Leslie, forthcoming). Mile End in particular has become a popular site because the rents are cheaper than those of the Plateau (the more established design district) and because the local community economic development corporation has helped to develop a fashion design incubator, LABCreatif, where young designers can share space, machinery, and information (Klein, Tremblay, and Bussières 2007). In interviews, designers also note that while these areas are close to Chabanel and the support services based there, they are not as 'corporate' or 'stifling' as the commercially oriented Chabanel. Rather, the bohemian character of the neighborhoods make them more conducive for aesthetic innovation. Thus, the geography of this design segment reflects a physical—as well as cultural—separation from Chabanel.

The Institutional Setting: Public and Non-profit Initiatives

Because of the historical signficance of the industry and the symbolic significance of fashion as a cultural product, local and provincial governments and trade organizations have introduced a series of intiatives to encourage restructuring and a greater orientation to design, marketing, and export promotion. At the local level, an important development was the establishment of the Montreal Fashion Network in 1998, a non-profit organization uniting several sector associations. Concerned about the survival of the industry in the face of trade liberalization, the network launched a biannual fashion show, Montreal Fashion Week, in 2001 to increase the visibility of Quebec's fashion products and position Montreal as a fashion capital by showcasing designers' collections to buyers and the media. The show is now organized by the marketing agency Sensation Mode, which also organizes an annual outdoor fashion show for the general public along a major downtown shopping street (Leslie and Rantisi 2006).

Another important institution is the municipal government's Fashion Commissioner, a position with a mandate to support and promote the industry. The commissioner works closely with trade organizations and other levels of government, and key activities include a website that announces all industry events and news, events that allow for knowledge exchange and networking among industry actors, and support for local initiatives such as Fashion Week, LABCreatif, and the redevelopment of Chabanel. In addition to the Fashion Commissioner, there is a Design Commissioner, who is responsible for promoting the design sector more generally and who successfully made a bid to designate Montreal as a UNESCO Design Capital in 2006. This commissoner sponsors exhibitions and open houses that feature local fashion designers.

Local initiatves are also complemented by provincial government programs. One such program is a designer tax credit progam that offers subsidies for the costs of hiring a professionally trained designer; both mass-market producers and independent designers have taken advantage of these subsidies (Leslie and Rantisi 2006; personal interviews). Another provincial program, Montreal Collections, seeks to market Montreal clothing producers to US buyers by sponsoring booths and fashion shows at major US trade shows, sponsoring ads in fashion publications, and hosting a website (Rantisi, forthcoming). In addition to these programs, in 2008 the provincial government introduced a three-year strategy, Pro-Mode, which stipulates funding for individual enterprises that are adapting their business models as well as funding for existing provincial and local programs and workforce development initiatives. (www.mdeie.gouv.qc.ca)

The Future of Clothing in Montreal

New strategies on the part of firms and an institutional context that encourages industrial upgrading and promotion suggest that Montreal clothing is charting a new path to development. However, like most transitions from 'old' to 'new', the path is neither smooth nor certain. One major challenge to restructuring, for example, is the legacy of a traditional manufacturing model that privileges commercial over aesthetic considerations, leading to the designer–manufacturer divide discussed above. Despite initatives such as the designer tax credit, designers complain that manufacturers continue to view design as a cost rather than value added and that there is little appreciation or understanding of the design process. Yet the few cases in which designer–manufacturer collaborations have worked point to benefits for both: greater production capabilities and distribution and marketing support for designers and distinctive products for manufacturers (personal interviews). This suggests the need for programs that move beyond the provision of subsidies and offer educational and networking opportunities for the two segments.

Montreal must also contend with the fact that as a regional city, it lacks the marketing and distribution channels found in Toronto. Most of the major retail buyers and the fashion media are based in Toronto, and Toronto has a larger Fashion Week, even though more clothing production occurs in Montreal. Events like Fashion Week are helping to build a local marketing infrastructure; however, the Toronto advantage is likely to persist, suggesting that co-ordination across the city-regions may be the way for both cities to benefit from regional specializations.

While some of Montreal's historical legacies present challenges to the process of adaptation, other features are helping to distinguish the local industry and establish a niche for Montreal within the global fashion hierarchy. Montreal's biculturalism, for example, has become a marker of distinction and a selling point for the industry. Promotional materials and press releases regularly mention how Montreal producers offer a North American business sense with a 'European flair' (e.g., http://www.touchofme.com/en/profile.html). Moreover, while there is a mix of francophones and anglophones in both segments, the designers who garner the most media visiblity— Denis Gagnon, Phillip Dubuc, and Marie St Pierre—are francophone, and this further emphasizes the French influence on the style of Montreal clothing. Montreal's biculturalism also means that producers are regularly exposed to different cultural traditions, which contributes to design innovation (Stolarick and Florida 2006; Rantisi and Leslie, forthcoming).

The other key issue influencing the prospects for renewal is workforce development. The new realities of the industry imply a new sets of skills—skills that are oriented to advanced technology and service-oriented activities (e.g., computer-aided design or logistics software system). While the historic focus of design schools is on the aesthetic dimensions of the industry (sketching, draping, history), this is starting to change, and a recent initiative by the Apparel Human Resource Council to co-ordinate a roundtable of fashion design schools and industry representatives is an important step forward. At the same time, programs are needed to support and retrain the thousands of workers who are losing their jobs as a consequence of outsourcing (Shragge and Siddiqi 2008). There are some cases where the demand for assembly skills still exists, as in the case of the designer segment or the firms competing on the basis of quick replenishment, but employers may not know how to locate such workers, particularly with the decline of Chabanel as a production pole and the increasing dispersal of industry actors. A new metropolitan spatial division of labour is leading to a spatial mismatch and the growing dislocation of traditional manufacturing workers and presenting new challenges for labour mobility and training.

Conclusion

Two intertwined stories have been told about Montreal. On the one hand, a story of stability: the principal historical, cultural, and industrial elements that constitute Montreal's identity and shape the way in which it is moving forward have remained relatively constant over the past century. Its industrial structure and internal geography of economic activity have changed only very slowly over the past 30 years. On the other hand, a story of change: Montreal's primacy in Canada's urban hierarchy has disappeared, suburbs are growing fast, and within traditional sectors (such as clothing) and employment centres (such as Anjou and Chabanel), important and fundamental changes are taking place in the skills required and in the way business is conducted in the face of global trends and technological change.

These stories are strongly connected: the internal geography of employment remains the same, yet buildings are renovated, modified, and put to different uses. Old industrial premises that used to house large factories now house numerous designers and small manufacturers, often in the same or similar—yet virtually unrecognizable—economic sector. Montreal's industrial structure changes very slowly, but the production processes, types of skills, and industrial organization of even the most traditional sector, clothing, are being revolutionized.

There is an understandable tendency for current discourse to focus on these changes: buzz, creativity, innovation are all very popular concepts constantly recurring in policy documents and academic work on cities. These concepts correspond to an important dimension of urban economies and need to be understood, but there is a danger that their popularity obscures the deeper, slower, yet fundamentally important structural and cultural contexts within which this buzz takes place.

Perhaps because Montreal has an evident cultural heritage and a strong industrial history, it is easier to identify the longer-term trends that shape its evolution, and this chapter is an attempt to illustrate how change and up-to-date adaptation to rapidly altering circumstances can nevertheless be compatible with long-term structural stability and slow evolution.

Stolarick and Florida (2006) point out the role that Montreal's heritage plays in today's economy. They discuss the importance of bilingualism and multiculturalism for creative inspiration and also the way in which this multiculturalism can serve as a testing ground for multiple markets. They highlight some of the art/design and technical collaborations that arise from a mixed economy with advanced manufacturing capability such as Montreal's (e.g., Ubisoft, Bombardier, Cirque du Soleil). Yet while these are important effects of Montreal's biculturalism and of its creative economy, the negative effects that the English–French divide continues to have on the city's management, the corporatist business climate, and the persistent low wages that characterize its economy are darker sides that should not be ignored.

Notwithstanding the path dependency and glacial pace of change that have been a theme of this chapter, change at a macro scale can and does happen: from time to time exogenous factors can open up new opportunities or create new problems. Over the past 30 years, Montreal has been adapting to the fact that it is no longer Canada's primary metropolis, and to some extent this has held Montreal's economy back. Its industries have constantly been adapting and reinventing themselves, but this has not been sufficient to compensate for its shrinking hinterland and loss of financial and managerial dominance (Polèse and Shearmur 2004). In many ways, this period of adaptation has come to a close: Montreal has (almost) accepted its status as Canada's second city and is increasingly taking pride in its cultural and economic particularities.

Thus, Montreal's economy may be charting a new course, one that builds upon its recent employment performance and shift toward qualified occupations to begin generating higher incomes for its inhabitants by fully exploiting their capacity to design, invent, and implement original goods and services. At the very least, many Montrealers hope that this is what the future holds.

References

Anon. 2005. *Profile: Peerless Clothing. Wear? Canada!* Ottawa: Canadian Apparel Federation, p. 13.

Bourhis, R. 2008. 'The English-speaking communities of Quebec: Vitality, multiple identities and linguicism'. In R.Y. Bourhis, ed., *The Vitality of the English-Speaking Communities of Quebec: From Community Decline to Revival*. Montreal: CEETUM, Université de Montréal.

Brault, S. 2002. 'How should Montreal define its cultural policy?' *Culture Montreal* 26 August. http://www.culturemontreal.ca/positions/020826ed_whichpolicy.htm.

Campbell, F., C. Ramey, E. Pungello, J. Sparling, and S. Miller-Johnson. 2002. 'Early childhood education: Young adult outcomes from the Abecedarian Project'. *Applied Developmental Science* 6 (1): 42–57.

Carr, J., and R. Feiock. 1999. 'Metropolitan government and economic development'. *Urban Affairs Review* 34 (3): 476–88.

Cloutier, L. 2007. 'Vêtement : 40 000 travailleurs toujours actifs'. *La Presse* 8 June. http://lapresseaffaires.cyberpresse.ca/.../01-673717-vetement-40-000-travailleurs-toujours-actifs.php.

Coffey, W. 1994. *The Evolution of Canada's Metropolitan Economies*. Montreal: Institute for Research in Public Policy.

Coffey, W., and M. Polèse. 1993. 'Le déclin de l'empire montréalais : regard sur l'économie d'une métropole en mutation'. *Recherches sociographiques* 34: 417–37.

———. 1999. 'A distinct metropolis for a distinct society? The economic restructuring of Montreal in the Canadian context'. *Canadian Journal of Regional Science* 22 (1/2): 23–40.

Coffey, W., and R.Shearmur. 2001a. 'Intrametropolitan employment distribution in Montreal, 1981–1996'. *Urban Geography* 22 (2): 106–29.

———. 2001b. 'The identification of employment centres in Canadian metropolitan areas: The example of Montreal, 1996'. *Le géographe canadien* 45 (3): 371–86.

Coté, M., and C. Séguin. 2009. 'Les grandes villes : locomotives du développement économique. Quitter le peloton de queue'. Synthèse du 34ième Congrès de l'ASDEQ. http://www.asdeq.org/congres/pdf/2009/Synthese-Cote-Segin-Congres-ASDEQ-2009.pdf.

Economist, The. 2009. 'Municipal corruption in Canada'. 25 June.

Fraser, M. 1987. *Quebec Inc: French-Canadian Entrepreneurs and the New Business Elite*. Toronto: Key Porter Books.

Friede, E. 2005. 'Chabanel still has lots of life'. *The Gazette* 5 June: A1.

Germain, A., and C. Poirier. 2007. 'Les territoires fluides de l'immigration à Montréal ou le quartier dans tous ses états'. *Globe* 10 (1): 107–20.

Germain, A., and D. Rose. 2000. *Montreal: The Quest for a Metropolis*. Chichester: Wiley.

Jedwab, J. 2004. *Vers l'avant : l'évolution de la communauté d'expression anglaise du Québec*. Ottawa: Commissariat aux langues officielles.

King, M. 2008. 'Garment sector hanging by a thread'. *The Gazette* 7 March. http://www.canada.com/montrealgazette/news/business/story.html?id=c51f66be-f87c-4fa6-8ab0-44a24e875d00&k=81914.

Klein, J.-L., D.-G. Tremblay, and D. Bussières. 2007. *Community-Based Intermediation Organisations and Social Innovation: A Case Study in Montreal's Apparel Sector*. Paper presented at the annual meeting of the Innovation Systems Research Network, Vancouver, 3–4 May.

Legault, J. 2009. 'Sabia meeting shows how much Quebec Inc. has changed'. *The Gazette* 10 April. http://www.montrealgazette.com/news/Sabia+meeting+shows+much+Quebec+changed/1483643/story.html.

Leslie, D., and N.M. Rantisi. 2006. 'Governing the design sector in Montréal'. *Urban Affairs Review* 41: 309–37.

Levine, M. 1990. *The Reconquest of Montreal: Language Policy and Social Change in a Bilingual City*. Philadelphia: Temple University Press.

Lewis, R. 2000. *Manufacturing Montreal: The Making of an Industrial Landscape, 1850 to 1930*. Baltimore: Johns Hopkins University Press

McNicoll, C. 1993. *Montréal : une société multiculturelle*. Paris: Belin.

Marsan, J.-C. 1990. *Montreal in Evolution*. Montreal and Kingston: McGill-Queen's University Press.

———. 2000. 'L'art d'enfoncer les portes ouvertes'. *Le Devoir* 28 November. http://archives.vigile.net/00-11/fusions-marsan.html.

Maughan, C. 2008. 'Facelift for slumping Chabanel; $17-million plan; City project aims to revitalize district's declining economy'. *The Gazette* 7 August: A7.

Milligan, K. 2005. 'Subsidizing the stork: New evidence on tax incentives and fertility'. *Review of Economics and Statistics* 87 (3): 539–55.

Milstein & Co. Consulting Inc. 2007. *Apparel Strategic Benchmarking Report and Planning Toolbox: Manufacturer's Report*. Prepared on behalf of the Apparel Human Resource Council. Montreal.

Ministère des affaires culturelles. 1992. *La politique culturelle du québec : notre culture, notre avenir*. Québec: Gouvernement du Québec.

Morgan, D., and P. Mareschal. 1999. 'Central city/suburban inequality and metropolitan political fragmentation'. *Urban Affairs Review* 34 (4): 578–95.

Nelson, A., and K. Foster. 1999. 'Metropolitan governance structure and income growth'. *Journal of Urban Affairs* 21 (3): 309–24.

Paytas, J. 2001. *Does Governance Matter? The Dynamics of Metropolitan Governance and Competitiveness*. Pittsburgh: Carnegie Mellon Center for Economic Development.

Pomfret, R. 1981. *The Economic Development of Canada*. Toronto: Methuen.

Prémont, M.-C. 2000. 'L'opposition farouche au projet Harel : ou l'art d'éviter la question difficile'. *Le Devoir* 23 November: A7.

Rantisi, N.M. Forthcoming. 'Cultural intermediaries and the geography of designs in the Montréal fashion industry'. In G. Rusten and J. Bryson, eds, *Industrial Design and Competitiveness: Spatial and Organization Dimensions*. Hampshire, UK: Palgrave Macmillan.

Rantisi, N.M., and D. Leslie. Forthcoming. 'Creativity by design? The role of informal spaces in creative production'. In T. Edensor, D. Leslie, S. Millington, and N. Rantisi, eds, *Spaces of Vernacular Creativity: Rethinking the Cultural Economy*. London: Routledge.

Savitch, H., and R. Vogel. 2004. 'Suburbs without a city: Power and city-county consolidation'. *Urban Affairs Review* 39 (6): 758–90.

Sancton, A. 2000. 'Amalgamations, service realignment and property taxes: Did the Harris government have a plan for Ontario's municipalities?' *Canadian Journal of Regional Science* 23 (1): 135–56.

Shearmur, R., and W. Coffey. 2002a. 'Urban employment sub-centres and sectoral clustering in Montreal: Complementary approaches to the study of urban form'. *Urban Geography* 23 (2): 103–30.

———. 2002b. 'A tale of four cities: Intra-metropolitan employment distribution in Toronto, Montreal, Ottawa and Vancouver. *Environment and Planning A* 34: 575–98.

Shearmur, R., W. Coffey, C. Dubé, and R. Barbonne. 2007. 'Intrametropolitan employment structure: Polycentricity, scatteration, dispersal and chaos in Toronto, Montreal and Vancouver, 1996–2001'. *Urban Studies* 44 (9): 1713–38.

Shearmur, R., and B. Motte. 2009. 'How are central cities and suburbs bound? A study of Montreal's intra-metropolitan economy'. *Urban Affairs Review* 44: 490–524.

Shragge, E., and Y. Siddiqi. 2008. 'The empire's new clothes: Where does boutique capitalism leave Montreal's garment workers?' *Education for Development Magazine* October.

Stolarick, K., and R. Florida. 2006. 'Creativity, connections and innovation: A study of linkages in the Montréal region'. *Environment and Planning A* 38: 1799–1817.

Terral, L., and R. Shearmur. 2008. 'Desserrement de l'emploi et diffusion de sa croissance dans la région métropolitaine de Montréal'. *Espace géographique* 37 (1): 16–31.

Notes

1. The third founding culture, that of the First Peoples, while it strongly influenced Montreal's initial development, particularly with respect to inland exploration and to the balance of power between English and French colonial settlements, has had far less influence since the Industrial Revolution.

2. The violent activities of fringe elements of the FLQ (Front de Libération du Québec) are sometimes erroneously pointed to as signs of the violent nature of the struggle as a whole. There are two recorded kidnappings and one murder associated with the FLQ (there are currently about 30 murders each year in Montreal): these are terrible events, but from a wider perspective, it is the peacefulness and civility of this struggle that makes it truly remarkable. Trudeau's invocation of the War Measures Act in response to the FLQ kidnappings—and the subsequent jailing of about 500 suspected FLQ members without charge—turned out to be very divisive.

3. Unless otherwise stated, the spatial unit referred to in this chapter is the census metropolitan area: statistics for 1971, 1981, 1991, 1996, and 2001 are

based on the 1991 boundaries, those for 2006 on the 2006 boundaries. All data, unless otherwise stated, are derived from special Statistics Canada census tabulations. Our 1971–2001 series is consistent, and we also possess consistent 2001–6 data. In order to maintain compatibility between the 1971–2001 data and the 2006 data (despite minor boundary changes and changes in classification systems), 2006 numbers are imputed from 2001 numbers by applying the 2001–6 growth rate of the closest matching variable to 2001 data.

4. Immigrants obtain 'points' if they speak French. Furthermore, once landed, their children must attend French-language schools, and the working language in all but the smallest establishments is French. These laws have been instrumental in re-establishing French as Montreal's dominant language and in supporting the thriving francophone cultural industry (Levine 1990; Bourhis 2008). Obviously, the language criteria make Quebec more attractive to French-speaking immigrants, but this does not fully compensate for the processes just described.

5. It should be noted that some fusions also occurred on the South Shore, and similar forced fusions occurred in Quebec City: this undermines the rhetoric claiming either that it was only anglophones who refused the fusions (opposition also emanated from francophone suburbs and in Quebec City) or that this was a provincial plot to strip only anglophones of their municipal governments. The rationale for these forced fusions remains opaque.

6. The methodology applied for identifying employment poles is described and discussed in Coffey

and Shearmur 2001b. The only new type of pole introduced here is the 'residential' pole, described in the text. In this chapter, 1996 tracts are combined with 1996 and 2006 census data to re-identify the poles, which differ slightly (because of changes in census tracts and the use of more recent data) from those identified for 1996 and 2001. It is felt that a detailed methodological section would not be appropriate in this chapter.

7. This section draws on interviews with industry actors conducted by Norma Rantisi and research assistants Mia Hunt and Jason Blackman from 2004 to 2008. More than 70 interviews were conducted with manufacturers, designers, fashion promoters, trade association representatives, and government officials.

8. In some cases, firms are not even performing preassembly functions or providing direction in terms of the styles or fabrics to be used but rather are co-ordinating the logistics of transferring finished products. These firms function strictly as import/export firms and thus are not considered 'manufacturers' for the purposes of the discussion here.

9. An important exception here are the high-end mass-market producers, which tend to be more design-intensive than their moderately priced mass-market counterparts and thus sell to a mix of specialty boutiques and department stores but not discount stores. These producers can still be differentiated from the independent designers, since the designer is an employee in the former and must still conform to the constraints of high-volume production (personal interviews with high-end producers).

9 Ottawa-Gatineau: Capital Formation

Caroline Andrew, Brian Ray, and Guy Chiasson

Introduction

For many if not most Ottawa-Gatineau[1] residents, a visualization of the city's economy would probably be dominated by the neo-gothic Parliament Buildings that sit on the escarpment above the Ottawa River or the stark modernist office buildings occupied by various federal government ministries. For some residents, their image of the economy might well accord considerably more importance to the suburban office campuses of high-technology firms that built Ottawa's once thriving and now largely tarnished Silicon Valley North reputation (Mosco and Mazepa 2003; Shavinina 2004). In fact, one of the challenges in writing about Ottawa's economy and employment geographies at this moment in time is the unresolved future of the city's high-technology sector given the recent demise of Nortel, the communications giant whose collapse has sent ripples of unemployment and uncertainty throughout the city's workforce (Martin Prosperity Institute 2009). Indeed, one of the most apt visualizations of Ottawa's post-industrial economy may well be vacant suburban office space, a material manifestation of the fact that an economy built on people with exceptional education, rather than brawn, is still vulnerable to the cyclical churnings of advanced capitalism.

Without doubt, however, few would visualize Ottawa's economy in terms of the manufacturing industries that have been at the core of employment in most urban economies in the twentieth century (Lewis 2000; Hohenberg and Lees 1985). Manufacturing and processing, transportation and warehousing, and construction, while far from absent, have not dominated Ottawa's industrial base since World War II and perhaps even World War I (Keshen 2001). In many ways, World War II was a critical watershed moment marking a transformation in Ottawa's national significance and the nature of employment in the city. As Keshen succinctly argues:

> It was during the Second World War that Ottawa experienced its most significant leap forward toward becoming a *government town*—an occurrence due not only to the massive demand for personnel from war-generated departments such as National Defence and Munitions and Supply, but also to the federal government's commitment by 1944 to increase social welfare programmes and implement a Keynesian-style, planned economy . . . hundreds of 'dollar a year' men—experts in fields such as machinery, science, law, accounting and statistics—moved to Ottawa to administer the war effort and to plan the post-war economy. Equally

critical and far more visible were the thousands of women who came to fill low-grade civil service posts (2001, 390).

After the British decision in 1857 to establish Ottawa as the permanent seat of government, the city's economy steadily moved away from once-dominant wood-based industries. But with the administrative demands of World War II, Ottawa truly became a services-based economy. On the Quebec side of the metropolitan area[2] (Gatineau), forest-based manufacturing remained a driver of the economy for some time after World War II. However, by the 1970s the pulp and paper mills that had dominated the urban landscape had already moved to the rural periphery of the region (Chiasson, Blais, and Boucher 2006), and Gatineau was quickly integrating itself into the federal service–driven economy (Beaucage 1994). As such, Ottawa-Gatineau is now an interesting case study for examining the qualities and characteristics of a city-region in which services rather than goods dominate employment and have done so for a number of decades.

A recent report by the Martin Prosperity Institute (2009) tends to confirm the status of Ottawa-Gatineau as a city where post-industrial employment occupies a very important place. Borrowing from Richard Florida's famous concepts about human creativity and economic growth (Florida 2005), the report shows how the Ottawa-Gatineau CMA ranks very high among North American cities in terms of the importance of the 'creative class', a category of workers 'who are paid to think for a living'.

> The Creative Class makes up 40.9% of Ottawa-Gatineau's labour force. This equates to just below 256,000 workers, from a workforce of approximately 627,000. Among Ottawa-Gatineau's peers only San Jose has a larger share of its workers in the Creative Class at 44.1% . . . The fact that Ottawa-Gatineau is only slightly below San Jose is a testament to how competitive it is in this dimension (Martin Prosperity Institute 2009, 9).

This branding of Ottawa-Gatineau as a successful post-industrial city with a creative workforce should not downplay the importance of the federal government in the regional economy and job market. Despite periodic efforts by regional elites to repackage Ottawa and Gatineau as technological cities (Andrew 2002), the federal government's action continues to shape the location of employment and who performs these jobs. This can be seen as the creation of a *national capital* out of the established urban form of a once-industrial town. The federal government, through agencies such as the National Capital Commission (NCC),[3] exerts an important influence on urban development and where people work.

The federal government is also an employer of a very significant portion of the regional workforce. Decisions by the government on the location of its offices have had a distinctive effect on where people work. As will be discussed, these decisions have at times led to suburbanization of work but at other times have anchored jobs in the downtown core. Federal hiring policies—namely, gender-based affirmative action practices and a large number of jobs now classified as bilingual imperative—have had an undeniable effect on the profile of workers in the public service. In both cases, important efforts have been made since the 1970s by agencies of the bureaucracy to promote women in the public service and bilingualism. As a result, the percentage of francophones and women in the overall public service has risen, and more recently, the percentage of women in the executive ranks has risen significantly (Treasury Board of Canada 2000). Considering the importance of federal jobs in the total workforce, this in turn has ripple effects both on the workforce of the city as a whole and the characteristics of work and workers in employment nodes across the CMA.

The types of industries and jobs that people occupy in Ottawa-Gatineau, how they have changed, and the complexities of where people work lie at the core of this chapter. We will also examine how the characteristics and locations of work in this city have been influenced by intersections between politics and public policy, in turn influencing the qualities of employment in specific parts of the city. In this respect, we pay particular attention to language, gender, and the intra-urban mobility patterns of Ottawa and Gatineau residents to highlight the complex qualities of work in post-industrial Ottawa. The city may or may not be an exemplar of the processes that construct post-industrial urbanism, but it is most certainly Canada's best example of a city where employment geographies, at their very root, are a function of the location demands (or lack thereof) of (public) services rather than goods production. In the first section, we present the characteristics of the Ottawa-Gatineau workforce as a whole. The second section adopts a geographical perspective in order to give a better sense of how the patterns of post-industrial work intersect with different spaces in the city. In the last section, we examine the patterns of work in four employment nodes across the city.

Industrial and Occupational Characteristics among Ottawa-Gatineau Workers

In the early twentieth century, secondary manufacturing, processing, transportation, warehousing, and construction industries in cities such as Montreal and Toronto held a prominent position, employing a very large and often growing segment of the workforce (Gad 2004; Linteau 1992). Ottawa-Gatineau, on the other hand, saw manufacturing employment in wood-based industries decline to secondary importance as the civil service became the city's principal employer (Beaucage 1994). For instance, within a short 10-year span, civil service employment grew from 4191 to 10,091 workers between 1911 and 1921 (Keshen 2001). But in many ways, it was World War II that set in motion major changes in the qualities of employment that would make Ottawa post-industrial long before other Canadian cities. During World War II, 24,000 new civil service jobs were created, and the city truly did become a one-industry town, that industry being public administration (Keshen 2001, 387).

Examining the distribution of employment across industrial sectors, it is clear that by 1961 public administration was extraordinarily significant in Ottawa-Gatineau and that its employment characteristics differed markedly from those of the nation. In Ottawa-Gatineau, 33.3 per cent of workers were employed in public administration, while the same was true for only 7.5 per cent of Canada's workforce (Table 9.1). Moreover, on a national level, 21.7 per cent of the labour force was employed in manufacturing industries, but in Ottawa-Gatineau only 10.6 per cent of workers were employed in this sector.[4] In fact, by 1961 almost three-quarters of the Ottawa-Gatineau workforce was employed in some type of service industry, compared to only 47.2 per cent of all Canadian workers. The proportion of workers employed in the services-producing sector both nationally and in Ottawa-Gatineau continued to grow steadily. It is clear, however, that in the early postwar decades, Ottawa-Gatineau was already well established not only as a city dominated by public-sector employment but one in which goods production was quickly declining in significance.

Simply describing the distribution of workers across industrial sectors, however, camouflages a more complex story about the characteristics of the workforce. In Ottawa-Gatineau, a far greater proportion of women than men have been employed in the services-producing sector. Even in 1961, almost 90 per cent of employed women were employed in services-producing industries compared to 65.8 per cent of men (Table 9.2). By 2006, the proportion among women had grown to 94 per cent, whereas among men only 78.4 per cent were employed in services. It is important to note that only recently has a greater proportion of women than men found employment in the relatively well-paid and secure public administration sector—in 2006, 22.8 per cent of employed women worked in public administration compared to 19.8 per cent of men. Over time, the proportion of women employed in key sectors of the service economy, such as business services and health and social assistance services, has steadily increased. Notably, the proportion of individuals employed in services to business, which includes services provided to governments, has steadily increased over time.

The rise of services-producing industries in Ottawa has coincided with the significant decline of the goods-producing sector, and in this sector too the changes have been experienced differently by women and men (Table 9.2). The number of women employed in goods-producing industries has never been large—only 2927 in 1961 and 7720 by 2006. However, traditionally a large proportion of men have found employment in this sector, particularly in manufacturing. There has been a significant decline in the absolute number of men employed in manufacturing, especially when compared to the national average. In many ways, the new century marks the last gasp of the manufacturing sector in Ottawa-Gatineau, and the decline has affected men much more directly than women.

One of the important changes in Ottawa's labour force over time is the declining proportion of both women and men who work directly in public administration (Table 9.2). More than 30 per cent of both sexes worked in this sector in 1961 and 1971, but by 2001 the proportion had dropped to 18.1 per cent among men and 20.2 per cent among women. Since 2001 there has been a slight increase in the proportion of individuals employed in this sector, particularly among women, but it is still well below the levels found in the 1960s and 1970s. This may well be attributable to governments outsourcing services to private businesses, and certainly the increasing importance of business services speaks to this trend. It must be noted, however, that the absolute number of people, especially women, employed in public administration did increase dramatically over time and is strongly indicative of the Keynesian role the federal government pursued after World War II. Even today, the federal government continues to grow in size—in the short five-year period between 2001 and 2006, the proportion of women and men employed in public administration grew to 21.4 and 18.8 per cent, respectively, and the increase was far greater than for the workforce overall. Although

Table 9:1: Proportion of Employed Workers by Industrial Sector, Canada and Ottawa-Gatineau CMA, 1961–2006

Industry	CANADA				OTTAWA-GATINEAU CMA			
	1961	1981	2001	2006	1961	1981	2001	2006
Services Producing Sector								
Public Administration	7.5%	7.5%	5.8%	5.8%	33.3%	27.3%	19.1%	21.3%
Finance, Insurance, & Real Estate	3.5%	5.3%	5.3%	5.4%	4.5%	5.5%	4.4%	4.3%
Information and Cultural Industries	1.4%	1.4%	2.7%	2.5%	1.6%	2.4%	3.7%	3.0%
Retail and Wholesale Trade	15.3%	16.8%	16.2%	16.2%	13.6%	13.9%	12.6%	13.2%
Business Services	1.5%	4.1%	10.3%	11.1%	1.7%	5.9%	15.0%	14.3%
Accommodation and Food Services	3.7%	5.6%	6.7%	6.7%	3.6%	5.4%	5.9%	6.1%
Education Services	4.1%	6.5%	6.6%	6.8%	4.0%	7.7%	6.8%	7.0%
Health and Social Assistance Services	4.8%	7.3%	9.7%	10.2%	4.9%	7.2%	9.1%	9.9%
Arts, Entertainment, and Recreation Services	0.6%	1.1%	2.0%	2.1%	0.5%	1.3%	2.0%	2.2%
Other Services	4.8%	4.1%	4.8%	4.9%	5.8%	4.9%	4.5%	4.8%
Services Sector Subtotal	47.2%	59.7%	69.9%	71.6%	73.6%	81.4%	83.1%	86.1%
Goods Producing Sector								
Secondary Industries								
Manufacturing	21.7%	18.9%	14.0%	11.9%	10.6%	6.8%	7.7%	4.6%
Construction	6.7%	6.4%	5.6%	6.3%	7.0%	4.4%	4.6%	4.9%
Transportation, Warehousing, and Utilities	8.0%	6.5%	5.7%	5.7%	5.2%	4.5%	3.9%	3.7%
Secondary Industries Subtotal	36.3%	31.8%	25.3%	23.9%	22.8%	15.7%	16.2%	13.2%
Primary Industries	14.0%	7.0%	4.7%	4.5%	1.0%	1.4%	0.7%	0.7%
Goods Sector Subtotal	50.4%	38.8%	30.1%	28.4%	23.9%	17.0%	16.9%	13.9%
Unspecified	2.5%	1.5%	0.0%	0.0%	2.6%	1.5%	0.0%	0.0%
Total	**100.0%**	**100.0%**	**100.0%**	**100.0%**	**100.0%**	**100.0%**	**100.0%**	**100.0%**

Table 9.2 Proportion of Female and Male Workers by Industrial Sector, Ottawa-Gatineau CMA, 1961–2006

Industry	Women				Men			
	1961	1981	2001	2006	1961	1981	2001	2006
Services Producing Sector								
Public Administration	30.8%	25.7%	20.2%	22.8%	34.5%	28.5%	18.1%	19.8%
Finance, Insurance, & Real Estate	6.6%	7.1%	5.2%	4.9%	3.4%	4.2%	3.6%	3.8%
Information and Cultural Industries	2.1%	2.4%	3.1%	2.6%	1.4%	2.3%	4.2%	3.4%
Retail and Wholesale Trade	13.0%	13.3%	13.5%	12.5%	13.9%	14.3%	11.8%	13.9%
Business Services	1.9%	5.3%	12.2%	11.5%	1.6%	6.4%	17.6%	17.0%
Accommodation and Food Services	5.6%	6.2%	6.2%	6.4%	2.6%	4.8%	5.6%	5.9%
Education Services	7.4%	10.4%	9.2%	9.5%	2.4%	5.6%	4.4%	4.7%
Health and Social Assistance Services	11.3%	12.4%	15.0%	16.2%	1.8%	3.1%	3.6%	3.9%
Arts, Entertainment, and Recreation Services	0.4%	1.1%	1.9%	2.2%	0.6%	1.5%	2.0%	2.2%
Other Services	10.3%	6.4%	5.1%	5.6%	3.7%	3.7%	3.9%	3.9%
Services Sector Subtotal	89.4%	90.4%	91.7%	94.1%	65.8%	74.4%	75.0%	78.4%
Goods Producing Sector								
Secondary Industries								
Manufacturing	5.3%	4.0%	4.7%	2.5%	13.2%	9.0%	10.4%	6.5%
Construction	0.6%	0.9%	1.1%	1.2%	10.2%	7.1%	8.0%	8.5%
Transportation, Warehousing, and Utilities	1.7%	2.2%	2.0%	1.8%	6.9%	6.3%	5.6%	5.6%
Secondary Industries Subtotal	7.7%	7.1%	7.8%	5.5%	30.2%	22.4%	24.1%	20.6%
Primary Industries	0.2%	0.6%	0.5%	0.4%	1.4%	1.9%	1.0%	1.0%
Goods Sector Subtotal	7.9%	7.8%	8.3%	5.9%	31.7%	24.3%	25.0%	21.6%
Unspecified	2.7%	1.8%	0.0%	0.0%	2.5%	1.3%	0.0%	0.0%
Total	100.0%	100.0%	100.0%	100.0%	100.0%	100.0%	100.0%	100.0%

public sector employment is still very important, by 2006 it is no longer possible to accurately characterize Ottawa-Gatineau as a city of bureaucrats, even if the city's reputation seems steadfastly wed to this stereotype (Mosco and Mazepa 2003).

An appreciation of the qualities of the workforce in Ottawa-Gatineau certainly demands that attention be paid to gender, but it is also important to recognize the differences in employment characteristics between residents on one side of the Ottawa River versus those on the other. In earlier decades, a slightly higher proportion of workers residing in Ottawa than in Gatineau worked in public administration, but today the opposite is true, with a larger proportion of workers residing in Gatineau working in this sector (22.3 per cent compared to 18.6 per cent among Ottawa residents) (Table 9.3). This is especially true among women in that just over one-quarter of the workforce living in Gatineau is employed in public administration, compared to less than one-fifth in Ottawa. The large proportion of Gatineau women who work in public administration is an important component of the region's workforce given that they are much more likely to be employed in federal government clerical positions designated as bilingual imperative. The everyday lives of these Gatineau-based women can be rendered quite difficult if their job moves to a suburban employment node, since the Gatineau bus system (Société de transport de l'Outaouais [STO]) only serves downtown Ottawa. Commutes that extend beyond that point require a switch to Ottawa's public transit system (OCTranspo), and this complicates and lengthens their work day.

Quite clearly, the services-producing sector, led by public administration, business services, and retail and wholesale trades, dominates Ottawa-Gatineau's industrial base. While the city's industrial composition provides a good overview of the workforce, it is also important to consider the kinds of jobs performed in order to round out an appreciation for the qualities of employment. In order to maintain the comparability of data, we could only meaningfully trace the occupation characteristics of Ottawa-Gatineau's labour force back to 1991. We have, however, tried to capture differences in skill requirements and social status associated with particular occupational groups by dividing them into two categories: 1) professional and specialists and 2) support and technical.

In 2006, a markedly greater proportion of workers in Ottawa-Gatineau than in the nation overall were employed in management, business (support and technical), natural sciences (professional and specialist), and social sciences (professional and specialist) occupations (Table 9.4). The starkest difference is in professional and specialist occupations in the natural sciences in which 9.6 per cent of Ottawa-Gatineau's labour force found employment compared to only 4.3 per cent of workers nationally. This concentration of highly skilled workers in the natural sciences speaks to both the importance of the private-sector high-technology industries (Silicon Valley North) and the demand for similarly skilled workers by various federal government agencies and departments. In contrast, several occupations are significantly under-represented in the region, most notably operators, assemblers, and labourers working in manufacturing, processing, and utilities. There is also a significant gap between the number of operators and skilled people working in trades, transportation, and equipment in the CMA compared to the national average. Again, the occupation characteristics of the labour force in Ottawa-Gatineau underline the degree to which employment is dominated by services rather than goods production.

The proportion of workers in most occupations has remained relatively stable since 1991. For example, there has been little change in the proportion of individuals working in management, sales, manufacturing, processing, and utilities occupations between 1991 and 2006. As a proportion of all workers, individuals employed in manufacturing as operators, assemblers, and labourers did decrease by one-half per cent, and this represents a loss of 1150 workers or 10.6 per cent of the labour force in this field (Table 9.5).

The change in the number of women and men working in particular occupations is quite revealing of the ways in which Ottawa-Gatineau's labour force has been changing (Table 9.5). Taking management occupations as an example, there was a 33.6 per cent increase in the number of women working as managers in Ottawa-Gatineau but a −4.7 per cent decrease among men. The growth in the number of women working in some occupations is particularly noteworthy. For instance, women working as business professionals and specialists grew by 105.6 per cent between 1991 and 2006 in Ottawa-Gatineau, far outstripping the growth in the number of women at the national scale working in this field (72.2 per cent) and that for men in the metropolitan area (64.4 per cent). Similarly, the growth in the number of women working as professionals and specialists in the social sciences in Ottawa-Gatineau has been greater than the national average, while the growth rate for men was much more modest. Largely reflecting the increasing importance of the high-technology sector in Ottawa-Gatineau between 1991 and 2006, the number of women and men working as professionals and specialists in the natural sciences increased significantly. It is important to note that the increase among women was from a relatively small number of individuals in 1991 (7595), and in 2006 this field is still overwhelmingly dominated by men (44,985 versus 14,905 women).

It is also important to note that this period of time also has been marked by sharp declines in the number of women and men working in some occupations. For example, there was a −12.1 per cent decline in the number of women working in business support and technical occupations in Ottawa-Gatineau, compared to a small increase nationally (0.5 per cent) (Table 9.5). Although more men were working in business support and technical occupations in 2006 than in 1991, the increase in Ottawa-Gatineau lagged well behind the national average. The number of women working as operators, assemblers, and labourers in manufacturing, processing, and utilities also declined in Ottawa-Gatineau, while the number of women working in these occupations nationally slightly increased. Among men there was a sharp decline in the number working in manufacturing and processing occupations in Ottawa-Gatineau (−13.5 percent), although nationally there was a slight increase (3.0 per cent).

Employment in Ottawa-Gatineau has changed in recent decades, with a greater share of individuals finding employment in services-producing industries. In particular, employment in professional and specialist occupations has grown proportionately and numerically, particularly among women. The goods-producing sector, which has not constituted a large part of Ottawa-Gatineau's economy for most of the twentieth century, declined both in terms of the number of people directly producing goods or working as low-skill labourers and helpers in the trades, transportation, and equipment sector. To a considerable degree, Ottawa-Gatineau exemplifies an economy in which services (both high- and low-skill) play an extraordinarily important role, and in this sense the city exemplifies the qualities of a post-industrial economy.

Table 9.3 Proportion of Female and Male Workers by Industrial Sector, Ottawa and Gatineau, 1981 and 2006

Industry	Ottawa						Gatineau					
	1981			2006			1981			2006		
	Total	Women	Men	Total	Women	Men	Total	Women	Men	Total	Women	Men
Services Producing Sector												
Public Administration	27.8%	25.3%	29.9%	18.6%	11.2%	18.0%	25.3%	27.5%	23.8%	22.3%	25.8%	19.0%
FIRE	5.8%	7.4%	4.4%	4.3%	4.7%	3.9%	4.5%	5.8%	3.5%	3.1%	3.8%	2.5%
Information and Cultural Industries	2.5%	2.4%	2.5%	3.0%	2.6%	3.5%	2.0%	2.4%	1.7%	2.9%	1.1%	4.5%
Retail and Wholesale Trade	13.9%	13.4%	14.3%	12.3%	11.6%	13.1%	13.9%	12.9%	14.7%	11.7%	11.0%	12.3%
Business Services	6.7%	59%	7.3%	19.6%	18.8%	20.4%	3.3%	3.1%	3.5%	17.7%	15.9%	19.3%
Accommodation and Food Services	5.5%	6.1%	4.9%	5.6%	5.7%	5.5%	5.3%	6.5%	4.3%	5.6%	6.1%	5.2%
Education Services	7.7%	10.2%	5.8%	6.3%	8.4%	4.3%	7.7%	11.2%	5.2%	6.8%	9.5%	4.3%
Health and Social Assistance Services	7.3%	12.7%	3.0%	9.0%	14.6%	3.5%	6.7%	11.3%	3.3%	9.4%	15.6%	3.8%
Arts, Entertainment, and Recreation Services	1.3%	1.1%	1.4%	1.9%	2.0%	1.9%	1.4%	1.1%	1.6%	2.3%	2.2%	2.4%
Other Services	4.7%	6.1%	3.6%	4.5%	5.3%	3.7%	5.3%	7.2%	3.9%	4.1%	4.7%	3.5%
Services Sector Subtotal	83.2%	90.8%	77.1%	85.2%	93.0%	77.7%	75.4%	89.0%	65.4%	85.8%	95.6%	76.8%

Goods Producing Sector

Secondary Industries

Manufacturing Industries	6.0%	3.8%	7.7%	4.3%	2.6%	6.0%	9.8%	4.7%	13.6%	4.0%	1.6%	6.2%
Construction Industries	4.0%	1.0%	6.4%	4.0%	1.1%	6.8%	5.8%	0.7%	9.6%	6.2%	1.1%	10.9%
Transportation, Warehousing, and Utilities	4.3%	2.3%	6.0%	5.9%	3.0%	8.6%	4.9%	1.8%	7.2%	3.5%	1.4%	5.3%
Secondary Industries Subtotal	14.3%	7.0%	20.0%	14.1%	6.7%	21.4%	20.5%	7.2%	30.4%	13.7%	4.1%	22.4%
Primary Industries	1.4%	0.7%	2.0%	0.7%	0.4%	0.9%	1.1%	0.5%	1.5%	0.6%	0.3%	0.8%
Goods Sector Subtotal	15.7%	7.8%	22.0%	14.8%	7.0%	22.3%	21.6%	7.6%	32.0%	14.2%	4.4%	23.2%
Unspecified	1.1%	1.4%	0.9%	0.0%	0.0%	0.0%	2.9%	3.3%	2.7%	0.0%	0.0%	0.0%
Total	**100.0%**	**100.0%**	**100.0%**	**100.0%**	**100.0%**	**100.0%**	**100.0%**	**100.0%**	**100.0%**	**100.0%**	**100.0%**	**100.0%**

Table 9.4 Proportion of Employed Workers by Occupation Classification, Canada and Ottawa-Gatineau CMA, 1991–2006

Occupation	Canada			Ottawa-Gatineau cma		
	1991	2001	2006	1991	2001	2006
Management	9.7%	10.4%	9.7%	12.3%	12.9%	11.1%
Business (professionals & specialists)	2.7%	3.6%	3.7%	3.0%	4.3%	4.7%
Business (support & technical)	16.2%	14.2%	14.2%	21.4%	16.6%	16.6%
Natural Sciences (professionals & specialists)	2.7%	4.2%	4.3%	6.0%	10.5%	9.6%
Natural Sciences (support & technical)	2.3%	2.3%	2.3%	2.6%	2.4%	2.1%
Health (professionals & specialists)	2.6%	2.7%	2.8%	2.7%	2.7%	2.8%
Health (support & technical)	2.4%	2.6%	2.8%	2.0%	2.1%	2.5%
Social Sciences (professionals & specialists)	1.7%	2.1%	2.2%	3.5%	4.3%	4.9%
Education (professionals & specialists)	3.9%	3.7%	4.0%	4.3%	4.2%	4.5%
Social Sciences (support & technical)	0.9%	1.1%	1.2%	0.9%	1.1%	1.2%
Arts, Culture, Recreation, Sport (professionals & specialists)	1.0%	1.2%	1.3%	2.1%	2.2%	2.5%
Arts, Culture, Recreation, Sport (technical & support)	1.4%	1.6%	1.7%	1.6%	1.5%	1.8%
Sales and Service Supervisors	0.6%	0.8%	0.8%	0.6%	0.8%	0.8%
Sales	8.4%	7.8%	8.3%	7.6%	7.2%	7.8%
Services	15.4%	16.0%	15.8%	14.9%	14.4%	15.2%
Trades, Transportation, & Equipment (supervisors & contractors)	1.5%	0.8%	0.7%	1.2%	0.7%	0.5%
Trades, Transportation, & Equipment (operators & skilled trades)	11.3%	11.2%	11.4%	7.7%	6.9%	7.2%
Trades, Transportation, & Equipment (labourers & helpers)	2.7%	2.1%	2.4%	1.8%	1.2%	1.5%
Primary Industries Employment	5.2%	4.3%	3.8%	1.4%	1.1%	1.2%
Manufacturing, Processing, & Utilities (supervisors)	0.5%	0.6%	0.5%	0.2%	0.2%	0.1%
Manufacturing, Processing, & Utilities (operators, assemblers, & labourers)	7.0%	7.0%	6.0%	2.0%	2.6%	1.5%
Total	**100.0%**	**100.0%**	**100.0%**	**100.0%**	**100.0%**	**100.0%**

Table 9.5 Change in Absolute Number of Workers by Occupational Classes and Gender, Ottawa-Gatineau and Canada, 1991–2006

Occupation	All Workers		Women		Men	
	Ottawa-Gatineau CMA	Canada	Ottawa-Gatineau CMA	Canada	Ottawa-Gatineau CMA	Canada
Management	7.0%	18.0%	33.6%	46.5%	-4.7%	6.0%
Business (professionals & specialists)	85.0%	65.2%	105.6%	72.2%	64.4%	56.1%
Business (support & technical)	-8.3%	3.8%	-12.1%	0.5%	2.9%	14.7%
Natural Sciences (professionals & specialists)	87.5%	88.3%	96.2%	94.8%	84.7%	86.4%
Natural Sciences (support & technical)	4.1%	18.7%	25.0%	51.3%	-9.9%	12.8%
Health (professionals & specialists)	22.1%	26.1%	21.2%	25.9%	25.5%	26.5%
Health (support & technical)	45.8%	41.0%	55.0%	45.0%	19.7%	24.1%
Social Sciences (professionals & specialists)	63.7%	53.0%	113.7%	87.3%	27.3%	24.1%
Education (professionals & specialists)	24.3%	21.6%	29.1%	31.6%	16.2%	6.4%
Social Sciences (support & technical)	58.0%	62.7%	74.3%	77.6%	23.8%	28.5%
Arts, Culture, Recreation, Sport (professionals & specialists)	40.7%	52.1%	61.1%	58.6%	14.7%	44.1%
Arts, Culture, Recreation, Sport (technical & support)	32.5%	46.5%	49.9%	49.2%	16.8%	43.5%
Sales and Service Supervisors	60.4%	49.8%	63.7%	70.1%	57.6%	29.1%
Sales	20.3%	17.8%	21.3%	21.2%	19.0%	13.0%
Services	20.8%	22.2%	34.5%	31.1%	8.8%	11.5%
Trades, Transportation, & Equipment (supervisors & contractors)	-50.8%	-41.8%	-16.4%	6.2%	-52.5%	-43.6%
Trades, Transportation, & Equipment (operators & skilled trades)	10.0%	19.5%	0.9%	20.1%	10.6%	19.5%
Trades, Transportation, & Equipment (labourers & helpers)	-3.3%	6.6%	43.4%	33.1%	-5.4%	4.2%
Primary Industries Employment	1.9%	-11.7%	-7.0%	-12.6%	4.1%	-11.5%
Manufacturing, Processing, & Utilities (supervisors)	-16.5%	0.6%	-3.8%	33.4%	-17.5%	-4.4%
Manufacturing, Processing, & Utilities (operators, assemblers & labourers)	-10.6%	2.5%	-3.3%	1.4%	-13.5%	3.0%
Total	**18.4%**	**18.6%**	**24.0%**	**25.0%**	**13.5%**	**13.3%**

It is important to recognize the substantial suburbanization of employment that has occurred in the metropolitan region. Especially in Ottawa, the emergence of a services-based, high-technology economy has been accompanied by employment suburbaniza-tion. The move of jobs to the suburbs has a substantial history in this city, and in many ways where people should or might work within this metropolitan area has been the focus of a number of experiments led largely by one of the region's two major employ-ers: the federal government and, more recently, the high-technology firms. In the next section of this paper, attention shifts to the qualities and characteristics of the primary employment centres in the city, with an emphasis on the role of suburbanization.

Ottawa-Gatineau's Geography of Employment

As is the case for most large cities, people work almost everywhere in Ottawa-Gatineau. From domestics working in private residences to corner stores to big-box stores and shopping malls, employment is far from being confined to the CBD or a few suburban nodes. In fact, there are several parts of the metropolitan area where almost no one lives (for example, the Central Experimental Farm and Greenbelt) but hundreds of people work. Nevertheless, employment in Ottawa is concentrated in several different nodes throughout the metropolitan area, most notably in the CBD (Figure 9.1). In this section, our attention turns first to describing employment nodes and the character-istics of employment in both the downtown and suburbs. We argue that where people work in this city reflects the ways in which several factors have interacted over time: federal public policy and politics, enthusiasm for urban planning experimentation, and the space/accessibility requirements of many industries that have located in this low-density city.

The dual effects of an absolute shortage of office space and urban planning visions for the capital have been instrumental in encouraging suburbanization of federal govern-ment office employment in Ottawa. As early as 1950, the Federal District Commission's *Plan for the National Capital* (also known as the Gréber Report) emphasized the need to decentralize several government departments, in part to relieve traffic congestion heading toward the city's core and a belief in the social value of campus-like workplaces for employees (Bellamy 2001). The enthusiasm for office suburbanization is difficult to deny when looking at where people work in the metropolitan area (Figure 9.1), although the CBD remains very important. In 2006, approximately 66,480 people, or 12.3 per cent of all people with a regular place of employment that is not their home, worked in Ottawa's downtown core (Table 9.6). It must be noted that Ottawa's CBD is somewhat unusual in that it is a relatively small geographic area extending south from the Ottawa River to Gloucester Street and from the Rideau Canal on the east to Bronson Street on the west and comprises only one census tract. It is also unusual for the fact that a sub-stantial concentration of workers lies just across the Ottawa River from the CBD in the Hull sector of Gatineau. More than 17,000 people work in this dense cluster of office buildings that is adjacent to, but physically divided from, the CBD.[5]

The suburbanization of employment, however, is undeniable, with six concentra-tions in suburban Ottawa standing out (Figure 9.1). Several of these suburban nodes, such as Tunney's Pasture and Baseline/Woodroffe in the west end, Confederation

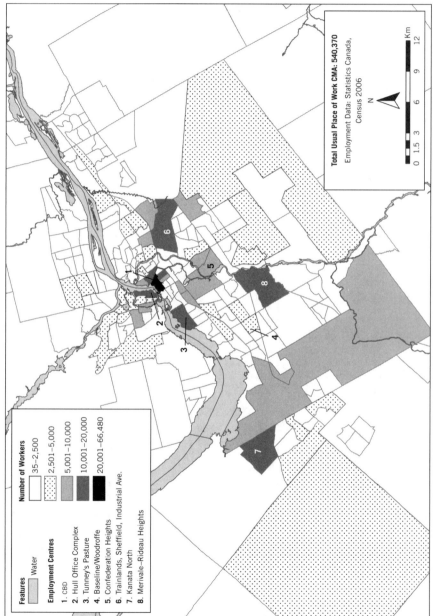

Figure 9.1 Primary centres of employment concentration, Ottawa-Gatineau CMA, 2006.

Table 9.6: Gender, Language, Industry, Occupation, and Education Profiles for the Major Employment Centres in Ottawa-Gatineau, 2006.

	1	2	3	4	5	6	7	8
Usual Place of Work								
Population	66,480	17,070	16,390	3,190	12,480	28,180	17,085	15,225
				Total				
				Per Cent				
Male	43.0	45.1	40.1	48.4	52.7	70.4	62.7	62.7
Female	55.9	56.9	54.9	59.9	51.6	47.3	29.6	37.3
Language								
English	50.1	31.1	52.7	73.5	55.0	55.3	63.5	62.7
French	35.1	57.0	31.1	12.2	26.7	28.3	9.3	17.8
Other	14.8	11.9	16.2	13.8	18.1	16.2	24.2	19.5
Employment by Industry								
Agricultural, Forestry, Fishing and Hunting	0.1	0.0	0.1	0.0	0.0	0.2	0.0	0.3
Mining and Oil and Gas Extraction	0.0	0.0	0.1	0.0	0.0	0.0	0.1	0.3
Utilities	0.0	0.0	0.0	0.3	0.0	0.1	0.1	1.1
Construction	0.5	0.3	1.6	0.5	0.2	3.6	1.1	9.1
Manufacturing	0.5	2.1	3.1	0.8	2.0	4.5	34.4	10.8
Wholesale Trade	0.5	0.1	0.5	0.9	0.5	6.4	12.0	8.7
Retail Trade	1.6	0.8	4.4	0.8	5.3	4.7	1.6	11.7
Transportation and Warehousing	1.6	0.2	0.2	0.0	19.6	14.2	0.3	3.7
Information and Cultural Industries	4.6	1.9	0.5	2.7	0.5	0.7	3.9	6.4
Finance and Insurance	6.7	0.5	1.7	1.7	1.4	1.6	1.3	2.3
Real Estate and Rental and Leasing	1.1	0.6	0.5	0.8	0.5	1.2	0.6	2.0
Professional, Scientific, and Technical Services	11.3	3.8	5.0	5.2	3.7	4.4	28.3	13.4
Management of Companies and Enterprises	0.0	0.0	0.0	0.0	0.0	0.1	0.0	0.2
Admin, Support, Waste Mgmt, and Remediation	4.5	3.3	2.7	2.0	2.6	4.2	8.2	5.1
Educational Services	1.6	0.8	1.1	2.7	2.3	2.9	0.9	1.9
Health Care and Social Assistance	0.8	2.1	3.7	12.9	5.0	31.6	1.6	2.1
Art, Entertainment, and Recreation	1.0	3.3	0.2	1.4	0.9	2.0	0.4	1.3
Accommodation and Food Services	3.9	2.8	2.7	1.3	2.4	3.1	3.7	2.6
Other Services (except Public Admin.)	3.7	1.4	1.3	2.4	2.9	4.5	0.9	4.6
Public Administration	56.0	78.9	70.5	63.3	50.0	10.1	0.6	12.3

Employment by Occupation

	1	2	3	4	5	6	7	8
Management occupations	15.0	13.2	12.1	8.6	14.8	9.4	15.0	14.8
Business, Finance and Administrative occupations	38.1	37.6	29.3	30.3	30.7	22.1	16.5	22.8
Natural and applied sciences and related occupations	12.6	15.3	20.7	19.4	26.3	5.6	48.1	17.1
Health occupations	0.6	1.4	2.8	10.7	3.0	20.7	1.0	1.5
Social science, education, gov. service and religion	14.5	15.9	16.9	15.0	6.8	6.3	3.0	4.4
Art, culture, recreation and sport	6.2	6.1	4.2	3.8	2.8	2.4	1.3	4.3
Sales and service occupations	10.7	7.6	9.9	9.6	12.5	14.1	8.5	15.5
Trades, transport and equipment operators and related	1.8	1.6	2.7	2.0	2.0	17.3	2.7	15.2
Primary industry	0.2	0.3	0.2	0.3	0.2	0.5	0.1	0.6
Processing, manufacturing and utilities	0.2	1.1	1.2	0.0	0.8	1.6	5.8	3.8

Education

	1	2	3	4	5	6	7	8
High School Certificate								
Male	16.9	13.7	17.5	15.6	17.9	29.0	13.0	24.6
Female	23.5	24.8	29.9	24.1	25.2	18.6	24.0	30.6
College, CEGEP								
Male	17.9	18.4	19.5	29.7	24.9	21.2	22.7	25.2
Female	21.0	18.6	19.7	19.6	27.1	30.4	23.9	24.7
University								
Male	28.9	27.5	27.6	28.6	30.4	11.4	34.5	17.2
Female	28.7	23.4	24.9	27.0	21.7	21.9	25.0	18.3
Master's degree								
Male	16.5	17.3	12.9	9.4	8.8	4.7	13.6	6.5
Female	11.2	12.1	13.0	8.6	5.9	6.4	6.6	6.4
Doctorate								
Male	2.9	2.2	4.7	0.8	1.6	2.2	2.0	1.5
Female	1.1	0.6	2.2	0.8	0.9	1.3	0.7	0.8

Employment Centres:
1 = CBD
2 = Hull Office Complex
3 = Tunney's Pasture
4 = Baseline / Woodroffe
5 = Confederation Heights
6 = Trainlands, Sheffield, Industrial Ave
7 = Kanata North
8 = Merivale / Rideau Heights

Heights in the south, and Trainlands, Sheffield, Industrial Avenue in the east, are linked to the city's bus transitway system and therefore are highly accessible to workers using public transportation. In contrast, other districts such as Merivale–Rideau Heights and Kanata North are more easily reached by private transportation. Of all the suburban employment nodes, the Trainlands, Sheffield, Industrial Avenue district is largest in geographic size and number of workers. People are also employed in a wide range of occupations and industries, and this diversity does distinguish Trainlands from the other employment nodes. Of the 28,180 people who work in the Trainlands district, only 10.1 per cent are employed in public administration, which is quite low relative to other districts. A large proportion of workers, however, do work in the health care and social assistance,[6] wholesale trade, transportation and warehousing, construction, and manufacturing sectors.

Where people work, of course, is strongly related to the type of industry and work being performed. Given the importance of the federal government as the city's major employer, the geography of work has been strongly influenced by government decision-making about where to build and lease office space. Public administration employment is extremely important in Ottawa-Gatineau (135,055 workers), and people are employed in several different nodes throughout the metropolitan area (Figure 9.2). The CBD, however, remains the core of public administration employment, with 37,215 people or 56 per cent of the workforce being employed in this sector. Other public sector workers are found in a number of inner suburban districts such as Tunney's Pasture and Confederation Heights. Federal government office suburbanization began in earnest just after World War II and was spurred on by a desire to control automobile congestion in the city's core and a critical shortage of office space throughout the metropolitan area. The Confederation Heights complex was built at the intersection of a planned but never constructed major roadway to the south and Heron Road, a busy east-west suburban arterial roadway. Today, Confederation Heights is highly accessible by car or public transit, and major government departments such as Revenue Canada, the Canadian Broadcasting Corporation, Canada Post, and Public Works have offices there. The Confederation Heights district is truly a product of postwar federal government planning and development when there was considerable interest in locating a number of offices out of the downtown core in order to decrease commuting pressures and quickly respond to the need for office space. However, locations of public administration employment today are also found in recently constructed industrial parks, such as Merivale–Rideau Heights, where the federal government has played almost no role in their development, and these areas are characterized by an array of different industries.

A quite different geography of employment emerges when other sectors of the economy are examined. For example, among individuals working in the professional, scientific, and technical services industrial sector (44,860), which includes engineers and other professionals associated with Ottawa's high-technology sector, the CBD retains importance as an employment node (8365 workers) (Figure 9.3). Most certainly, the federal government does interact with this sector. However, many of the suburban districts where there is a strong concentration of public administration employment are relatively insignificant sites of employment for these workers. Much more important are a number

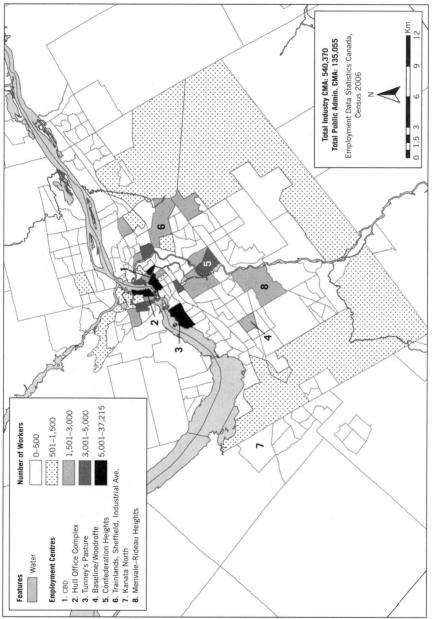

Figure 9.2 Place of work for individuals employed in the public administration industrial sector, Ottawa-Gatineau, 2006.

Features

Water

Employment Centres

1. CBD
2. Hull Office Complex
3. Tunney's Pasture
4. Baseline/Woodroffe
5. Confederation Heights
6. Trainlands, Sheffield, Industrial Ave.
7. Kanata North
8. Merivale–Rideau Heights

Number of Workers

0–500
501–1,500
1,501–3,000
3,001–5,000
5,001–37,215

Total Industry CMA: 540,370
Total Public Admin. CMA: 135,055
Employment Data Statistics Canada, Census 2006

N

Km
0 1.5 3   6   9   12

of west-end suburban areas and most especially the Kanata North district (7885 workers). In fact, 28.3 per cent of all workers in Kanata North work in the professional, scientific, and technical services sector, compared to only 11.3 per cent in the CBD.

Although manufacturing industries employ a much smaller number of workers (26,805) than either public administration or professional, scientific, and technical services, these workplaces are highly suburbanized, both in Ottawa and Gatineau (Figure 9.4). The CBD and areas around the downtown core are inconsequential as sites of employment in manufacturing industries. In contrast, the Trainlands, Merivale–Rideau Heights, and Kanata North districts, as well as a number of areas in Gatineau that are in close proximity to expressways, remain important sites of manufacturing. It must be noted, however, that this map simply indicates where people work in manufacturing or processing industries; not all of these workers in fact are directly engaged in producing goods. As such, the map does somewhat overstate the importance of goods production.

From the end of World War II through to the present, the federal government has consistently built office complexes or rented space in suburban areas. Throughout the same postwar decades, the government has also built, purchased, or rented buildings in Ottawa's downtown and in the Hull sector of Gatineau. Place du Portage and Les Terrasses de la Chaudière office complexes in Hull constitute one of the largest concentrations of office employment in close proximity to the downtown core (discussed in detail below). The Hull Federal Office Complex is the only major employment destination for individuals who live in Ottawa but work in Gatineau—7725 workers or 45.3 per cent of all workers in this district live in Ottawa (Figure 9.5). More than twice as many people live in Gatineau but work in the CBD (16,155), but they make up less than a quarter of the workforce in the CBD. One of the important characteristics of employment in the metropolitan area is that a large number of Gatineau residents work in both downtown and suburban districts of Ottawa, but apart from in the Hull Federal Office Complex, relatively few Ottawa residents work in Gatineau.

The geography of work also strongly reflects the region's linguistic duality and the federal government's bilingualism policies for the public service.[7] The proportion of workers whose home language is French exceeds the metropolitan average (32.7 per cent) in all areas of Gatineau, which is to be expected given the substantial francophone history of the Quebec portion of the region. The metropolitan average for French home-language workers is also exceeded in Ottawa's CBD and several inner and suburban areas on the east side of Ottawa. The eastern suburbs of Ottawa historically have had a relatively substantial francophone population, and therefore it is to be expected that francophones would also work there. The number of people working in the east-end areas, however, is relatively small compared to that in other parts of the metropolitan area.

The relatively high proportion of francophones working in the CBD is an important indicator of the efforts made by the federal government to create bilingual work environments in government ministries (Hudon 2009). It also underlines the reality of interprovincial commuting for people living in Gatineau. Importantly, the proportion of people working in the CBD whose home language is English (50.1 per cent) is just below the metropolitan average (50.8 per cent), and therefore the 'dominance'

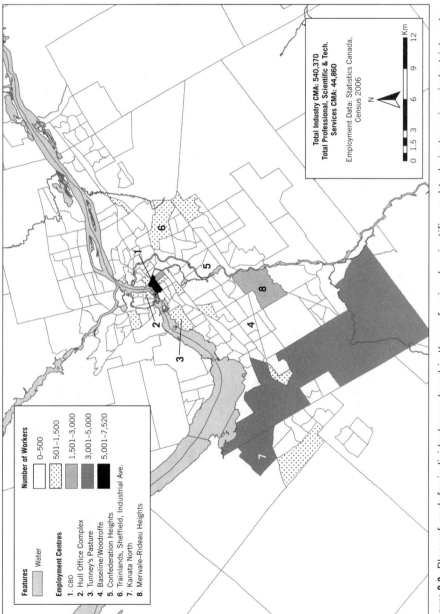

Figure 9.3 Place of work for individuals employed in the professional, scientific, and technical services industrial sector, Ottawa-Gatineau, 2006.

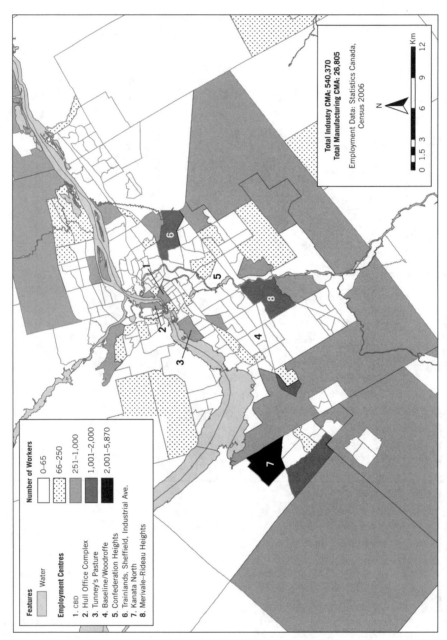

Figure 9.4 Place of work for individuals employed in the manufacturing industrial sector, Ottawa-Gatineau, 2006.

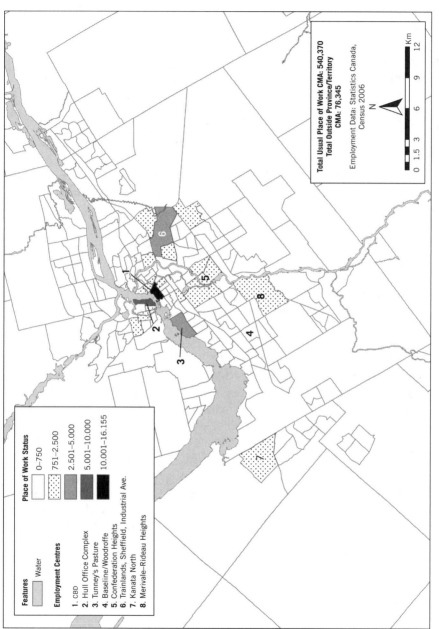

Features

Water

Employment Centres

1. CBD
2. Hull Office Complex
3. Tunney's Pasture
4. Baseline/Woodroffe
5. Confederation Heights
6. Trainlands, Sheffield, Industrial Ave.
7. Kanata North
8. Merivale–Rideau Heights

Place of Work Status

0–750
751–2,500
2,501–5,000
5,001–10,000
10,001–16,155

Total Usual Place of Work CMA: 540,370
Total Outside Province/Territory
CMA: 76,345

Employment Data: Statistics Canada,
Census 2006

N

Km
0 1.5 3   6   9   12

Figure 9.5 Work locations for individuals who live in a province that is not their province of work, Ottawa-Gatineau, 2006.

of francophones must not be overstated. It must, however, be noted that the proportion of workers whose home language is English exceeds the metropolitan average in all of Ottawa's suburban employment nodes, including those dominated by public sector employment. Finally, there are a few areas on the east side of the city where the proportion of anglophones and francophones exceeds each group's metropolitan average. These areas are unique in terms of the workforce being fairly evenly split between anglophones and francophones. Although data about home-language use among workers are interesting and suggestive of the linguistic composition of workplaces, they provide little insight into the work locations of people who use *both* French and English in their professional lives. We suggest, however, that the bilingualism requirements associated with employment in the federal government means that the geography of bilingualism is strongly associated with the location of major sites of public administration employment.

The seemingly simple movement of jobs to suburban locales belies considerable complexity with regard to the type of work performed in particular places, intricate commuting patterns between Ottawa and Gatineau residents, and language use. In short, the qualities and characteristics of work are highly variable across the metropolitan area, even if most people are employed in the services-producing sector. As the following short profiles of several employment centres demonstrate, intersections between government policy, language, and urban planning experiments have in no small measure contributed to the creation of distinct employment landscapes in this post-industrial city where services have long held a dominant position over goods production.

Ottawa-Gatineau's Employment Centres

The kinds of work people do, who does the work, where they work, and in which language(s) in Ottawa-Gatineau has been influenced by a number of employment policies and urban planning experiments. While it is impossible to catalogue in great detail all the various projects that have occurred over the postwar decades, we have chosen to profile four areas of the city to provide a sense of the social and political complexity that underlies the geography of work in this post-industrial city: the CBD, Tunney's Pasture, the Hull Federal Office Complex, and Kanata North (Table 9.6). Our selection of these areas is strategic in that they highlight the ways in which urban development associated with the city's service-producing industries has waxed and waned between the downtown core and suburbs over the past half-century. Our sample also regroups employment nodes where the federal government is the main employer and places where it is almost absent (e.g., Kanata North).

Making a Modern CBD

It is difficult to underestimate the degree to which Ottawa was prepared for the considerable growth in office employment associated with World War II or the enhanced implication of the federal government in the country's economic and social development in the postwar decades. As Keshen notes, 'Not long after the war started, practically every square foot of federal government office space was crammed with civil servants' (2001, 388). Between 1939 and 1945, the federal civil service increased from 12,000 to almost 36,000 employees (Bellamy 2001, 447), and given the shortage of office space, 14 three-storey, white clapboard 'Temporary Buildings' were built to accommodate office workers. While most of these buildings were located in and around the downtown, a few were built near Dow's Lake and are some of the earliest examples of office suburbanization. The demand for office space did not disappear with the end of the war, and the downtown streets of Ottawa were almost completely transformed in the postwar decades by the development of hastily constructed modernist office buildings. While a few pre-war buildings of architectural and historical significance remain on Wellington and Sparks streets, the CBD is mostly composed of postwar office towers that accommodate the largest concentration of workers in the metropolitan area. The boom in office-tower construction in the downtown core did create a great deal of functional space but with little architectural merit. As Cohen (2007, 180–1) has argued,

> What has emerged beyond Parliament Hill is a shapeless, featureless skyline . . . to accommodate a growing bureaucracy . . . [the federal government] has thrown up monstrosities designed by the *cum laude* graduates of the School of Brutalism. . . . The catalogue of the soulless includes . . . the clutch of anonymous office blocks with bland food courts in their basements. It isn't that these siege-works are ordinary; it is that they are awful.

The architecture, built form, and streetscapes of downtown Ottawa have received substantial criticism over the years, but the population working in the downtown core is very much indicative of the qualities and characteristics of a post-industrial city (Table 9.6). The workforce is strongly dominated by women (55.9 per cent), a characteristic that the CBD shares with a number of suburban office clusters with substantial public sector employment. The population working in the CBD also has a relatively high level of education—less than a quarter of women and one-fifth of men have only a high school certificate. An almost equal number of women and men working in the CBD have a Bachelor's degree (26.7 and 28.9 per cent respectively). However, a substantially larger proportion of men (16.5 per cent) than women (11.2 per cent) hold a Master's degree. Overall, the workforce in the downtown core is quite well-educated. The suburban office clusters, especially those with substantial public sector employment, also have highly educated workforces. None of the suburban areas, however, quite match the CBD in this respect. Even in Kanata North, the home of Ottawa's high-technology sector, the proportion of women and men with a Master's or doctoral degree lags behind that of the CBD. A substantially higher proportion of men working in Kanata North (34.5 per cent) than in the CBD (28.9 per cent), however, do hold a Bachelor's degree.

The characteristics of employment in terms of industries and occupations also distinguish the CBD from other employment nodes in the metropolitan area. The CBD is dominated by public administration (56 per cent), followed by professional, scientific, and technical services (11.3 per cent) and finance and insurance (6.7 per cent). Public administration is a very important sector in the downtown core, although it is less dominant than in the federal government's Hull and Tunney's Pasture office complexes. The downtown also shares with the Kanata North and Merivale–Rideau

Heights suburban areas the fact that the professional, scientific, and technical services sector is relatively large. The kinds of occupations workers hold in the CBD are also strongly indicative of the dominance of service employment—business, finance, and administrative occupations dominate the CBD's employment profile (38.1 per cent), as they do the Hull Federal Office Complex, Tunney's Pasture, and Confederation Heights districts. Jobs in the social sciences, education, government services, and religion are also quite important in the downtown core, as well as in the same suburban areas previously mentioned. The CBD lags well behind Kanata North in terms of employment in the natural and applied sciences occupations (12.6 versus 46.1 per cent of workers), but clearly the federal government remains active in various forms of research-related employment in the downtown core and a number of suburban areas in which it is the dominant employer.

The CBD is the single largest employment node in the metropolitan area, and demand remains strong for downtown office space. In 2008, rental rates in the downtown increased by 12.5 per cent, driven largely by federal government leasing demand. In turn, vacancy in the downtown fell to only 2.3 per cent (City of Ottawa 2009, 40–1). The federal government's office space requirements continue to be very strong—in December 2008 a Request for Information (RFI) was issued to lease up to 360,000 square metres of office space in the metropolitan Ottawa-Gatineau area by 2011. A number of new office buildings are also under construction in the core, including two 19-storey buildings—one on Kent Street and the other at the corner of Slater and O'Connor streets—and the federal government will be the major tenant in each. At least five new downtown towers are also in the planning phase, all of which would contribute to maintaining the importance of the downtown core (City of Ottawa 2009).

Unlike other Canadian downtown cores where office developments have incorporated some housing, relatively little of this mixed-use activity has occurred in Ottawa. Until relatively recently, most rental apartment and condominium development occurred adjacent to, but not in, the city's CBD. To the extent that mixed-use developments with a substantial residential component are characteristic of the downtown cores in post-industrial cities, Ottawa is not a particularly strong example. This is beginning to change as two new mixed-use office, retail, and residential buildings have been completed in the CBD and another is in the planning stages—Hillside, to be located between Sparks and Queen streets near Metcalfe. These developments should have a substantial influence on the character of the CBD, which has long been regarded as a dead zone during evenings and weekends.

Tunney's Pasture: A Planned Move to the Suburbs

Tunney's Pasture represents the first example of a suburban satellite campus model of federal government buildings. It was developed following the Second World War when Ottawa clearly became a government town. This district reflects the aesthetic vision of Jacques Gréber and the political vision of then–Prime Minister William Lyon Mackenzie King. King brought Gréber to Canada before the war and again after the war to finish his plan for the national capital. Gréber's plans are based on decentralization and the placing of government buildings in park-like environments in proximity to the suburban residential areas that were growing up. This vision is strongly influenced by Le Corbusier, concentrating buildings in open green spaces and designing scenic roadways through these greenfields (Udovicki-Selb 2001).

King saw the planning of Ottawa as creating a capital worthy of the Canada emerging from the war (Gordon and Donald 2007). It was part of his plan to develop a sense of Canadian nationalism, albeit marked by his own cautious character. The plan certainly indicated a transformation of the rough and tumble lumber town into the respectable national capital. Population expansion also led to the City of Ottawa amalgamating surrounding municipalities during the postwar period and as a result, the territory of Tunney's Pasture became part of the City of Ottawa. By 1947, the National Capital Planning Agency authorized the Department of Public Works to expropriate land for federal buildings, and Tunney's Pasture became federal land. This land had been designated for industrial development by Nepean (the municipality, now amalgamated with Ottawa, where Tunney's Pasture is located), but the federal government's desire to house the new functions of government and the increased numbers of civil servants won the day. The Gréber Plan had identified two buildings for Tunney's Pasture, both symbolic of the desired entry of Canada into modernity: the Dominion Bureau of Statistics and the National Film Board. The second was never built, but Statistics Canada is both one of the pivots of Tunney's Pasture and one of the symbolic centres of the modern state.

Tunney's Pasture was first envisaged in the Beaux-Arts style of Jacques Gréber: low-rise buildings set back from the roadway so as to emphasize green spaces and park-like surroundings. The work environment was to be suburban in atmosphere, peaceful and even bucolic. There was to be plenty of parking space and even two baseball diamonds. The first buildings of the 1950s were low-rise (Statistics Canada main building, Finance, and a variety of health-related buildings), and this style continued to be employed during the 1960s (National Defence and Public Archives). By the mid-1960s and the 1970s, however, high-rise buildings were added across the site (Brooke Claxton, Jeanne Mance, R.H. Coats, Jean Talon), thus breaking with the initial unified architectural statement.

Tunney's Pasture today has more female than male employees, consistent with the pattern we described for the CBD (Table 9.6). In 2006, it had 16,390 employees, and 54.9 per cent of them were women. The female workforce is highly educated—Tunney's Pasture has the highest percentage of women with a doctorate of all employment nodes with significant public sector employment (2.2 per cent of the female employees, as compared to 4.7 per cent of the male employees, also the highest percentage of all the federal centres).

At the present time, the Tunney's Pasture site is to be renovated, indicating that the initial architectural vision is seen as outmoded but also that this initial vision was not actually followed, with the addition of a number of high-rise buildings. The central Health Canada building (the Brooke Claxton Building) is considered as having some architectural merit, but most of the other high-rise buildings are seen as less distinguished. The low-rise, car-dominated, park-like setting was the postwar vision, but by the 1970s the federal government was questioning this vision. Tunney's Pasture continued to be an important concentration of federal employment, but attention shifted back to the centre of the urban area, in proximity to the CBD but on the other side of the river.

Downtown Hull: Working for a Bilingual Canada

The federal government employment complex in downtown Hull represents the same mixture of federal government political will and evolving notions of good urban design. By the 1960s and 1970s, the suburban vision of Le Corbusier had been replaced by a more urban sensitivity, and federal planners began to favour a recentralization of federal buildings in proximity to the Parliament Buildings, seen as the symbolic centre of the region.

The federal government's Hull office complex would not exist had it not been for the political will of Pierre Elliott Trudeau, who became prime minister in 1968. Trudeau's policy in the capital region was a mixture of territorialized social justice aims and tough intergovernmental politics in that he came to power intending to ensure that the Quebec side of the river would share in the benefits of the federal government presence and, at the same time, that the independence of Quebec would become as difficult as possible (Couture 1996). Ably backed by his Quebec lieutenant, Jean Marchand, who was responsible for the National Capital Commission, Trudeau had the provincial premiers and prime ministers declare in 1968 that the National Capital Region was composed of Ottawa-Gatineau, thus establishing a base of equality between the two sides of the river. The story of the establishment of the federal government buildings is a complex one involving the progressive deindustrialization of Hull and Gatineau (Chiasson and Simard 2007), the efforts of local elites to enlist the interest of the Quebec and/or federal governments to support local development, and the political battles between the Quebec government and the federal government (Andrew, Bordeleau, and Guimont 1981). This led to expropriations, by both the federal and the Quebec governments, and the transformation of downtown Hull from a working-class residential neighbourhood to a development serving office workers (Andrew, Blais, and DesRosiers 1975). The agglomeration of high-rise office towers that makes up Place du Portage and Les Terrasses de la Chaudière is a strong material statement about the important ways in which Gatineau's economy and workforce have transitioned from manufacturing to services since World War II—a bureaucracy's offices, not sawmills, dominate the city's landscape.

Trudeau was not only interested in increasing federal government employment on the Quebec side, he was also interested in the aesthetic and symbolic equality of the Ontario and Quebec parts of the capital region. In this sense, he followed in the footsteps of Laurier and King in wanting to mark the National Capital Region personally. Trudeau's contribution to Gatineau was the Museum of Civilization, which faces his other major architectural initiative, the National Gallery, located immediately across the river in Ottawa. Indicative of the important role played by all kinds of services in the local economy, the Museum of Civilization is now the second most important tourist site in the region (the Casino du Lac Lemay, also located in Gatineau, is the leading attraction). The location of the Museum of Civilization in Quebec has shifted tourist destinations in the capital region to some extent, although the Ontario side still dominates, since it still has the most tourism sites overall (Chiasson and Andrew 2009).

The federal employment sites on the Quebec side of the river also include the National Archives building in the Gatineau sector. The Achives' location could be seen as a return to an earlier suburbanization era, although it is probably better explained as an exception based on the exigencies of equalizing federal employment in what were at that time the separate municipalities of Hull and Gatineau.

The Hull Federal Office Complex has, as its Ottawa CBD counterpart does, more female than male employees (Table 9.6). In 2006, 56.9 per cent of the workers were female, an even higher percentage than in Tunney's Pasture. Of the overall number of employees, 57.2 per cent indicated that they speak French at home, 32.1 per cent English, and 10.6 per cent some other language—an almost complete reversal of the distribution among Tunney's Pasture workers (54.1 per cent English, 31.3 per cent French, and 14.6 per cent other). The female employees have slightly lower levels of education than do those in Tunney's Pasture, but the differences are not considerable.

As we suggested earlier, the federal government buildings do not blend easily into the suburban built stock of postwar Gatineau. Moreover, the transformation of the working-class industrial community of nineteenth-century Hull into the public sector, service-oriented city of twenty-first-century Gatineau has been uneven. Gatineau has one of the lowest high school graduation levels in Quebec, which suggests that an important share of the local population would have difficulty obtaining public sector employment. The commercial development around the federal office buildings, for the most part, does not comprise local companies, and therefore the benefits of a federal office presence for the local population are once again muted.

As stated earlier, the set of federal government decisions involved in creating the Hull Federal Office Complex represents a turning away from the decentralization trend exemplified by the suburban campus complexes of Tunney's Pasture, Confederation Heights, and Kanata North. While the Hull office towers do not directly contribute to reinforcing employment in the CBD, they do reinforce the material and symbolic significance of the urban core for the federal government in building the National Capital Region.

The Kanata North District: High-Technology Suburbanization on the Edge

Our fourth case study focuses on the high-technology cluster located in the former municipality of Kanata in the western suburbs of Ottawa. Since its initial moments in the 1940s, the high-tech sector's growth in Ottawa has been impressive although not very steady. Growth in the 1990s was such that many authors (e.g., Shavinina 2004) were quick to identify Ottawa as 'Silicon Valley North' and to consider it 'one of Canada's foremost technology clusters' (Madill, Haines, and Riding 2004). In the context of the regional economy, the high-tech sector's growth has at times even challenged the supremacy of the federal government, at least in terms of job numbers. Wellar and Novakowski (2007, 66) report that 'at the peak of the high-technology boom in 2000, high-tech employment reached 85,000 and for the first time exceeded federal employment on the Ottawa side of the Ottawa-Gatineau CMA.' With the crash of 2001 and the more recent demise of Nortel, the number of high-technology jobs has plummeted, and since the end of hiring freezes in the early to mid-1990s, federal public service jobs have begun to grow again both numerically and as a proportion of all employment. The importance of high-technology in the regional labour force has declined, but this sector remains an important part of the regional economy and has a major imprint on the region's geography of work.

The development of Ottawa's high-technology sector has been well documented (see, for example, two fairly recent books: Shavinina 2004 and Novakowski and Tremblay 2007). Relatively little attention, however, has been given to the particular geography of high-technology employment within the region. It is well known that the high-technology sector has been heavily concentrated in the western part of Ottawa, more precisely in industrial parks in and around the western portion of the Greenbelt. Gordon and Donald (2007, 105), in their 'local geography of high technology', maintain that 'the overall pattern of employment in the late 1990s is strongly skewed to the southwest of the NCR for all the high-technology sectors except software consulting, which is concentrated near the Ottawa downtown.'

Like our other cases (the CBD, Tunney's Pasture, and the Hull Federal Office Complex), Kanata North is a major hub in the geography of work of the National Capital Region (NCR), with a very significant portion of the high-technology employment. In 2006, more than 17,000 people worked in that area. The creation of this node, however, owes much to the federal government's investments in technology for purposes of national defence and industrial development, as well as its efforts in planning the national capital. Wellar and Novakowski (2007, 67) argue that 'the high-tech industry is considered "home-grown," with its origins primarily related to the presence of federal government agencies (National Research Council, Department of National Defence) and the associated spin-offs involving each type of high-tech firm.' In other words, the federal government gave the initial impetus to the high-technology sector in Ottawa by implementing its own high-technology facilities in the postwar era. For instance, the National Defence Department located its research lab in the western Greenbelt, while the National Research Council set up shop in the eastern part of the Greenbelt (Novakowski and Tremblay 2007). The location of these public labs in the Greenbelt meant not only that public servants would work and most likely live in the suburbs but also that there was sufficient land for private spin-off companies to develop and grow in adjacent industrial parks (Wellar and Novakowski 2007).

The decision to locate federal research labs in the eastern and western suburbs can be seen as a small part in a broader plan that intended to transform Ottawa (and Gatineau) into a modern, functional, and beautiful capital city. This decentralization of federal employment was a major thrust of the fore-mentioned 1950 Gréber report.

> The 1950 *Plan for the National Capital* proposed a greenbelt to surround the new suburbs, with satellite communities to absorb future growth, similar to the 1946 Greater London Plan. While the greenbelt was to provide visual relief and some recreational amenities to the suburban residents, it was also intended as a site for the airport and research facilities that required vast tracts of land and would benefit from a non-urban location. The greenbelt became home to the Defence Research Board, the Communications Research Center, the Agricultural Research Center, the Defense Proving Grounds and other federal facilities (Gordon and Donald 2007, 99–100).

In the Gréber report, the Greenbelt was to function as a divider between the city and its rural surroundings (Gauthier and Mevellec 2010). Decentralization of federal buildings and jobs was intended to occur within the boundaries of the Greenbelt (Gordon 2001). Following the path set by the Gréber Report, in the 1950s the federal government did decentralize a number of its offices in the suburbs near the Greenbelt while the National Capital Commission promoted industrial parks in that area (Gordon and Donald 2007). Before too long, in contradiction to Gréber's plan, urban sprawl extended over the limit of the Greenbelt (Gauthier and Mevellec 2010), and private promoters turned rural townships (e.g., March) into satellite communities and industrial parks (e.g., Kanata) to accommodate companies and employees (Gordon and Donald 2007).

The origins of the high-technology cluster in Kanata lie in the federal government's planning decisions for the region, but recent developments have been led largely by the private sector. Even though there are still a number of federal research labs, Kanata North district's high-technology employees are now mostly in private companies with markets extending well beyond the needs of the federal government. In 2006, a minuscule 0.6 per cent of employment in the Kanata North district was in public administration (Table 9.5), while 28.3 per cent of workers were employed in professional, scientific, and technical services industries (the highest percentage in any of the region's nodes).

As we mentioned earlier, feminization of jobs is a major trend in Ottawa-Gatineau as a whole. Kanata North, however, departs significantly from this trend (Table 9.6). Among the 17,000 people who work in the area, 70 per cent are male. Although it might be tempting to conclude that the feminization of jobs is attributable to the public sector's employment equity policies, the low proportion of women working in Kanata North remains exceptional even when compared to other districts in the city with considerable private-sector employment. For instance, in the Trainlands, Sheffield, Industrial Avenue district where only 10 per cent of jobs are in public administration, women constitute 47 per cent of the labour force. The striking absence of gender parity in Kanata North may be attributed to the continued dominance of men in fields such as engineering and computer science. It also confirms that the feminization of the labour force has a specific geography, even in places with a sizeable proportion of high-skill service-sector jobs that are dispersed across a metropolitan area.

The Kanata North district is also distinctive in terms of the home languages of its workers. First, there is a remarkably small number of francophones working in this area. Of the 17,000-plus workers, only 1600 indicate French as their only home language, and only another 95 indicate that French is one of two or more languages used at home. In part, this can be explained by the fact that very few workers from Gatineau cross the border to work in the western part of Ottawa but also by the fact that franco-Ontarians have, in the postwar period, tended to live in the opposite end of the Ottawa region because of the presence of Catholic churches and institutions there.

On the other hand, companies in the Kanata North area employ a large number of people whose home language is neither French nor English. Twenty-six per cent of workers, or 3360 people, indicate home languages other than French and English, and another 100 workers indicate other languages in combination with an official language. This makes Kanata North by far the most linguistically diverse node in our sample of districts. The difference is quite striking, since in none of the other districts did the percentage of workers indicating 'other' as home language surpass 20 per cent. In comparison, the CBD, Tunney's Pasture, and the Hull Federal Office Complex have very low proportions of workers whose home language is not English or French—less than 15 per cent. In Kanata North, a very significant portion of these workers identified Chinese as their home language. In fact, there are more Chinese speakers (615) than Italian, Portuguese, Polish, German, and Spanish speakers combined. Nevertheless, Chinese individuals represent less than 15 per cent of all workers whose home language is 'other'. This confirms that the workforce in this district is linguistically very diverse, even among non-official language groups.

Our analysis underlines the very distinctive traits that make Kanata North stand out relative to other parts of Ottawa's post-industrial landscape and economy. Kanata North has a strikingly low proportion of female workers and relatively few franco-phones. On the other hand, linguistically and ethnically, this district is extremely diverse. These distinctive traits might be explained, at least in part, by the predomin-ance of private sector jobs in the area. Areas that have more public administration jobs have shown greater feminization and a stronger francophone presence, in part because of employment equity and language policies pursued by the federal public service. The trends observed in Kanata are also magnified by the specificities of the high-technology sector. Many companies in Kanata North are involved in high-end research, innovation, and development that require a very specialized workforce. Such workers tend to be very mobile, and despite local efforts to promote workforce develop-ment (Paquet, Roy, and Wilson 2004), many firms recruit employees internationally. This makes the high-tech sector less permeable to local labour market trends in terms of both gender and linguistic diversity.

With its downturn in fortunes in the early years of the twenty-first century, the weight of the high-technology sector in the regional economy has diminished. Public service jobs once again drive employment trends in the region, but the city's high-technology clusters remain important. If creativity is a source of economic growth in the decades to come, to a very large extent the region's future is tightly bound to the small and mid-sized firms that still remain after the collapse of industry giants such as Nortel.

Conclusion

We have portrayed Ottawa-Gatineau as a city that has been dominated by employ-ment based on the production of services rather than goods for many decades. Yet as a post-industrial city, it is an unusual hybrid because investments by private corpora-tions and the federal government have played an influential role in shaping the type of work performed and where people work in the region. In terms of the sheer number of people employed in service-based industries, the high proportion of highly educated and 'creative' individuals, and the large proportion of women working in all indus-tries and across the job spectrum, Ottawa-Gatineau is an exemplar of post-industrial urbanism. The role played by the federal government, as an employer and driver of commercial development, undercuts the idea that Ottawa-Gatineau can be seen as just another post-industrial city. We have tried to show how the presence of the federal government has had significant effects on who performs various kinds of work and where people work. Because the federal government is a major employer, its gender equity and bilingualism policies have influenced both the number of women working in various employment nodes and the number of francophones and anglophones working on the opposite side of the Ottawa River from where they live. It is also important to emphasize that women working in several of the suburban employment nodes are very well-educated and enjoy good wages and job security primarily because they are working in unionized public sector employment. These suburban centres of employment are not the kind of back-office enclaves of feminized and relatively low-wage work found in many American cities (England 1993; Nelson 1986).

As our case studies have shown, the federal government's planning initiatives have at times favoured the suburbanization of jobs, but for functional, symbolic, and political reasons, the federal government has continued to invest in office space in the downtown core. As Sassen (1991) has argued, other major post-industrial cities such as New York and London have also seen considerable investment in downtown commercial buildings, but we would argue that the investments in Ottawa-Gatineau's urban core have been guided by somewhat different motivations and logic than those in other cities. Given the city's status as a national capital, symbolic considerations remain important, as do politics associated with an urban space that straddles the Quebec–Ontario border.

On the other hand, as the Kanata North case study illustrates, private investors who are not implicated in the symbolic production of the capital have overwhelmingly directed their investments to the suburban fringe of the metropolitan area. As such, they have created a low-density, car-oriented employment landscape that bears a strong resemblance to the suburban industrial parks of other North American cities (Soja 2000). The type of work undertaken in the Kanata North district is distinctly different from that found in the other major employment nodes, and the characteristics of the labour force also depart from metropolitan trends in important ways. Ottawa's high-technology node is the one place where men are still the vast majority of employees. It is also the location that most directly links Ottawa to an international labour market of talent. As the home languages of workers indicate, a significant number of employees are relatively recent immigrants from Asia. Although an important portion of Ottawa-Gatineau's population is foreign-born (18 per cent), it is only in Kanata North that this portion of population really stands out as an important segment of the labour force. In terms of gender and language, the post-industrial 'creative class' of Kanata North is different from that of other districts where public sector employment is significant.

At the present time, the role played by Kanata North in linking Ottawa-Gatineau into international circuits of post-industrial economic growth, stagnation, and decline is of uppermost concern to entrepreneurs and politicians. The demise of Nortel and other companies, once the core of Ottawa-Gatineau's knowledge economy, strongly underlines the fact that this city is not completely insulated by the federal government's investments and employment practices. As the unravelling of the old high-technology economy in Kanata North continues, and the reinvention of the industry and district is in its infancy, the qualities and geography of post-industrial work in Ottawa-Gatineau over the next decade may well come to resemble that of an earlier period when the city really was a 'government town'.

References

Andrew, C. 2002. 'Ottawa : le regard touristique'. *Teoros* 21 (1): 32–5.

Andrew, C., A. Blais, and R. Desrosiers. 1975. 'L'information sur le logement public à Hull'. *Recherches sociographiques* 16 (3): 375–83.

Andrew, C., S. Bordeleau, and A. Guimont. 1981. *L'urbanisation : une affaire*. Ottawa: Ottawa University Press.

Beaucage, A. 1994. 'From manufacturing to services'. In C. Gaffield, ed., *History of the Outaouais*, 481–524. Quebec City: Institut québécois de recherche sur la culture.

Bellamy, R. 2001. 'The architecture of government'. In J. Keshen and N. St-Onge, eds, *Ottawa: Making a Capital*, 433–65. Ottawa: University of Ottawa Press.

Chiasson, G. 2007. 'A cross-border high-tech cluster: A view from Gatineau'. In N. Novakowski and R. Tremblay, eds, *Perspectives on Ottawa's High-Tech Sector*, 225–42. Brussels: Peter Lang.

Chiasson, G., and C. Andrew. 2009. 'Modern tourist development and the complexities of cross-border identities within a planned capital region'. In Robert Maitland, ed., *City Tourism: National Capital Perspectives*, 253–63. CABI.

Chiasson, G., R. Blais, and J. Boucher. 2006. 'La forêt publique québécoise à l'épreuve de la gouvernance : le cas de l'Outaouais'. *Géocarrefour* 81 (2): 113–20.

Chiasson, G., and E. Simard. 2007. *Le renouvellement des rapports entre la ville et sa region : le cas de l'Outaouais*. Rapport remis au Ministère des affaires municipales et des régions.

City of Ottawa. 2009. *Annual Development Report 2008*. Ottawa: City of Ottawa, Infrastructure Services and Community Sustainability Department.

Cohen, A. 2007. *The Unfinished Canadian*. Toronto: McClelland and Stewart.

Couture, C. 1996. *La loyauté d'un laic : Pierre Trudeau et le nationalisme canadien*. Paris: l'Harmattan.

England, K.V.L. 1993. 'Suburban pink collar ghettos: The spatial entrapment of women?' *Annals of the Association of American Geographers* 83 (2): 225–42.

Florida, R. 2005. *The Flight of the Creative Class: The New Global Competition for Talent*. New York: HarperBusiness, HarperCollins.

Gad, G. 2004. 'The suburbanization of manufacturing in Toronto, 1881–1951'. In R. Lewis, ed., *Manufacturing Suburbs: Building Work and Home on the Metropolitan Fringe*, 143–77. Philadelphia: Temple University Press.

Gauthier, M., and A. Mevellec. 2010. 'La ceinture de verdure de la capitale nationale du Canada : une frange périurbaine en transformation'. In M. Dumont and E. Helier, eds, *Les nouvelles périphéries urbaines*, 153–64. Rennes: Presses universitaires de Rennes.

Gordon, D. 2001. 'Weaving a modern plan for Canada's capital: Jacques Gréber and the 1950 plan for the National Capital Region. *Urban History Review* 29 (2): 43–61.

Gordon, D., and B. Donald. 2007. 'Unanticipated benefits: The role of planning in the development of the Ottawa region technology industries'. In N. Novakowski and R. Tremblay, eds, *Perspectives on Ottawa's High-Tech Sector*, 91–116. Brussels: Peter Lang.

Hohenberg, P.M., and L.H. Lees. 1985. *The Making of Urban Europe 1000–1950*. Cambridge, MA: Harvard University Press.

Hudon, M.-E. 2009. *Les langues officielles dans la fonction publique : de 1973 à aujourd'hui*. Ottawa : Service d'information et de recherche parlementaires, Bibliothèque du Parlement.

Keshen, J. 2001. 'World War Two and the Making of Modern Ottawa'. In J. Keshen and N. St-Onge, eds, *Construire une capitale/Ottawa—Making a Capital*, 383–410. Ottawa: University of Ottawa Press.

Lewis, R. 2000. *Manufacturing Montreal: The Making of an Industrial Landscape, 1850–1930*. Baltimore: John Hopkins University Press.

Linteau, P.-A. 1992. *Histoire de Montréal depuis la Confédération*. Montreal : Les Éditions du Boréal.

Madill, J., G. Haines, and A. Riding. 2004. 'A tale of one city: The Ottawa technology cluster'. In L. Shavinina, ed., *Silicon Valley North: A High-Tech Cluster of Innovation and Entrepreneurship*, 85–117. Amsterdam: Elsevier.

Martin Prosperity Institute. 2009. *Ottawa's Performance on the 3Ts of Economic Development*. Rotman School of Management, University of Toronto.

Mosco, V., and P. Mazepa. 2003. 'High tech hegemony: Transforming Canada's capital into Silicon Valley North'. In L. Artz et al., *The Globalization of Corporate Media Hegemony*, 93–112. New York: SUNY Press.

Nelson, K. 1986. 'Labor demand, labor supply and the suburbanization of low-wage office work'. In A. Scott and M. Stroper, eds, *Production, Work and Territory*, 149–71. Winchester, MA: Allen and Unwin.

Novakowski, N., and R. Tremblay, eds. 2007. *Perspectives on Ottawa's High-Tech Sector*. Brussels: Peter Lang.

Paquet, G., J. Roy, and C. Wilson. 2004. 'Ottawa's TalentWorks: Regional learning and collaborative governance for a knowledge age'. In L. Shavinina, ed., *Silicon Valley North: A High-Tech Cluster of Innovation and Entrepreneurship*, 311–29. Amsterdam: Elsevier.

Sassen, S. 1991. *The Global City: New York, London, Tokyo.* Princeton, NJ: Princeton University Press.

Shavinina, L, ed. 2004. *Silicon Valley North: A High-Tech Cluster of Innovation and Entrepreneurship.* Amsterdam: Elsevier.

Soja, E.W. 2000. *Postmetropolis: Critical Studies of Cities and Regions.* Oxford: Blackwell.

Treasury Board of Canada Secretariat. 2000. *Employment Statistics for the Federal Public Service.* Ottawa: Treasury Board of Canada Secretariat.

Udovicki-Selb, D.F. 2001. 'The elusive faces of modernity: Jacques Gréber and the planning of the 1937 Paris World Fair'. *Urban History* Review 29 (2): 20–35.

Wellar, B., and N. Novakowski. 2007. 'Local government's record of assessing the impacts of the high-tech industry on Ottawa's land use–transportation relationship'. In N. Novakowski and R. Tremblay, eds, *Perspectives on Ottawa's High-Tech Sector,* 61–90. Brussels: Peter Lang.

Notes

1. Ottawa-Gatineau is divided by the Ottawa River, and consequently the metropolitan area bridges two provinces—Ottawa is located in Ontario and Gatineau in Quebec. Until the provincial governments amalgamated municipalities in 2001 (Ottawa) and 2002 (Gatineau), the Ontario and Quebec sections of the metropolitan area were fragmented into a number of municipalities, most of which were suburban. Given that the municipal amalgamations are relatively recent, it is not uncommon that people continue to refer to the old municipal names. This is especially confusing in the new city of 'Gatineau' where the name of one of the suburban municipalities (Gatineau) now designates the entire Quebec section of the metropolitan area. Many people also continue to refer to the urban core of the new amalgamated city, where the greatest concentration of employment is found, as 'Hull'. To avoid confusion, we have decided to use the name by which the region is currently known.

2. In this essay, the metropolitan area is based on Statistics Canada's definition of a census metropolitan area (CMA), and metropolitan area and CMA are used synonymously.

3. The predecessor of the NCC was the Federal District Commission, created in 1927 by Prime Minister Mackenzie King to develop and implement an overall urban plan for the capital city and surrounding region.

4. It is important to recognize that not all workers employed in manufacturing industries are in fact directly processing or fabricating goods; many workers in fact provide services to manufacturing industries.

5. We refer to this area as the Hull Federal Office Complex. It comprises several sets of large office buildings that were constructed in the 1970s and early 1980s and architecturally have much more in common with the CBD than with surrounding areas. The complex comprises Place du Portage Phases I through IV (Phase I was opened in 1973) and les Terrasses de la Chaudière. It should be noted that the definition of the CBD used to compare cities in Chapter 5, which includes all census tracts contiguous to the CBD tract that have employment levels greater than 5000 and net in-commuting, differs from the one used in this chapter.

6. The Ottawa Hospital General Complex and Canadian Forces Health Services Centre are located in this district.

7. For the Ottawa-Gatineau metropolitan area overall, 50.8 per cent of workers indicate that English is the language they use most often at home, while 32.7 per cent indicate that their home language is French.

10

The Greater Toronto Region: The Challenges of Economic Restructuring, Social Diversity, and Globalization

Larry S. Bourne, John N.H. Britton, and Deborah Leslie

Introduction

During the past half century, the Toronto region has become the country's pre-eminent metropolis, its dominant economic engine and innovation milieu, as well as its principal gateway to the rest of the world. It is also a node, as Chapters 2 and 3 have demonstrated, in the continental North American urban system. The region's economy is not only large, roughly 20 per cent of the nation's, but exceptionally diverse.[1] It has relative strengths in automobile manufacturing, information and communication technologies, financial services, business and professional services, education, health, culture, and new media. That robust economy, combined with the region's role as the country's largest financial centre and major destination for recent immigrants to Canada, has produced a consistently high rate of population growth and employment generation, as well as enhanced wealth and a relatively highly-ranked quality of life.[2] Depending on how it is defined, the greater Toronto region is the fifth or sixth largest metropolitan economy in North America and one of the continent's fastest growing mega-regions.

The sustainability of these conditions, however, has been increasingly challenged by a range of economic factors and social changes, all deepened by the latest economic recession and global financial crisis. These challenges include widespread economic restructuring and intense international competition for new investment and skilled human capital and the application of new technologies. These external pressures have coincided with the accumulating social consequences of high levels of immigration and unprecedented ethno-cultural change; by concentrated poverty and spatial polarization; by jurisdictional conflicts over property development, land use, and transportation investments; by uncertainty with respect to strategic planning and municipal budgets; and by intergovernmental tensions over who (if anyone) is managing the city region.

Viewed over the longer term, three of the defining features of the region have been rapid population and economic growth, dramatically increased cultural diversity, and the immense geographical spread of urban development. The Toronto census metropolitan area (CMA) had fewer than 2.1 million residents and 800,000 jobs in 1961 and ranked second to Montreal in the Canadian urban hierarchy. By 2006, it had more than tripled to over 5.5 million people and over 3.0 million jobs (Table 10.1). Most of that growth in recent decades has been driven by immigration rather than by domestic migration. During the same period, the geographical extent of the urbanized area and the regional labour market increased by more than six times, in the process enveloping small towns and cities in the region. In each recent decade, the Toronto region added more than 800,000 people (and corresponding numbers of jobs, housing units, infrastructure, and services), the equivalent of constructing a city as large as metropolitan Winnipeg or Quebec City every 10 years.

This scale of growth has produced a pattern of development that is simultaneously dispersed and concentrated, economically integrated but politically fragmented. It is an evolving form that challenges conventional wisdom on how cities are organized and undermines the ability of governments and other agencies of civil society and economic governance to respond in a coherent and effective fashion. All three traditional spatial processes—suburbanization, inner suburban decline, and core area revitalization (and gentrification)—have played a prominent role. These processes, in turn, have been accompanied by institutional conflicts driven by a fragmented political structure and a lack of inter-municipal cooperation. The latter are reflected in the difficulties of co-ordinating infrastructure provision and in meeting the challenges posed by economic changes, fiscal deficits, and increasing social inequality. One further outcome is that there is no consensus on the identity of the city or the region, since these identities are perceived from different vantage points. Thus, it is difficult to construct a single narrative that does justice to the highly varied and distinctive local economies, cultures, and landscapes that make up this expanding region. There are, in fact, many different 'Torontos' co-existing within a shared urban space.

Despite these tensions, and the uncertainties inherent in globalization (and fluctuating exchange rates), there are immense opportunities for economic development and innovation in the region provided by a pool of highly skilled labour, clusters of leading-edge firms, strong proactive civic organizations and institutions, and an unmatched resource of cultural diversity and talent. The city and the region also have, by North American standards at least, relatively high-quality public infrastructure and strong public transit systems—the TTC and the regional GO commuter system[3]—which have considerably shaped both urban form and economic development. The former has contributed to a relatively viable inner city, the latter to suburban spread and dispersion.

Still, the future form, character, and economic viability of this self-proclaimed global region have yet to be defined: will it emerge as a global financial and ICT centre or a postmodern centre of international culture, or will it settle in as a second-order city region within the North American urban system? Or, at worst, will it become a post-industrial Detroit North? Whatever scenario is considered, the future will most certainly be the combined outcome, as it has been in the past, of many often contradictory forces of change and of a myriad of struggles among many different actors and agents of change.

Approach

This chapter explores recent structural changes in the Toronto region, with particular focus on three themes: the overall scale and magnitude of growth; transformative shifts in the regional economy (in employment, occupations, investment) by sector

Table 10.1 Population Growth and Redistribution, Greater Toronto Area (GTA) and Hamilton, 1961–2006

Population by Municipality (in 000s)

Year	Metro* Toronto	Peel	York	Durham	Halton	Total (GTA)	GTA Change (in 000s)	Hamilton CMA**
1961	1620	110	110	150	110	2110	–	348.2
1971	2090	260	170	210	190	2920	810	410.9
1981	2140	490	260	280	250	3420	500	427.3
1991	2280	730	500	410	310	4240	820	470.2
2001	2481	989	729	507	375	5082	840	511.6
2006	2503	1159	892	561	439	5554	472	528.5

Municipality as % of Total GTA Population					Suburban % of Total	
1961	76.9	5.3	5.3	7.2	5.3	23.1
1971	71.6	8.9	5.8	7.2	5.3	28.4
1981	62.6	14.3	7.5	8.2	7.3	37.3
1991	53.8	17.3	11.8	9.7	7.3	46.1
2001	48.8	19.5	14.4	10.0	7.4	51.2
2006	45.1	20.9	16.1	10.1	7.9	54.9

Source: Statistics Canada. *Census of Canada.*

* Former Metro Toronto became the new City of Toronto after amalgamation, January 1998.
** Excludes the City of Burlington (164,415 in 2006), which is part of the Hamilton CMA but is also included in the Halton region in the standard definition of the GTA.

and type of industrial cluster; and the changing form and structure of the region. We also give attention to parallel but interrelated processes underlying the dramatic, indeed unprecedented, social and ethno-cultural transformation—the emergence of a new multi-level social geography—and to devising policies for economic renewal, land-use planning, and governance that are deemed appropriate to the expanding scale and diversity of the region. The following section sets the stage by defining the area of interest and outlining the region's recent population growth and current governance structures. In section three, we provide an overview of the metropolitan region's space-economy, the changing structure and geographies of employment and investment, incorporating the impacts of the region's changing social morphology—again notably through immigration—on employment and regional labour markets. Section four focuses on how that economy is organized into industrial clusters, with representative examples from the ICT, biotech, financial services, and automobile sectors. Each sector, each economic cluster, and each local district has its own story to tell.

The concluding section highlights some of the challenges, conundrums, and opportunities this narrative presents for the Toronto region's economy and quality of life over the next few decades. As part of this mandate, we focus on examples of the recent policy initiatives intended to manage the growth trajectory of the economy, to encourage technological innovation, to enhance both global competitiveness and local social cohesiveness, and to design (or redesign) the physical form, density, and environmental sustainability of the region.

We do not, given space limitations, spend much time on the historical development of the region, although we recognize the importance of context (e.g., in terms of the city's position in the Canadian urban hierarchy) and the path dependency of the region's current economic structure. There is, fortunately, an extensive background literature that documents the city's nineteenth-century (Careless 1980) and early twentieth-century (Lemon 1985) growth histories, as well as substantive treatments of the evolution of regional governance (Frisken 2007), finance (Slack and Bird 2004), politics (Boudreau, Keil, and Young 2009) and planning (Neptis Foundation 2006; Filion 2007), the changing social structure (Bourne 2000; Anisef and Lanphier 2003; Walks and Maaranen 2008), the historical development of infrastructure provision (White 2003), and the origins of the sprawling suburbs (Sewell 2009). We do not intend to repeat this information here but to build on it. While there are surprisingly few systematic overviews of the long-term evolution of the entire regional economy, numerous studies document recent trends and current economic issues while offering differing visions of—and multiple strategies for—change, several of which we cite at appropriate points in the text.

Setting the Stage: Boundaries, Governance, and Fragmentation in the Region

There is no single or widely accepted definition of the geographical extent and boundaries of the Toronto region. To some readers, the term Toronto conveys the image of the downtown core or the old pre-amalgamated central city (700,000), to others it is the post-1999 amalgamated city, the former Metro (2.6 million). But for research and planning purposes, the most commonly used boundary is the Statistics Canada definition of the CMA (5.5 million; Figure 10.1). The latter is the spatial unit for which most statistical data are available and is defined based on commuting to work and thus on the assumption that it represents an integrated labour market. Unfortunately, the Toronto CMA is now seriously under-bounded as economic integration increases within the larger region. Increasingly common in the media and government publications is use of the Greater Toronto Area (GTA 5.6 million), an unofficial amalgam of municipalities that combines the City of Toronto and the four regional government units (Halton, Peel, York, and Durham) that surround—and are functionally integrated with—the city but are politically separate from it (and from each other).

Boundaries do matter for economic development, especially in this region. Unlike most other metropolitan areas in Canada, where the definition of the CMA captures most of the built-up or urbanized region and most of the urban labour market, in southern Ontario and particularly in the Toronto region this is not the case. The Toronto CMA is not an isolated and free-standing economic region but is surrounded by and closely integrated with several nearby cities and metropolitan areas—notably

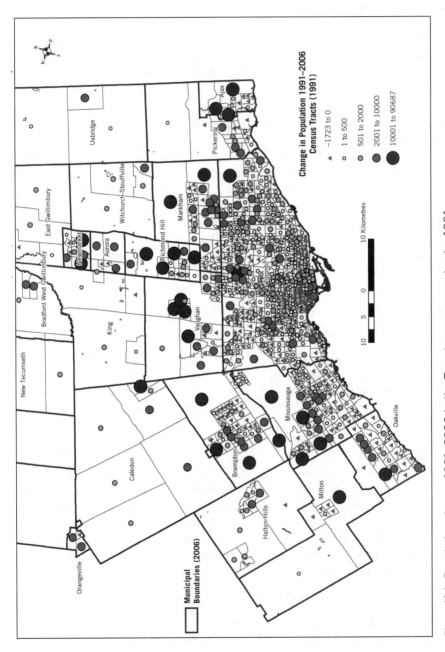

Figure 10.1 Population change 1991–2006 in the Toronto CMA census tracts, 1991.

(Source: Statistics Canada, Census 1991 and 2006.)

Oshawa and Hamilton but also Barrie, Guelph, and Cambridge. Toronto, in effect, is the urbanized core of a much larger economic region and a highly urbanized regional landscape. The point here is that whatever boundary is chosen substantially influences any analyses undertaken and any attempt to characterize the nature of the region, its changing economy and life styles, and the policy challenges it faces.

In this context, we use different spatial definitions depending on the processes and issues under discussion and the availability of data on the economy. For most purposes, we use CMA data, but where possible we define a broader region, the GTA+H; this is the urbanized core consisting of the Greater Toronto Area and Hamilton (an area that is frequently approximate for data purposes by combining the Toronto, Oshawa, and Hamilton CMAs). The 2006 population of the GTA+H was about 6.1 million. An even broader definition is the Greater Golden Horseshoe (the GGH), a region that extends from Oshawa and Peterborough in the east to Kitchener-Waterloo and Guelph in the west, to St Catharines–Niagara in the south.[4] This region is both multi-nodal in its physical form and multi-scalar in its economic structure and houses more than 7.5 million people. The GGH is largely an abstraction; nonetheless, it is the region that the provincial government now argues is necessary for its latest regional planning, transportation, and infrastructure initiatives (Ontario 2006; Filion 2007). In one sense at least, it offers a perspective on, and a vision of, the longer-term future size and form of the Toronto region, and we address these initiatives in the concluding section.

Governance

It is worth reiterating that the current structure of institutions and government in the Toronto region is not well understood outside the region (or even inside) and that it shapes the region's form, character, and growth prospects, including the geography of its economy. The GTA, despite common perceptions, is not a unit of government. It is instead an amalgam of distinctly separate municipalities: the City of Toronto and the four two-tier regional governments (Halton, Peel, York, and Durham) surrounding the city.[5] In effect, there is no overall regional government, or even regional institutions, encompassing the entire GTA and definitely not for the broader GGH, and there is certainly no economic strategy for either region (OECD 2009; GTR Economic Summit 2009). To the extent that any formal governance structures now exist for the region, they are provided by the provincial government and its various ministries and special boards (e.g., the Ontario Municipal Board for resolving land-use conflicts and Metrolinx, the new provincial transportation agency for the region).

This fragmented structure of government is awkward for any purpose, and to some observers it is almost dysfunctional.[6] Although the region has worked reasonably well to date as an ad hoc set of governments, fragmentation does have other direct and often negative effects that are relevant here in our discussion of the economy. It significantly inhibits political initiatives and strategic planning for economic development and innovation at the regional scale, while at the same time it encourages inter-municipal competition for new industries, investments, and tax revenues. The uneven expenditure burden between the amalgamated City and the newer suburbs, especially for welfare, social services, and housing, limits the City's fiscal flexibility. It makes it difficult, as examples, for the City to be entrepreneurial, to set competitive commercial tax rates, to maintain social services (some mandated by the province), and to enhance the public realm and built environment.[7] In parallel, political fragmentation facilitates, if not accelerates, the decentralization of jobs, since new growth tends to migrate to lower-cost municipalities, and over the long term it contributes to structural inefficiencies in economic activity and infrastructure provision and to the emergence of a discontinuous urban fabric. The collective failure to address the issue of regional governance and tax sharing in the decades since the initial metropolitan government was established (1954) may seem surprising given the historical record (Bourne 2001; Frisken 2007). The creation of Metro is widely viewed as one of the most innovative and successful experiments in the governance of any metropolitan region on the continent. This success has not been replicated in the decades since for the expanded urban region.

The Changing Metropolitan Space-Economy

The economy of the Toronto region, like that of most other large metropolitan regions, is not only massively complex and diverse but is in constant flux as new firms and jobs replace older versions while others relocate outside the region or the country. That economy, in theory at least, includes the sum of all goods produced and services provided, as well as all transactions and flows (e.g., in transportation and communications) that link together the varied sectors and widely dispersed parts of the region. It also encompasses all expenditures in the consumption sector and all transactions in the residential and non-residential property markets, as well as the circulation of capital embedded in the built environment and infrastructure. Based on traditional measures, the GDP for the GTA in 2009 amounted to more than $290 billion, with more than 180,000 businesses employing more than 3 million workers in 80,000 different work locations. In this section, we use primary data on employment and occupations as surrogate measures for overall economic activity, since comprehensive data on other indices (e.g., total output, investment, capital flows) are not available at the city or metropolitan scale.

Employment Structure and Change

Perhaps the most succinct way of conveying a sense of the overall structure of the regional economy, for the GTA+H, is through the distribution of employment by sector from the 2006 census (Table 10.2).[8] Depending on the specific industrial classification used, the largest sector of employment is business services, with 494,000 jobs, and second, manufacturing with 440,000 jobs. Indeed, the Toronto region has a relatively large proportion of manufacturing jobs for a major tier-two global city (see Chapter 2). The next largest sector is retailing (325,000), followed by financial services (279,000), health (263,000), and education (191,000). These totals can of course be modified (aggregated or disaggregated) in various ways to illustrate the particular importance of individual sectors or sub-sectors. Combined, business services and financial services provide more than 954,000 jobs, manufacturing (plus transportation, utilities, and construction) more than 695,000, services to consumers more than 678,000, and the public sector more than

Table 10.2 Employment by Place of Work: Greater Toronto Region, 1971–2006 and 2001–6 (for the Toronto, Hamilton, and Oshawa CMAs)*

Employment Sector	1971 (in 000s)	2006 (in 000s)	Change 1971–2006 (in 000s)	Growth Rate 1971–2006 %	Change 2001–6 (in 000s)	Growth Rate 2001–6 %
Blue collar	564.7	695.0	130.3	23.1	–30.4	–4.2
Primary	15.1	18.5	3.4	22.5	–0.6	–2.5
Transport & utilities	78.4	152.2	73.8	94.1	5.7	3.9
Construction	88.3	84.2	–4.1	–4.6	5.3	6.7
Manufacturing	382.9	440.1	57.2	14.9	–40.8	–8.5
Services to business	291.6	854.4	662.8	227.3	66.4	7.5
Wholesale	78.4	181.6	103.2	131.6	13.0	7.7
Financial	94.1	279.2	185.1	196.7	23.6	5.3
Business service	119.1	493.6	374.5	314.4	29.8	6.4
Public service	245.8	567.2	321.5	130.8	72.7	14.7
Education	90.1	190.8	100.7	111.8	25.9	15.7
Health	81.5	262.7	181.2	222.3	35.2	15.5
Public administration	74.2	113.7	39.5	53.0	11.5	11.3
Services to consumers	291.4	678.4	390.0	133.8	40.2	8.2
Leisure	15.5	51.8	36.3	234.2	5.1	10.9
Food & lodging	46.8	165.9	119.1	254.5	8.7	5.5
Personal service	61.4	136.0	74.6	121.5	13.1	10.6
Retail	167.7	324.7	157.0	93.6	13.3	4.3
Totals	**1393.5**	**2895.1**	**1501.6**	**107.8**	**147.9**	**5.4**

Source: Statistics Canada, Census 1971 and 2006.

* For 1166 census tracts in the three CMAs.

568,000 (Simmons, Bourne, and Kamikihara 2009). The regional economy then, at this scale at least, is not simply diverse but surprisingly balanced.

Over the period of 1971 to 2006, not surprisingly given the rapid growth of the region, all sectors of employment grew. More than 1.5 million jobs were added to the region. The most rapidly growing sectors are in business services, services to consumers (e.g., leisure, food), health, and financial services. Some of this employment growth reflects a response to simple population growth and the increase in household and community income over the past few decades, while growth in other sectors, such as business and financial services, in contrast, is primarily driven by the needs of the production sector and limited by the intense pressures of global competition.

Among Toronto's industries, the most responsive to inter-regional and international competition are manufacturing, professional services, scientific and technical services (PST), finance and insurance (FIRE), and information (IT) and cultural industries. Individually, these industries are large and relatively concentrated in the Toronto region: each employed more than 100,000 in the region and recorded location quotients of 1.0 or greater. Interestingly, none of these industries grew by more than the regional average (9.4 per cent) over the last census period. One explanation for slower employment growth, in addition to the effects of global competition, is that these sectors were able to innovate in their production processes (ICT especially), to increase productivity (without increasing job numbers), or to outsource their operations to firms in other regions or abroad. Manufacturing manifested the greatest changes among these industries, as we discuss later, with a decline in employment of more than 6 per cent.[9]

In contrast, the industries that grew most rapidly over this period, and that were also of a scale (>100,000 employees) sufficient to warrant close attention here, were all activities that have largely responded to regional demands rather than global markets. They include health and education, both part of the public sector, as well as transportation and warehousing, construction, and accommodation and food (the latter likely a joint response to tourism and income growth). These same sectors are also especially vulnerable to local cyclical downturns. Whether growth in any or all of these sectors, especially those in the public sector, is sustainable in the near future, given the pressures on municipal and provincial government revenues, is open to continuing debate.

How do employment trends in the region's occupational structure compare with those of its industrial structure? In terms of occupations, the Toronto region's workforce has two unequal parts: just over 40 per cent is employed in sales and services, trades and equipment operations, and manufacturing and utilities. Some of these jobs involve skilled trades, but the majority are low-skilled, putting them in a weak situation in the current recession. The larger part of the workforce (57 per cent) is in white-collar jobs where a higher level of education is usually required. This applies to managerial, financial, and administrative jobs, as well as to jobs in health, natural science, education, government, and art, culture, recreation, and sport. All of these occupations have location quotients (LQs) above 1.0, indicating relative strength in Toronto, except for occupations in health. In this sense, Toronto has fewer health workers than one might expect given its population size and the number of specialized hospitals and medical facilities. The most localized of these occupational groups in Toronto is the arts, culture, and recreation category. While this group employs only 3.0 per cent of the regional labour force, it is clearly of importance symbolically and strategically in contemporary urban planning and policy discourse (Gertler 2000; Florida 2002, 2007; Wolfe 2009).

The Changing Geography of Employment

The geography of employment in the region is as varied as its industrial and occupational structures. Figure 10.2 maps the distribution of employment for 2006 based on census place-of-work data. As expected, jobs are widely distributed throughout the region but with marked concentrations in selected locations, notably the downtown core (with 415,000 jobs) and the region around the airport (198,000 jobs). Other smaller nucleations stand out throughout the older and newer suburbs, some at redevelopment sites around TTC (e.g., North York City Centre) and GO stations, others at major shopping centres (Yorkdale, Vaughan Mills at 401 and 407) and suburban town centres (e.g., Scarborough Town Centre, Mississauga City Centre), others in office parks, and still others in large industrial estates (e.g., Oakville). Other large industrial nodes are in the northwest part of the region, particularly in Vaughan, Mississauga, and Brampton, and to the east in Oshawa, nodes that are often linked to highway access and the availability of suppliers. Employment in retail and consumer services, in contrast, is primarily market-driven and thus tends to follow the pattern of population growth (and decline), while business and financial services are relatively concentrated in the downtown core and inner city as well as in a few expanding suburban nodes.

Economic activity in the region is obviously highly volatile over both space and time. Analyses of the overall pattern of employment change in the Toronto region for the most recent census period, 2001–6 (Figure 10.3), indicate considerable variation among individual sectors throughout the entire region (Simmons, Bourne, and Kamikihara 2009). Manufacturing employment, as in other cities, has continued to shift away from the city centre and is increasingly suburban and decentralized. The overall pattern of employment change is dominated by suburban growth, especially in the manufacturing, communications, and transportation sub-sectors and in services to consumers. The downtown core has been more or less stable in total number of jobs but of course not in the composition of these jobs. In contrast, most areas of the inner city and older (early postwar) suburbs show widespread employment losses. Even some of the older nodes of office development from the 1960s (e.g., Don Mills, Eglinton-Yonge) have shown employment declines.

There are, on the other hand, some rather dramatic increases in employment in specific inner-city districts and within particular clusters of activity. To the west of the downtown core, for example, employment has been growing in the arts, professions, and media sub-sectors as part of what is often called the new economy of the inner city. King-Spadina, Queen West, and Liberty Village are three well known examples. The latter we highlight in a later section. Overall though, the City of Toronto's population and labour force has been growing much more slowly than that of the GTA. Between 2001 and 2006, for example, the city's labour force grew by only 1.9 per cent, while the GTA grew by 17.2 per cent (City of Toronto 2008a).

Despite the continued shift in activity and employment to the suburbs, the City of Toronto continues to represent the core of the region's economy, accounting for about 44 per cent of its labour force (City of Toronto 2008a). Most of the recent increase in the city's labour force between 2001 and 2006, however, occurred in only two sectors—education and health care—while the manufacturing labour force, as shown above, decreased significantly. Location quotients reveal that employees in the City of Toronto are more likely than other GTA and Canadian workers to work in information and cultural industries, finance and insurance, real estate, professional, scientific, and technical services, management enterprises, and administrative services (City of Toronto 2008a).[10] Cultural industries are still largely concentrated in the central core and inner city, and the City of Toronto remains the dominant centre for a number of

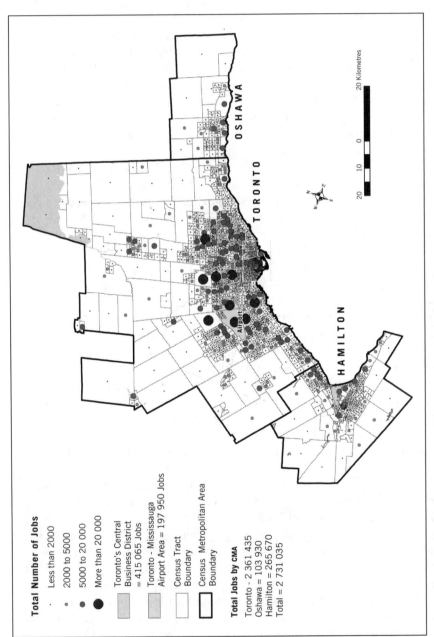

Total Number of Jobs

- · Less than 2000
- · 2000 to 5000
- ● 5000 to 20 000
- ● More than 20 000

Toronto's Central
Business District
= 415 065 Jobs

Toronto - Mississauga
Airport Area = 197 950 Jobs

Census Tract
Boundary

Census Metropolitan Area
Boundary

Total Jobs by CMA

Toronto - 2 361 435
Oshawa = 103 930
Hamilton = 265 670
Total = 2 731 035

Figure 10.2 Jobs by census tract place of work, 2006, Toronto, Oshawa, Hamilton CMAS.

(Source: Statistics Canada. Place of Work Custom Data, Census 2006. Table 97C0060, 20% Sample Data.)

Note: Data excludes persons with no usual place of work.

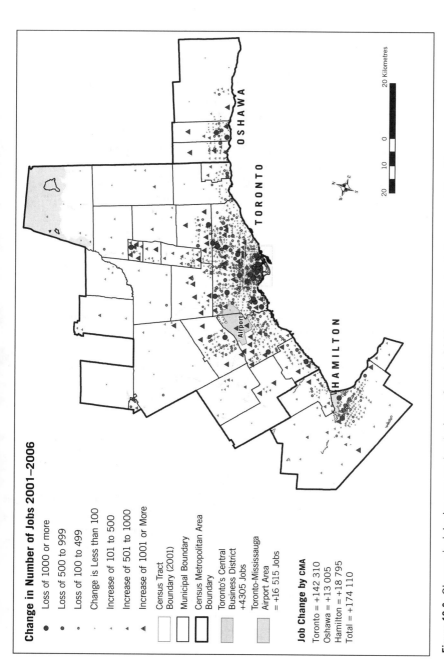

Figure 10.3 Change in jobs by census tract place of work, 2001–6: Toronto, Oshawa, and Hamilton CMAs.

(Sources: Statistics Canada. Place of Work Basic Cross-Tabulations, Census 2006, Semi-Custom Table 97C0037, 20% Sample Data; Statistics Canada. Place of Work Custom Data, Census 2006, Table 97C0060, 20% Sample Data.)

Note: Data excludes persons with no usual place of work.

cultural industries in Canada, including film and television, theatre, dance, food, art, architecture, design, and music.[11]

Transformation of the Downtown Core: Jobs and People

The region's downtown core provides a particularly interesting case study of the intersection of all of these trends. Table 10.3 summarizes the occupational structure of jobs in the core area—defined here as the city's central planning district—for 2006 as well as changes over the recent census period 2001–6. As expected, the economy of the downtown specializes in particular activities and occupations relative to the rest of the city and the CMA. The core has higher proportions of workers in managerial occupations, in business and finance, natural and social sciences, health, education and government services, and in art, culture, recreation, and sport. It has relatively fewer jobs in sales and services, distribution and transportation, and manufacturing and utilities, activities typically located in suburban areas. Over time, that level of functional specialization has tended to increase, reflecting the changing role of the downtown core in the regional economy.

Three temporal trends in the economic structure of the downtown area stand out in this ongoing transformation. One is the relatively high rate of growth of employment in health, education, government, and arts and culture. This trend follows the national experience of continued expansion in the public sector generally. It also attests to the critical importance of the public sector in maintaining employment levels in the core during a period when new commercial office construction was almost non-existent. Several smaller office buildings were converted to residential use, further reducing office space and associated employment. The second trend is the continuing rapid decline in the numbers and proportions working in manufacturing, utilities, and the trades, albeit measured from a relatively small base. These activities have been declining for decades and have now essentially vanished from the downtown core.

The other recent trend in the downtown economy is the relative decline in managerial occupations, for which there is no obvious or single explanation. In this instance, the decline in managerial positions in Toronto basically follows trends in the national economy, reflecting overall declines in white-collar employment and the difficulties facing high-tech industries and the manufacturing sector generally. These trends in turn mirror the local effects of global competition, the re-engineering of occupations and downsizing of management hierarchies in many large corporations, the dot.com bust beginning in 2000, and the economic shock of 9/11, as well as a host of other factors. Interestingly, in the Toronto case the recent decline of managerial occupations has occurred not only downtown but throughout the GTA, and in fact the rate of decline has been higher in the rest of the city than in the downtown core. The latter is likely the result of a reduction in middle management positions outside the core itself and may follow from the loss of some head-office functions.

Repopulating the Core

Despite widespread economic restructuring, the downtown core (and many other areas of the older city), have also witnessed a significant *repopulation process*. This has been

Table 10.3 The Changing Occupational Structure of Jobs in Toronto's Downtown (CBD) Core and the Toronto CMA, 2001–6

Occupations	2001 (in 000s)	%	2006 (in 000s)	%	Change 2001–6 (in 000s)	Change as % of 2001
Downtown (CBD) Total	410.7	100	415.1	100	4.3	1.0
Managerial	68.9	16.8	64.4	15.5	−4.5	−6.5
Business, finance, admin.	131.3	32.0	132.4	31.9	1.0	0.8
Natural & applied sciences	43.8	10.7	41.7	10.0	−2.1	−4.9
Health	21.6	5.3	24.1	5.8	2.5	11.4
Social science, education*	40.6	9.9	48.0	11.6	7.4	18.2
Art, culture, recreation	21.6	5.3	24.3	5.9	2.7	12.4
Sales and service	66.9	16.3	68.6	16.5	1.7	2.5
Trades, transportation	10.5	2.6	8.1	2.0	−2.4	−22.7
Primary industry	0.5	0.1	0.5	0.1	−0.1	−2.9
Manufacturing, utilities	4.9	1.2	2.9	0.7	−1.9	−40.2
Toronto CMA Total	2361.4	100	2503.7	100	142.3	6.0
Managerial	327.5	13.9	317.8	12.7	−9.7	−2.9
Business, finance, admin.	539.2	22.8	575.7	23.0	36.6	6.8
Natural & applied sciences	196.9	8.3	209.7	8.4	12.7	6.5
Health	101.0	4.3	118.9	4.7	17.9	17.7
Social science, education*	182.0	7.7	216.6	8.7	34.6	19.0
Art, culture, recreation	78.3	3.3	89.6	3.6	11.3	14.4
Sales and service	494.4	20.9	543.1	21.7	48.7	9.8
Trades, transportation	228.1	9.7	229.0	9.1	0.9	0.4
Primary industry	16.9	0.7	17.7	0.7	1.0	5.9
Manufacturing, utilities	197.4	8.4	185.7	7.4	−11.7	−5.9

Source: Statistics Canada, Census of Canada, 2001 and 2006. Place of work data, special tabulations (excluding those with no usual place of work).

Notes: The downtown is defined as the central area, stretching from the Don River to Bathurst and from the lake to the CPR tracks.
* Includes government service.

especially the case in the central area and waterfront districts. Unlike many older US cities, the population of the central city as a whole has continued to grow, albeit slowly. Despite a process of demographic thinning and an aging population that has reduced population numbers in most older and established neighbourhoods, the city's population has continued to grow and passed 2.6 million in 2006. The city's current Official Plan in fact calls for the addition of a further 500,000 people to the city's population over the next three decades, largely through residential intensification and infill. This is a very optimistic objective given economic circumstances, the region's changing demographic structure (e.g., an aging population), and the cumbersome and time-consuming process of redeveloping existing urban sites.

There are several nodes of very rapid population growth—for example, through planned intensification around subway stations (e.g., North York City Centre), in crowded suburban immigrant reception areas (e.g., Thorncliffe Park), and in areas of brownfield redevelopment along the former railway and port lands and within the downtown core itself (City of Toronto 2007b). The population of the downtown core passed 175,000 in 2006 and based on the number of new condominium units in the planning pipeline, could reach 250,000 within 10 years. Downtown living has become fashionable, especially for smaller, younger, and largely professional households. This process is, for the most part, an urban success story; it adds people and services, expands the tax base, improves local amenities, and enhances street life, usually without substantial displacement of existing populations. It also both mirrors and supports the retention of firms and the generation of new employment in the core and in districts surrounding the core.

Inner City Revitalization: The Case of Liberty Village

Liberty Village is one highly visible example of the kind of specialized inner-city district that has led the revitalization of areas on the periphery of the downtown core. Liberty Village is a 45-acre previously derelict district situated west of Toronto's downtown. The area was once home to a number of manufacturing plants assembling an eclectic mix of consumer goods, such as toys, electrical appliances, agricultural implements, carpets, furniture, and food. By the 1970s, however, the area witnessed the widespread closure of many industrial plants. In part this was related to deindustrialization and overseas competition and in part to technological obsolescence and rising costs. Some plants simply relocated to the suburbs to access cheaper land and improved transportation and logistics support, while others moved farther afield.

By the mid-1990s, however, the area started to go through a period of rapid transformation. It was close to downtown, was accessible by transit, and offered relatively low rents and other amenities. In effect, the area was rediscovered, initially as artists moved in and renovated older factories and warehouses, followed by more risk-averse developers and investors. The Village now contains a diverse array of 'creative' businesses, both old and new, including graphic, industrial, interior, and fashion design firms, as well as architects, artists, advertising agencies, music, film, and television companies, and new media start-ups. By 2008, there were more than 500 businesses in Liberty Village employing approximately 5000 workers. The area has been one of the fastest-growing employment districts in Toronto, with the number of employees growing by more than 30 per cent between 1995 and 2005 (City of Toronto 2006b). Over the same period, the surrounding area has witnessed an explosion of population growth through condominium construction and conversion, especially since 2001.

One of the intriguing features of new creative districts such as Liberty Village is the high level of inter-firm collaboration and networking (Catungal and Leslie 2009).

Employers are attracted to the diversity of industries in the area, which in turn provides possibilities for doing business with related firms. The variety of firms also creates the potential for cultural experimentation and for the diffusion of knowledge and best practices from one industry to another. The City of Toronto is currently investigating the feasibility of promoting similar 'creative hubs' elsewhere in the city. The challenge, in the case of Liberty Village and for other similar districts of new employment growth in the inner city, is in large part a consequence of their success: can the growth in jobs continue in the face of a tidal wave of new residential construction on its periphery and rising land values within the district? Already, a number of creative firms (particularly artists and photographers) and support services (such as arts organizations) have been displaced because of rising rents (Catungal and Leslie 2009).

Will residential uses and larger corporate creative industries outbid small creative firms for land and location, as has happened on portions of Queen West? The city has a clearly stated policy of protecting employment lands from residential (and commercial) invasions in places such as Liberty Village. This policy, although widely contested, offers some protection to Liberty Village at least in the short term, but the surrounding area has witnessed a displacement of employment and industrial uses, as well as low-income residents. A key problem is the fragmented approach to planning adopted by the city, which tends to focus economic development on small neighbourhoods and 'urban villages' rather than on larger scales.

Growing Socio-economic Polarization

One of the inevitable consequences of widespread economic restructuring, technological change, and global competition, combined with the ripple effects of demographic change, residential redevelopment, and gentrification, is heightened income polarization (Price and Benton-Short 2007; OECD 2009). While still low by international standards, numerous studies have shown that the Toronto region has been subject to increasing socio-economic polarization (ICF Consulting 2000; Bourne 2000; United Way 2004; Hulchanski 2007; Walks and Maaranen 2008). The process of social polarization also has a geographic component, as is evident in increasing inequalities among neighbourhoods within the city itself and between different parts of the region (Chapter 6 in this volume). The City of Toronto, for example, has concentrations of very low- and very high-income households, while the rest of the GTA has relative concentrations of middle- and high-income households, yet there are also emerging areas of relatively low and declining income even in the newer outer suburbs.

It is, however, the city's older inner suburbs of Etobicoke, North York, and Scarborough that have experienced the greatest decline in income. Many of the most disadvantaged populations now live in suburban neighbourhoods with concentrations of social housing and low-rent housing built in the 1960s and 1970s. In effect, poverty in the region was suburbanized and deconcentrated during this period—unlike in many US cities— through explicit public policy decisions on the location of public housing. Not surprisingly, suburban neighbourhoods characterized by falling relative incomes also tend to be areas of high unemployment and increasingly the home of many recent immigrants. Such areas are now the focus of the city's 'priority' neighbourhoods strategy, which targets public resources to those areas most in need. While the economic consequences of increased social and spatial polarization are difficult to measure, and many are likely to be felt only in the long term, most analysts agree that it discourages economic growth and inhibits social cohesion.

Summary

In sum, it is clear that the economy of the Toronto region has been undergoing a number of critical transformations in recent years, including a gradual shift toward service employment. In this regard, the region is not unique. Services of all kinds—including business and consumer services as well as the public sector—now constitute 70 per cent of employment in the region and generate more than 97 per cent of all employment growth (Simmons, Bourne, and Kamikihara 2009). The service sector, of course, is strongly bifurcated, in Toronto and nationally, providing both relatively high-income and markedly low-income occupations and employment opportunities. The shift to services then adds to the increasing level of income and employment inequalities.

Manufacturing employment in Toronto, in contrast, has been in relative decline. This follows a general trend across North America over the past several decades, beginning with the crisis of Fordist manufacturing systems and a consequent process of deindustrialization in the 1970s and continuing into the 1990s, when the Canada–US Free Trade Agreement and NAFTA were signed. Trade liberalization had an immediate and largely negative impact on many manufacturing sectors in Toronto, particularly on mature industries such as furniture and apparel, resulting in a substantial reduction in the number of firms and employees. While some industries have rebounded, North American cities like Toronto now find themselves at a competitive disadvantage in traditional mass production manufacturing, since operating and labour costs are increasingly expensive by global standards. However, a recent international assessment of competitive alternatives (KPMG 2010) puts Toronto in a strong overall position taking all locational costs into account, although the situation varies according to industry and is sensitive to exchange rate fluctuations.

Despite these challenges, manufacturing remains an important sector in the region, especially as a source of tradable (or export) products that bring income into the regional economy. Moreover, many services perform work directly related to the industrial economy (such as designing and advertising industrial products), and manufacturing remains important in terms of its contribution to regional GDP. Firms that have survived, however, have had to increase their use of skilled labour, pursuing a higher value-added and design-intensive strategy (Gertler 2003; Leslie and Reimer 2006). Successful firms have also had to become more flexible in terms of their use of technology, labour, and resources and by increasing their focus on export markets.[12] Nevertheless, in the opinion of the OECD (2009), negative aspects of structural economic change in Toronto are not being adequately offset by industrial innovation and increased productivity, and it suggests that weak private financing for innovation is part of the problem. This is reiterated by a recent Toronto Board of Trade study (2010).

The region is also increasingly characterized by a split locational pattern of employment, whereby the City of Toronto itself is focused on high-order knowledge-intensive businesses, professional and financial services, and cultural industries, including tourism and recreation. Post-secondary institutions and other research facilities are also concentrated in the city. In contrast, the outer areas of the GTA+H are characterized by a mix of high- and low-order services (including back office functions such as call centres), distribution, and warehousing and transportation, as well as manufacturing. These activities usually require, or seek, relatively inexpensive land and quick access to major highways and international airports.

The Organization of the Regional Economy: Clusters in the Toronto Economy

Any assessment of the Toronto region's economic performance and its possible future must go much deeper than mere recognition of the strong position of both secondary manufacturing and the service sector. Within both of these sectors, the Toronto region's real strength lies in the specializations that enjoy competitive advantages in the region and that have achieved substantial employment scale. These clusters of economic activity have grown in importance through the direct and indirect interdependence of firms within identifiable parts of the industrial spectrum, and they rely on the depth of skills and knowledge in the labour market and build on the platform provided by the educational and research infrastructure of the region. They also drive the region's inter-regional and international exports, since they are nodes in national and global industrial systems and supply chains, which generate transnational flows of components, producer services, and final products. In this section, we focus on a selection of key clusters that give distinctiveness to the Toronto economy. Here we draw heavily on the results of parallel research undertaken through organizations such as the Institute for Competitiveness and Prosperity (Milway et al. 2007), the Martin Prosperity Institute (MPI 2009), the Innovation Systems Research Network (ISRN) (Spencer at al. 2010; Wolfe 2009), the Toronto City Summit Alliance (2003; 2010), the Toronto Region Research Alliance, and the City of Toronto, among many others.

Clusters are vitally important to the economic health of the region. One illustration is that the proportion of employment in clusters accounts for a large portion of the variance (R-squared 0.45) in average income levels among Canadian cities (Spencer et al. 2010). Toronto has a distinct position in this system because it has the highest share (46.5 per cent) of employment contributed by clusters, confirming the economic advantages of both clustering and diversification. Although Toronto's clusters are substantial, their path dependence on earlier periods of industrial development is clear, as is the influence of supportive public investments (e.g., in higher education, hospitals) and the economic responses of corporations to political events and policy decisions (e.g., trade policy and innovation funding).

Various reports have listed as many as 12 prominent clusters of economic activity in the region.[13] There is, however, a wide variation among researchers in the actual specializations that have been identified. Although we have relied on the most recent and systematic Canada-wide analysis of metropolitan specializations in constructing Table 10.4, it is important to note that this summary excludes activities that others may deem important for the regional economy.[14] We review four clusters here in detail

(Table 10.4). The selection is intended to represent the intersection of two facets of economic life: whether activities are manufacturing or service-based and whether they are recognizably newer clusters based on the knowledge-intensive capabilities of the regional labour force or are activities with roots in the old economy but with continuing importance in the employment structure of the Toronto region.

Superficially, Table 10.4 gives the impression that only a modest share of the Toronto economy reflects the development of 'new' activities that have developed to the point of recognizable specializations and suggests that the region is heavily reliant on activities that have existed in varying forms for many decades. In reality, however, our dichotomies are only a starting point, and their use here points to the way that service and manufacturing activities are functionally closely intertwined. Even clusters with origins in the old economy need to be understood as major, innovative users of new technologies. The latter is true of the creative and cultural cluster whose origins are in conventional theatre production and the early years of motion picture production but whose current activities are recognized because of the penetration of information technology in animation, post-production activity, interactive gaming, and other new media entertainment products for the Internet.

The four clusters we examine in detail here are: 1) automobile production, an industry with roots in the region going back to the early twentieth century but which has continually succeeded in incorporating new technologies and is now restructuring once again; 2) financial services, which, although they predate the rise of Toronto's industrial economy, are now the epitome of information-intensive activities and are highly dependent on communication and information systems; 3) ICT, the archetype of interdependent technology-intensive manufacturing, software design, and service provision; and 4) the life sciences cluster, including biotechnology and biomedical production, which also relies on a highly research-intensive base and on the synergy between service and manufacturing companies in the health science fields; in this cluster, the private sector, university and hospital research teams, and public funding exhibit their closest relationships.

Toronto's Auto Sector

Automotive is the largest component of the durable goods manufacturing sector in the Toronto region, as well as in southern Ontario, and an industry that is especially vulnerable to intense international competition and technological change (DesRosiers 2007). Nonetheless, KPMG (2010) indicates that Toronto is cost-competitive with most US cities, although significantly lower costs in Mexico mean that auto parts firms in the Toronto region need to build their future on innovation and a highly skilled labour force.

The importance of this industry makes the Toronto and Ontario economies unique among Canadian cities and provinces.[15] Since at least the 1960s, the industry has also been an integral part of the North American industry, and its supply chains and markets are similarly continental. Despite its roots in the 'old economy', over the last four decades the auto industry has experienced phases of restructuring and modernization that have allowed it to remain a large, though shrinking, employer of labour in the Toronto region (Fitzgibbon et al. 2004). Jobs in the industry are localized in the outer suburban parts of the region and in nearby cities.

Table 10.4 The Toronto Economy: Cluster Characteristics

Cluster (n = 12)	Labour Force (in 000s)	LQ	% Change 2001–6
Old Economy Clusters			
Manufacturing:			
Textiles & apparel	28.5	1.28	−18.0
Food	86.8	1.12	13.9
Automotive	107.3	1.47	4.3
Plastics & rubber	101.4	1.55	−6.2
Services:			
Finance	277.2	1.61	8.1
Business services	354.6	1.33	11.8
Higher education	100.6	1.11	23.8
Logistics	204.9	1.52	9.2
New Economy Clusters			
Manufacturing:			
Biomedical	41.7	1.68	17.2
ICT	58.7	1.76	−14.2
Services:			
ICT	181.4	1.38	4.6
Creative & cultural	159.2	1.52	10.3

Source: Modified from Spencer et al. 2010.

LQ = location quotient (relative to the national economy).

Notes:
1. Clusters in bold are discussed in detail in the text.
2. Labour force figures above refer only to employment in the clusters, not to total employment as in Table 10.2.

The industry is a composite of auto assembly plants and auto parts suppliers.[16] Four major assembly plants are located in the greater Toronto region: Ford (Oakville), Chrysler (Brampton), GM (Oshawa), and Honda (Alliston). Three other assembly plants are located just to the west of the region in Ingersoll, Cambridge, and Woodstock (Toyota). Most parts manufacturers are similarly concentrated. For the Toronto CMA, total employment in auto manufacturing grew during the 1990s, peaked in 2005, and then declined to 59,000 in 2008.[17]

These declines reflect, as noted earlier, the severity of the current economic recession and mirror the reduced demand for vehicles produced by the North American Big Three auto manufacturers in the face of competition from Asian and European producers. Fortunately for the regional economy, both Honda and Toyota have retained (more or less) their levels of production and employment in southern Ontario. The current recession has clearly exposed the level of inefficiency and overcapacity in the industry in the Toronto region. Permanent reductions in production employment are expected, especially in the outer suburbs and smaller cities in the region, as are negative impacts on sales and related services. Widespread second-order multiplier effects are also expected throughout the region because of the vast input and output connections of the industry (MPI 2009).

The Financial Services Cluster

Recent trends toward deregulation and consolidation have strengthened this sector. Toronto now ranks as the third largest centre of finance in North America after New York and Chicago and twelfth among 46 international financial centres (Milway et al. 2007). Symbolically, and in terms of the quality of jobs generated, this position indexes Toronto's international functions and its stature as an aspiring 'global city'. Its position may have improved, owing to the prudent management of the banking industry, and its specializations reflect its centrality to the Canadian economy. As the prime financial centre in Canada, Toronto houses 29 per cent of Canada's total workers in financial services (LQ = 2.3) and is the head office location for many major financial institutions.[18]

After a period of rapid growth, Toronto's financial services cluster reached employment of 220,000 in 2007 (TFSA 2007). This is a knowledge-intensive sector, and firms have gained considerably from the relatively high quality of human capital and infrastructure available in the region. More than 50 per cent of the labour force in financial services have a university degree, compared with less than 31 per cent in all industries in the region, and another 25 per cent have other post-secondary qualifications.[19] Nevertheless, considerations of cost and access to qualified personnel are driving some firms, starting with IT services, to outsource functions overseas.

Illustrating the depth of interactions within and between clusters, financial services are a significant source of demand for a wide variety of ICT inputs (hardware, software, and service), and it has stimulated a large wireless (WiFi) service zone in the central area of Toronto. Moreover, Toronto-based banks and other financial service firms lead internationally in using ICT. The computer and communication services industry, another regional strength (LQ = 2.8), supplies computer services to financial services firms. The latter are also a substantial source of demand for many other business service inputs, including advertising and new media and legal, accounting, and educational and training services.

Information and Communication Technology (ICT)

ICT embraces the technology-intensive activities that continue to develop the means for manipulating, storing, and redistributing digital information. Canada's ICT sector was ranked sixth in competitiveness, reflecting its IT infrastructure (broadband access) and relatively high level of use by consumers, but was comparatively weak in business environment and R&D intensity. At the urban scale, Toronto's ICT cluster ranked eighth in North America in terms of employment but fifth in the degree of localization (City of Toronto 2007a). Canadian ICT is heavily concentrated in the

Toronto region—about 40 per cent of all electronics firms—while the remainder are distributed among smaller clusters in Ottawa, Montreal, Vancouver, and Calgary. ICT is distinctive among Toronto's clusters, since parts of it are in manufacturing (e.g., hardware production) and parts are in services, including software development and Internet services. By 2006, after recovering from the high-tech recession of the early 2000s, the Toronto region had more than 240,000 jobs in ICT.

Toronto first attracted multinational firms in ICT hardware, then subsequently in software and services. While some of these firms have established global R&D facilities or world product mandates in Canada, most multinationals focus on the domestic market, adapting their products to do so. By contrast, domestic firms, especially the larger ones, increasingly focus on external markets, although manufacturing firms in Toronto tend to focus on near-market innovation, which requires only modest levels of R&D (Britton 2004).

Despite the recent collapse of Nortel, Toronto's ICT prospects appear to be positive. The cluster has relative advantages of scale, cost, and talent in electronics as compared with many peer cities in Canada and in North America and Europe (KPMG 2010). It also prospers in digital media, which respond to both business and entertainment markets, because of relatively low costs combined with high levels of technical and media talent.

Life Sciences Cluster

Another leading-edge cluster in the Toronto economy, and a potential building block of the new economy, is the life sciences cluster. Like ICT, the life sciences industries are innovation-driven sources of employment growth (Gertler and Vinodrai 2009). One of the most recently developed components, biotechnology, is an area in which Canada ranks among the world leaders, although its R&D intensity is much lower than that of its comparators. Internationally, Toronto is recognized as having strong comparative cost advantages in specific fields such as biotechnology and clinical trials (KPMG 2010). Toronto leads the country in life science employment (51 per cent), although in Montreal (43 per cent) biotechnology appears better developed. Toronto has attracted a wide diversity of life science firms; employment of more than 43,000 is divided between medical equipment and assistive technologies (MAT), pharmaceuticals, and biotechnology.

The origins of this cluster lie in the history of life science innovations associated with the city's hospitals and the University of Toronto, which has spawned a strong concentration of medical research facilities, the fourth largest in North America. This development offers a clear illustration of the importance of co-ordinated public and private actions. It is a long-run outcome of intersecting research, teaching, clinical practice and manufacturing, and the professional and scientific services.[20] Recently, the Toronto Academic Health Science Network has acted as a framework for this cluster.

The latest addition to this cluster is the Medical and Related Sciences (MaRS) Discovery District, located in close proximity to the university's Faculty of Medicine and the city's largest hospitals. Its new research facilities have attracted international scholars and helped to formalize a variety of private, public, and academic connections in new ways. It acts as a focal point for firms and research initiatives from related industries, serves as an incubator, offers advisory services, and acts as an access point for start-up firms. The federal and provincial governments have funded MaRS in conjunction with private sector investments and the universities and hospitals in Toronto. How well this innovative model will work, especially during a period of investment restraint and service cutbacks, has yet to be determined.

Redesigning and Renewing the Toronto Region: Challenges, Policy Initiatives, and Conjectures on the Future

The Toronto region faces many of the same challenges that other large city-regions do. Most of these challenges are also now well known and need not be repeated here (Slack, Bourne, and Gertler 2003; Clark 2007; TD Economics 2007; OECD 2009; TCF 2009). Yet there are location-specific issues, or local manifestations of common issues, that reflect the specific history and current circumstances of the region. In this concluding section, we focus on the particular challenges facing the region in three broad areas— unemployment and industrial restructuring; culture, social polarization, immigration, and labour markets; the evolving form of urban development, congestion, and infrastructure provision—and on selected recent policy initiatives by different levels of government in different sectors.

Challenges and Opportunities

At no time in recent history have the challenges facing the region been more apparent than during current economic circumstances. There is both good news and bad, offering both challenges and opportunities. On the one hand, Canada's national banks, which cluster in Toronto, have been relatively stable and commercially successful compared to their international counterparts. On the other hand, as illustrated above, the global economic recession has triggered thousands of job losses in manufacturing, construction, and related sectors. Many of these jobs are unlikely to be replaced by similar jobs in other industries. There are also social and gender implications to these labour market shifts: these losses are typically in blue-collar jobs held predominantly by males yet providing incomes that are distinctly middle-class. Overall, the Ontario and Toronto regional economies have been especially hard hit because of their reliance on automobile assembly and parts.

With an unemployment rate for the region exceeding the national average (9.7 per cent versus 8.9 per cent, respectively, and more than 11.5 per cent in the city), there is widespread recognition that the regional economy will only resume its growth if there is a substantial change in the flow of new employment opportunities. Toronto's leading sectors and the industrial clusters described above ought to be the primary sources of robust adaptations to rapidly evolving markets. But will they be? In international terms, the financial services cluster has inherited the benefits of the prudent regulation of the country's banks, but augmenting its strength in other areas of finance could also be targets for strategic public policy. Any attempt to reduce the lag in the supply of venture capital, for example, would have major impacts on the region's emergent innovative firms (Creutzberg 2006; OECD 2009).

Some commentators have speculated that Toronto may be able to leverage its reputation for economic stability to establish itself as a major financial centre or global risk regulator. It is far more likely, however, that Toronto could become the location for a Canadian securities regulator, which would provide greater international visibility (GTR Economic Summit 2009). One problem in this regard is political: adding to the challenges is the commonly held local view that the federal government cannot be seen to favour Toronto as the site for strategic new investments, even if that location is the best choice in economic terms.

The most immediate sectoral issue flowing from the latest restructuring of the Toronto regional economy involves the auto cluster. On the positive side, the industry is large enough to command the financial attention (and assistance) of governments, and productivity innovations have allowed the industry to maintain a reasonably competitive position despite overcapacity in the industry. On the other hand, these innovations have limited Canadian origins, and competition from other countries as well as the locational drift of the industry will undoubtedly affect the industry negatively. The financial backing made available for the latest restructuring of the auto industry is only a short- to medium-term strategy, while the lack of interest and funding for firms undergoing comparably difficult adaptations in other manufacturing industries reflects the absence of systematic co-ordination between the different levels of government. As one indication of the severity of the economic downturn in manufacturing and associated industries, the federal government recently (2009) announced the establishment of a new regional development agency for all of southern Ontario, the country's traditional industrial heartland.[21] The symbolism of this new 'have-not' status is poignant.

A related critical challenge is attracting, developing, and retaining talent in an era of heightened interurban competition. A number of authors have written about the crucial role of the diversity and quality of place (i.e., amenity) characteristics of an urban region in attracting and retaining highly qualified workers (Florida 2002; Glaeser 2005; Clark et al. 2002). Others question the direction of causality in these arguments, suggesting that people follow jobs rather than amenities (Storper and Scott 2009). For these critics, production activities, employment opportunities, and supporting institutions are keys to attracting talent. There is, however, a growing consensus that both employment and amenities matter and that they may be mutually constitutive.

As noted earlier, Toronto ranks relatively high in surveys designed to measure the city's quality of life and economic opportunities. It is also one of the most ethnically diverse cities in the world, ranking very high in terms of the proportion of the population that is foreign-born—more than 50 per cent in the city and 47 per cent in the region as a whole in 2006. This diversity is both a challenge and an opportunity. In difficult economic circumstances, however, the region could experience a declining quality of life, a reduced rate of immigration, and perhaps a 'brain drain' as talented workers leave to pursue job opportunities and higher wages (or lower living costs) available elsewhere. This possibility points to the need to enhance the attractiveness of Toronto's knowledge-based and cultural clusters and to improve the quality of amenities and living spaces of the city and the region (Gertler 2003; Milway et al. 2007; Sewell 2009; OECD 2009).

Although still relatively low by international standards, especially American standards, the growing trend toward social, economic, and geographic polarization in the region threatens to undermine the region's social cohesion and quality of life (United Way 2004). There is a uniquely fine-grained diversity in Toronto's income distribution, overlain by a tapestry of ethno-cultural enclaves, but the overall picture is one of deepening polarization, as documented in Chapter 6 (and in Hulchanski 2007). This draws attention explicitly to the need to focus on ensuring that all groups are included within the new knowledge/creative economy. As argued earlier, the immigration issue is crucial: it is a prime factor in the recent increase in income inequalities, and it is an increasingly important component in maintaining the growth of Toronto's labour force. While the federal government sets international immigration levels, the costs are borne largely by the provinces and local municipalities. Ontario has a relatively low level of federal financial support for settlement assistance when compared to other provinces, despite recent improvements in assistance levels.

The Toronto region absorbs the overwhelming majority of immigrants to Ontario, and this poses serious challenges in terms of providing shelter and jobs and other essential services, particularly because the city is already confronting a lack of affordable housing. As in other parts of the country, many recent immigrants find themselves essentially 'deskilled'—working in jobs well below their qualifications (Bauder 2003; United Way 2004). If the region is to take advantage of the immense stock of human capital added through immigration, and retain existing immigrants, policies need to ensure that newcomers face fewer barriers to entry in regulated professions, that they have access to appropriate employment and skills training, and that they are incorporated into the region's business networks as well as into its political power structures (see TCF 2009; MPI 2009).

Recent Policy Initiatives

The response of all three levels of government to these issues, and to the combined pressures of globalization, technological change, and economic restructuring, has with a few exceptions been hesitant, uneven, and disjointed. Certainly, policy reports and discussion papers abound; indeed, the production of strategy plans has itself become a growth industry. The challenge is implementation. To further complicate the policy landscape, governments at all levels have instituted a range of sector-specific strategies to encourage employment growth, innovation, inward investment, and skills training within recognized clusters of activity, some of which have in fact received international awards. Here we examine only a sample of recent initiatives that follow directly from the themes raised in our earlier discussion.

An Integrated Economic Region?

In broad terms, it is reasonable to assert that there is no overall economic plan or strategy for the entire Toronto region, either for the GTA or for the much larger Greater Golden Horseshoe. Most of the 30 municipalities in the GTA region, as well as adjacent Hamilton, have their own economic development offices, and some, such as Barrie, Mississauga, and Markham, even have an explicit economic strategy (see Markham

2008). In fact, the principal economic strategy of most local governments, here as elsewhere, is to pursue whatever it takes to increase their tax revenues and employment base. Until the tax system is modified and the dependence of local governments on the property tax is reduced, this situation is unlikely to change. There is therefore a clear need for a rethinking of municipal finance and for identifying new modes of local government financing in Canada and in Toronto specifically (Slack and Bird 2004). While the new City of Toronto Act (2009) provides the city with marginally more fiscal flexibility and access to more funding sources, it does not address the fundamental issues of an over-reliance on the property tax and rising debt.

The recent GTR summit report, *Choosing our Future: An Action Plan for Economic Recovery* (GTR Economic Summit 2009) represents an important step—or call to action—at the regional scale, and it offers some suggestions, albeit vague, on possible initiatives and priorities. Specifically, it stresses the importance of region-wide thinking and intergovernmental policy co-ordination. The report makes the case that there is a need for one voice for the region instead of a multiplicity of agencies, offices, and groups representing industries, sectors, and associations of municipalities but not the regional economy as such. The recent OECD report on the regional economy and on governance issues makes essentially the same point (OECD 2009). The question of who would provide leadership in meeting these challenges, and who would offer the 'regional voice', remains unclear.

City Planning and Policy

Perhaps the most explicit policy initiatives in economic development have come from the City.[22] This is also not surprising, given its strategic position in and dominance of the regional economy and the continuing pressures on the City's economic base and the vitality of its revenue/cost balance sheet. These initiatives have encompassed almost every dimension of urban policy, from economic development and welfare provision, to neighbourhood stability and affordable housing, to green space preservation and transportation planning. The latter includes the recently released and very ambitious *Transit City* plan (2009). This plan proposes billions of dollars in expenditures over 25 years to expand and upgrade the city's subway, rapid transit, and streetcar lines and to renovate Union Station, with substantial support 'promised' from both federal and provincial governments. These plans, along with proposals for major improvements in waste, water, and energy systems, suggests that infrastructure investment will be a significant element of local economic development strategies in the future. Nevertheless, under current fiscal arrangements, infrastructure expenditures require the massive participation of the province. Provincial funding, however, despite commitments made for hosting the PanAm Games (in 2015), may be delayed or moved to an indefinite timeframe.

Looking back, however, it might be argued that a key policy initiative in reshaping the geography of development in older parts of the city was the initial *Central Area Plan* (1974) and its associated measures. These measures provided the flexibility in zoning that allowed for—and indeed encouraged—massive private residential and commercial investment and renewed population growth in the central area, as previously described. Another watershed was the further modification of typically single-use exclusionary zoning that permitted the transformation of industrial/warehousing space into residential lofts, design studios, and various entertainment sites.

It is possible to argue, although not without debate, that the city's political capital has been primarily expended on controlling footloose redevelopment and protecting neighbourhoods and rental housing. That local politicians see the city as an amalgam of neighbourhoods is perhaps not surprising, given that all councillors represent small wards and thus their priority is to protect their turf and their voters from change. It is also not surprising that neighbourhoods dominate the agenda in the city where Jane Jacobs spent so many productive years. The reduced emphasis on comprehensive planning models may also be seen as a manifestation of the shift toward neo-liberal models of urban governance, which privilege new scales of intervention such as the neighbourhood or urban 'village'. This very local focus invariably takes energy and resources away from city-wide initiatives in economic development and actively discourages civic entrepreneurship and inclusion.

One specific strand of innovative policy recognizes the crucial role of cultural industries and facilities in charting the city's future economic development. In 2003, the city released its *Culture Plan for the Creative City*, a long-term strategy designed to position Toronto as a leading international centre of culture and creativity (e.g., City of Toronto 2006a). This agenda was further extended in 2008, when 'Creative Toronto' was identified as one of four pillars in the mayor's *Agenda for Prosperity* (City of Toronto 2008b). A key priority in this agenda, as cited in the discussion of the Liberty Village example above, is to establish and strengthen creative districts throughout the city, again singling out particular neighbourhoods as worthy of economic development attention.

Over the same decade, there has clearly been a significant cultural renaissance—evident in massive investments by all three levels of government, as well as the private sector, in the expansion and refurbishment of the region's cultural icons, including institutions such as the Art Gallery of Ontario (AGO) and the Royal Ontario Museum (ROM). Aside from the immediate and direct impact of these investments on the economy, they also serve as an explicit recognition of the importance of culture and tourism, and growth of the public sector generally, as 'drivers' of the regional economy.[23] Since most of these renovations are now complete, however, it would be pure speculation to ask what (if any) investments will come next and whether financial limitations and mounting public deficits will allow for similar restorations of other community assets.

Regional Planning and Policy

In parallel, but at the broader regional scale, are the provincial government's most recent attempts to redefine and manage the form and growth dynamics of the entire region. These ambitious initiatives in growth management include the *Places to Grow* plan (Ontario 2006), intended to increase densities through intensification, to encourage nodal development and mixed uses, and to co-ordinate land-use, employment, and transportation decisions; the *Greenbelt Plan* (Ontario 2005), designed to limit urban sprawl and to protect rural and environmentally sensitive landscapes; and most recently the proposal by the new provincial Metrolinx agency, titled *The Big Move* (Metrolinx 2008), to develop a long-term transportation plan for the city and the region with a clear emphasis on transit investment—in the TTC and the GO commuter system—and based on smart growth principles.

These initiatives were stimulated, at least in part, by political concerns over continued rapid growth in the region, environmental degradation, increased congestion—no doubt a restraint on economic development—and the massive costs of new infrastructure provision. The latter costs are brought into sharp focus given the 2 to 3 million new residents that Ontario's Growth Secretariat and Ministry of Finance anticipate will be added to the region's population by 2036 (Ontario Ministry of Finance 2009). Even if these growth forecasts turn out to be exaggerated, which now seems likely, the challenge of absorbing growth of anywhere near this scale is certainly daunting.[24]

These are, on balance, impressive and laudable plans; indeed, they are perhaps the most expansive and far-reaching planning and transportation proposals for a region of this size anywhere in North America in recent years. In the Toronto case, they are arguably the first serious efforts at regional planning since the 1970s (Frisken 2007). They are not, to be clear, economic plans or strategies per se, yet they will obviously have substantial economic implications well beyond their first-order effect in directing future infrastructure investment decisions throughout the region. The underlying objectives of all three plans are certainly ambitious and more or less conventional: to increase the efficiency of urban growth; to encourage more compact urban forms and more locally integrated labour markets; to reduce commuting times and congestion costs; to reduce pollution, energy use, and environmental stress; and to encourage the clustering of new employment in areas with transit service.

These objectives are now part of the standard planning model for smart growth and sustainable urban development strategies that are being introduced almost everywhere. But are they realistic, in general and in this region? Although it is far too early to assess the impacts of these growth plans, there is concern that implementation will be slow and piecemeal. It is also not clear, for example, how or to what extent firms, local governments, and individuals will adhere to these new directives. To what extent can local governments dictate where employment growth will locate and where it cannot? Can governments choose between good jobs and bad jobs? And, of course, as noted earlier, success of any kind in this planning process depends on the consistent financial support of the provincial government and that government's political commitment to tighter regulation of the private land market.

It is not at all certain that sufficient financial resources will be there to implement the infrastructure proposals in the *Places to Grow* plan or the vast transportation improvements in the Metrolinx plan. Although all local and regional governments in the region are compelled to make their official plans compatible with the provincial growth plan and to protect employment lands for future use, beyond these legislative requirements there is considerable ambiguity and uncertainty. These concerns are not unique to the Toronto region, but the scale of the initiatives brings the questions into sharper focus.

Another related challenge for the region is also political: to reduce inter-municipal competition for development and for tax revenues and to improve intergovernmental communication. The Greater Toronto Region (GTR) Economic Summit report stressed the need to encourage more 'region-wide' policy thinking and to strengthen the 'sense of a region' among decision-makers—public, private, and civil society—and the general population. Part of this challenge is to improve policy co-ordination between the four levels of government in the region and among local municipalities—particularly between the core city and the rapidly expanding municipalities that surround it.[25] Can the municipal components of the GTA collaborate to compete globally and to address the social, economic, and development problems of the region?

Over the past several decades, the city has been declining in terms of its share of regional jobs, population, investment, and incomes. As has been argued, the economies of the city and the region are different and may well be becoming more so. And some suburban municipalities are following distinctly different development paths. In one sense, this diversity of economic activity is positive, and it reduces risk. Yet these municipalities are all intricately connected and mutually interdependent parts of the regional economy. They are, most obviously, integral parts of the same housing and labour markets and are dependent on the same transportation and infrastructure systems. It is therefore essential to forge a strong place-based identity and a broadly based, aggressive, and collaborative economic strategy for the region as a whole in order to attract and retain talent and investment and to stimulate employment opportunities appropriate for the global economic realities of the twenty-first century.

What the future holds for the economy of the Greater Toronto Area—a question raised in the introduction to the chapter—remains to be seen. Toronto's position as the dominant Canadian metropolis is unlikely to be threatened, but its pretensions to be a major player on the international stage are open to debate. Will the GTA become a major player in the global network of cities, and if so what sectors will drive this renewed growth? Will it become a centre recognized for its financial and risk management services, or as a newly reinvigorated manufacturing region, or as a knowledge-based economy combining IT, culture, education, research, and innovation? Or will it continue to live off revenues from the resource base, the staple products, of the rest of the country? Or will it be driven by some sector or industries that are not yet identified? Alternatively, the inevitable contraction of the auto industry, on which the regional economy heavily depends, will require substantial growth in other sectors to take up the slack. What happens if immigration levels decline, thus reducing population and labour force growth and the rate of expansion of consumer income and demand? The least plausible scenario is 'more of the same'.

References

Anisef, P., and M. Lanphier, eds. 2003. *The World in a City*. Toronto: University of Toronto Press.

Bauder, H. 2003. '"Brain abuse" or the devaluation of immigrant labour in Canada'. *Antipode* 35 (4): 699–717

Boudreau, J.-A., R. Keil, and D. Young. 2009. *Changing Toronto: Governing Urban Neoliberalism*. Toronto: University of Toronto Press.

Bourne, L.S. 2000. *People and Places: The Changing Social Character of the Toronto Region*. Toronto: Neptis Foundation.

———. 2001. 'Designing a metropolitan region: The lessons and lost opportunities of the Toronto experience'. In M. Freire and R. Stren, eds, *The Challenge of Urban Governance*, 27–46. Washington: World Bank.

Britton, J.N.H. 2004. 'High technology localization and extra-regional networks'. *Entrepreneurship and Regional Development* 16: 369–90.

———. 2007. 'Path dependency and cluster adaptation: A case study of Toronto's new media industry'. *International Journal of Entrepreneurship and Innovation Management* 7: 272–97.

Careless, M. 1981. *Toronto before 1918*. Toronto: Lorimer.

Catungal, J.P., and D. Leslie. 2009. 'Placing power in the creative city: Governmentalities and subjectivities in Liberty Village, Toronto.' *Environment and Planning A* 41 (11): 2576–94.

City of Toronto. 2006a. *Imagine Toronto: Strategies for a Creative City.* Toronto: City of Toronto.

———. 2006b. *Liberty Village Study Area.* Staff Report. Toronto: City of Toronto.

———. 2007a. *Benchmarking Toronto's Economic Performance.* Toronto: City of Toronto, Economic Development Office.

———. 2007b. *Living Downtown.* Research Bulletin. Toronto: City Planning.

———. 2008a. *Backgrounder. Release of 2006 Census Results: Labour Force, Education, Place of Work and Mode of Transportation.* Toronto: City of Toronto.

———. 2008b. *Agenda for Prosperity.* January. Toronto: Toronto Mayor's Economic Competitiveness Advisory Committee.

———. 2008c. *Creative City Planning Framework.* February. Toronto: City of Toronto.

———. 2009a. *Profile Toronto. Toronto Employment Survey 2008.* March. Toronto: City of Toronto, Economic Development Department.

———. 2009b. *Economic Indicators October 2009.* Toronto: City of Toronto, Economic Development Department.

Clark, Greg. 2007. *World Cities and Economic Development: Case Studies in Economic Strategies.* Toronto: City of Toronto.

Clark, T.N., R. Lloyd, K.K. Wong, and P. Jain. 2002, 'Amenities drive urban growth'. *Journal of Urban Affairs* 24: 493–515.

Creutzberg, T. 2006. *At a Crossroads: Strengthening the Toronto Region's Research and Innovation Economy.* Toronto: Toronto Region Research Alliance.

Deloitte. 2005. *City of Toronto: Economic Contribution of Toronto's Culture Sector.* April 2005.

DesRosiers, D. 2007. *Strategic Review of the Canadian Automobile Industry.* Toronto: Toronto Region Research Alliance.

DIAC (Design Industry Advisory Committee). 2004. *What Can 40,000 Designers Do for Ontario?* Executive Report. Toronto: DIAC.

Filion, P. 2007. *The Urban Growth Centres Strategy for the Greater Golden Horseshoe.* Research paper 2.9.3. Toronto: Neptis Foundation.

Fitzgibbon, S., J. Holmes, T. Rutherford, and P. Kumar. 2004. 'Shifting gears: Restructuring and innovation in the Ontario automotive parts industry'. In D. Wolfe and M. Lucas, eds, *Clusters in a Cold Climate*, 11–41. Montreal: McGill-Queen's University Press.

Florida, R. 2002. *The Rise of the Creative Class—and How It's Transforming Work, Leisure, Community and Everyday Life.* New York: Basic Books.

———. 2007. *Who's Your City.* Toronto: Random House.

Frisken, F. 2007. *The Public Metropolis: The Political Dynamics of the Toronto Region 1924–2003.* Toronto: Canadian Scholars Press.

Gertler, M. 2000. *The Regional Economy of Toronto.* Toronto: Neptis Foundation.

———. 2003. *Smart Growth and the Regional Economy.* Toronto: Neptis Foundation.

Gertler, M., and T. Vinodrai. 2009. 'Life sciences and regional innovation: One path or many? *European Planning Studies* 17: 235–61.

Glaeser, E. 2005. *Smart Growth: Education, Skilled Workers and the Future of Cold-Weather Cities.* Policy Brief PB-2005-1. Cambridge, MA: Harvard University Kennedy School.

Greater Toronto Region Marketing Alliance (GTMA). 2005. *Investing in the GTA: Automotive and Advanced Manufacturing.* Toronto: GTMA.

Greater Toronto Research Alliance (GTRA). 2005. *Toronto Region R&D—Quick Facts.* Toronto: GTRA.

GTR Economic Summit. 2009. *Choosing Our Future: An Action Plan for Economic Recovery.* Toronto: GTR Economic Summit.

Hemson Consulting. 2005. *The Growth Outlook for the Greater Golden Horseshoe.* Toronto: Hemson Consulting.

Hulchanski, D. 2007. *The Three Cities within Toronto: Income Polarization among Toronto's Neighbourhoods, 1970–2006.* Research Bulletin 41. Toronto: Centre for Urban and Community Studies, University of Toronto.

ICF Consulting. 2000. *Toronto Competes: An Assessment of Toronto's Global Economic Competitiveness.* Toronto: Toronto Economic Development.

Keil, R., and D. Young. 2008. 'Transportation: The bottleneck of regional competitiveness in Toronto'. *Environment and Planning C* 26: 728–51.

KPMG. 2010. 'KPMG's guide to international business location'. http://www.competitivealternatives.com.

Lemon, J. 1985. *Toronto since 1918: An Illustrated History.* Toronto: Lorimer.

Leslie, D., and S. Reimer. 2006. 'Situating design in the Canadian household furniture industry'. *The Canadian Geographer* 50 (3): 319–41.

Lu, V. 2009. 'Charting a route to recovery: In an unprecedented gathering, leaders pool ideas to steer region's economic engine out of recession'. *Toronto Star* 8 May.

Markham (Town of). 2008. *Markham's Economic Strategy 2009–2018: An Economic Development Blueprint for the Second Decade of the 21st Century.* Markham: Economic Development Department.

MPI (Martin Prosperity Institute). 2009. *Ontario in the Creative Age: Toward an Economic Blueprint.* Research Paper Series. Toronto: MPI, Rotman School of Management, University of Toronto.

Metrolinx. 2008. *The Big Move: Transforming Transportation in the Greater Toronto and Hamilton Area.* Toronto: Government of Ontario.

Milway, J., S. Nisar, C. Poole, and Y. Wang. 2007. *Assessing Toronto's Financial Services Cluster*. Toronto: Institute for Competitiveness and Prosperity.

Moore-Milroy, B. 2009. *Thinking Planning and Urbanism*. Vancouver: University of British Columbia Press.

Neptis Foundation. 2006. *A Commentary on the Province's Growth Plan for the Greater Golden Horseshoe*. Research paper 2.11.3. Toronto: Neptis Foundation.

OECD (Organisation for Economic Co-operation and Development). 2008. *Growing Inequality? Income Distribution and Poverty in OECD Countries*. Paris: OECD.

———. 2009. *Territorial Reviews, Toronto, Canada*. Paris: OECD.

Ontario, Government of. 2005. *Greenbelt Act*. Toronto: Queen's Printer.

Ontario. Growth Secretariat. 2006. *Places to Grow: Better Choices, Brighter Future: Growth Plan for the Greater Golden Horseshoe*. Toronto: Queen's Printer.

Ontario. Ministry of Finance. 2009. *Ontario's Population Projections 2008–2036*. Toronto: Queen's Printer.

Price, M., and L. Benton-Short. 2007. 'Immigrants and world cities: From the hyperdiverse to the bypassed'. *Geojournal* 68: 103–17.

Sewell, J. 2009. *The Shape of the Suburbs: Understanding Toronto's Sprawl*. Toronto: University of Toronto Press.

Simmons, J., L.S. Bourne, and S. Kamikihara. 2009. *The Changing Economy of Urban Neighbourhoods: An Exploration of Place of Work Data for the Greater Toronto Region*. Research Paper 219. Toronto: Cities Centre, University of Toronto.

Slack, E., and R. Bird. 2004. 'The fiscal sustainability of the Greater Toronto Area'. ITP Paper 0405. Toronto: Rotman School of Management, University of Toronto.

Slack, E., L.S. Bourne, and M. Gertler. 2003. *Vibrant Cities: Responding to Emerging Challenges*. Research Report 17. Toronto: Panel on the Role of Government in Ontario, Government of Ontario.

Spencer, G., T. Vinodrai, M. Gertler, and D. Wolfe. 2010. 'Do clusters make a difference? Defining and assessing their economic performance'. *Regional Studies* forthcoming.

Storper, M., and A.J. Scott. 2009. 'Rethinking human capital, creativity and urban growth'. *Journal of Economic Geography* 9: 147–67.

TCF (Toronto Community Foundation). 2009. *Toronto's Vital Signs 2009*. Toronto: TCF.

TD Economics. 2007. *The GTA: Canada's Primary Economic Locomotive in Need of Repair*. Toronto: TD Financial Group.

TFSA (Toronto Financial Services Alliance). 2007. *Talent Matters: Shaping Talent Strategies in a Changing World*. Toronto: TFSA Deloitte.

Toronto Board of Trade. 2010. *Towards a Global City: Scorecard on Prosperity*. Toronto: Toronto Board of Trade.

Toronto City Summit Alliance (TCSA). 2003. *Enough Talk: An Action Plan for the Toronto Region*. Toronto: TCSA.

———. 2010. *Time to Get Serious: Reliable Funding for GTHA Transit/Transportation*. Toronto: TCSA.

United Way. 2004. *Poverty by Postal Code*. Toronto: United Way.

Walks, A., and R. Maaranen. 2008. 'Gentrification, social mix and social polarization: Testing the linkages'. *Urban Geography* 29 (4): 293–326.

White, R. 2003. *Urban Infrastructure and Growth in the Toronto Region*. Toronto: Neptis Foundation.

Wolfe, D. 2009. *The Geography of Innovation: 21st Century Cities in Canada*. Toronto: Conference Board of Canada.

Notes

1. Toronto's contribution to total Canadian GDP is roughly equivalent to the combined contribution of New York, Chicago, Boston, and San Francisco to US GDP (City of Toronto 2008b).

2. The Economist Intelligent Unit, in a recent report, ranked the Toronto area fourth in terms of quality of life among world cities. Vancouver was first and Calgary sixth. In other studies, Toronto ranked #1 in a recent issue of *Foreign Direct Investment* for Best Quality of Life in North America and #2 in North America and #15 worldwide in a 2007 Mercer Human Resources Quality of Living survey (City of

Toronto 2008b). The World Intellectual Property Organization (WIPO) ranks Toronto seventeenth among global cities as a centre for innovation using patent data. Such rankings, of course, are open to debate and to frequent modification.

3. The TTC carries 1.6 million passengers per workday (475 million per year), the GO system 220,000 per workday. Suburban transit systems, although relatively small, add another 50 million per year.

4. The GGH is an exceptionally large region, covering 70,000 square kilometres, incorporating

several existing urban nodes and vast tracts of rural, recreational, and agricultural land. The GGH includes seven other metropolitan areas (CMAs) and more than 100 local municipalities.

5. The City of Toronto (like the adjacent and recently amalgamated City of Hamilton) is a single-tier municipal government that performs both local and regional functions, while the surrounding suburban regional governments are two-tier and contain 24 local municipalities that split responsibilities accordingly. To further illustrate the complexity, Mississauga is a lower-tier government despite its huge population size (710,000), which together with Brampton and Caledon report to upper-tier Peel region (population 1.1 million).

6. With 44 local councillors, one mayor, and no political parties, achieving a consensus on any major issue is difficult, and there is no one other than the mayor who speaks for the whole city.

7. The city's financial situation was made worse through the lingering effects of amalgamation in 1998 and, perhaps more importantly, by the simultaneous 'downloading' of several provincially mandated services (e.g., welfare, social housing, disability, and court costs) to local governments by the Ontario government of the day. Only in Ontario do local governments carry this burden. The current (2009) provincial government has agreed to 'upload' these costs by 2018, but in the meantime the negative impact on Toronto's finances has been substantial.

8. Employment locations are based on the 2006 Census place of work data.

9. Manufacturing is itself a complex sector, since its employment (370,000) is distributed unevenly over a number of distinctly different activities.

10. Location quotients (LQs) measure the relative concentration (in %) of employment in a city in a particular industry compared to the national average. Figures greater than 100 indicate that residents in a city are more likely to work in a specific industry than the Canadian average.

11. The city is the third largest centre of English-language theatre production in the world, following New York and London (City of Toronto 2006a). It is also the third largest centre of screen-based arts in North America (City of Toronto 2008c) and ranks third in North America, after New York and Boston, in terms of design employment (DIAC 2004). In addition, Toronto has grown faster than other North American centres, such as Montreal, Chicago,

San Francisco, Los Angeles, and New York, in terms of employment in creative industries (City of Toronto 2006a).

12. In 1981, the value of Ontario exports to the rest of Canada was equivalent to the value of its international exports. By the late 1990s, international exports were nearly three times those to the rest of the country. Most of those exports went to the United States.

13. Spencer et al. (2010) use labour force data by industry from the 2006 Census. We have modified this structure by incorporating information about the roots of Toronto's industrial structure.

14. Activities excluded by Spencer et al. (2010) are publishing and printing, aerospace, tourism, distribution services and transportation, which is sometimes combined with logistics.

15. The auto industry in southern Ontario has recently overtaken Michigan's as the leading North American regional producer.

16. Based on classifications NAICS 3361 and NAICS 3363, respectively.

17. In southern Ontario, the auto parts industry is the largest source of employment (87,000 jobs in 2008), while assembly plants add another 60,000.

18. The Toronto Stock Exchange, for example, is the third largest in North America and leads internationally in listings of mining and oil and gas firms. Toronto is also the location of the head offices of five Canadian-owned banks, six insurance companies, and the majority of foreign bank subsidiaries, as well as a high proportion of Canada's securities firms and mutual fund and pension fund management companies.

19. Although it is often stated that personnel in the Toronto region does not meet the university-level educational attainment of those in, for example, Boston, New York, San Francisco, and Chicago, in practice comparisons of educational systems and outcomes are difficult. At the same time, the relatively high proportion of workers with professional accreditations is a distinguishing feature of Toronto's labour force.

20. The association between Connaught Laboratories and the University of Toronto is an important part of this history of research and innovation.

21. The new agency, with an initial capital budget of $500 million and headquartered in Kitchener, is intended to stimulate innovation and investment in manufacturing and related industries.

22. The city's economic development arm, TEDCO, has recently been divided into two new agencies: Invest Toronto, which is intended to attract new investment, and Build Toronto, intended to market and develop its own real estate holdings.

23. Notable among these, in addition to the Art Gallery of Ontario and the Royal Ontario Museum, are upgrades to the National School of Ballet, the Royal Conservatory of Music, the Gardiner Museum, the new Opera House, and the (still under construction) TIFF Bell Centre.

24. The province of Ontario is currently (2010) undertaking a reassessment of the growth forecasts for the province and for the entire region (the GGH). See Hemson Consulting 2005 and the Ontario Ministry of Finance 2009. The original forecasts are likely to be revised downward.

25. In the Toronto case, the four levels are local, regional, provincial, and federal. One of the few examples of relatively effective cooperation between levels of government involves the creation and funding of Waterfront Toronto (2003), the agency responsible for redeveloping the waterfront. Nevertheless, intergovernmental bickering has continued to slow investment and to complicate longer-term planning, as it has done for decades.

11

'Heart of the New West'? Oil and Gas, Rapid Growth, and Consequences in Calgary

Byron Miller and Alan Smart

If Calgary is now the 'heart of the new west', as Calgary Economic Development's trademarked brand suggests, it is a heart that pumps oil. For the past half century, the story of Calgary's economic development has been one of transformation from a regional beef and agriculture centre to a major centre of the fossil-fuel industry, with global aspirations. That Calgary's contemporary economy is driven by the fossil-fuel industry is not in doubt. A 2006 study produced for the University of Calgary's Institute for Sustainable Energy, Environment and Economy estimates that by 2004 the oil and gas industry accounted for 53 per cent of Alberta's provincial GDP, counting both the direct and indirect (multiplier) effects of the industry (Mansell and Schlenker 2006, 29). Moreover, Mansell and Schlenker estimate that looking forward through 2013, the average annual contribution of the oil and gas industry to provincial GDP will be $87 billion per year, or 40 per cent of Alberta's GDP.

With Calgary as home to most of the head offices and many of the producer-service firms of the Canadian oil and gas industry, its fortunes have been tightly tied to this industry. In 1970, Calgary's population was just over 400,000. By 2009, the population of the Calgary census metropolitan area (CMA) was more than triple that—1,230,248 (Statistics Canada 2009)—making Calgary the fastest-growing of Canada's five largest urban economic regions. But Calgary's growth has been anything but steady. Given its resource-dependent nature, Calgary's economy exhibits classic boom and bust cycles (Markusen 1987) that track global changes in the price of oil. When global oil prices rise, the floodgates of capital investment in the oil patch open, job growth and in-migration accelerate, and wages, housing costs, and the cost of living generally take off. When global oil prices decline, Calgary's economy sputters. (Figure 11.1) Not surprisingly, the boom and bust cycles of the oil and gas industry govern the pace of Calgary's urban development and rural land conversion (Climenhaga 1997). Rapid shifts in the economy reverberate throughout the urban landscape (Ghitter and Smart 2009), with responses in real estate markets amplifying booms and reinforcing busts.

In this chapter, we consider the political economy of the processes shaping Calgary's growth and development, their relationship to increasingly scarce but variably priced fossil fuels, and the social transformation of the metropolitan region as waves of immigration and changes in economic structure shape the metropolitan area's ethnic and class composition as well as culture. The oil-fuelled boom and bust economy affects virtually every growth and development issue Calgary faces. During booms, labour is in short supply, wages rise—albeit unevenly—housing costs skyrocket, infrastructure

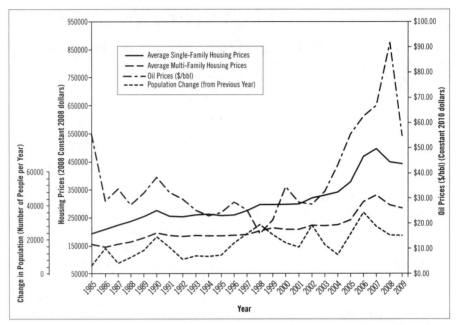

Figure 11.1 Relationship between crude oil price, population growth, and housing prices in Calgary.

becomes overburdened, and quality of life declines from the stresses of red-hot growth. During downturns, labour supply is less of an issue, but wages stagnate, revenues needed to provide basic infrastructure and services decline, and quality of life suffers not from the stresses of growth but from the state's seeming inability to carry out basic functions. Of course, the dynamics of the Calgary economy do not take place in a political vacuum. Calgary's growth and development both shapes and is shaped by political dynamics at the municipal and provincial scales, with significant federal entanglements added for good measure.

Oil and Gas, Labour, and the Calgary Economy

Calgary's economy and population grew rapidly from the early 1970s to the early 1980s as the 1973 OPEC oil embargo, the 1979 Iranian revolution, and the onset of the Iraq–Iran war in 1980 increased the average annual price for a barrel of oil almost five-fold, from $20.66 in 1972 to $98.52 in 1980 (all oil prices quoted in constant 2010 dollars).[1] During this period, Calgary saw a rash of nondescript skyscrapers built in its downtown to house the rapidly growing oil and gas industry. Mass produced suburbs grew rapidly too, typically at low automobile-dependent densities of 4.5 to 6.0 dwelling units (du) per acre (City of Calgary 2005). But even more quickly, the boom turned to bust. The recession of the early 1980s—widely attributed to rapidly rising global energy costs and contractionary monetary policies adopted by the US Federal Reserve Bank to squelch energy-driven inflation—led to a dramatic downturn in the price of oil. In addition, the

National Energy Program (NEP) set Canadian oil prices below those of the global market and raised taxes on the oil and gas industry, further dampening Alberta's growth. The drop in the price of oil was precipitous, hitting an average of $27.28 per barrel in 1988—only slightly above what it had been prior to the boom of the 1970s. Not surprisingly, Calgary's population growth was stagnant through most of the 1980s, plunging to less than 1 per cent in 1987 (City of Calgary 2008c). While the end of the NEP in 1986 provided some relief to Alberta's oil and gas industry, global oil prices fell lower still in the 1990s, bottoming out at $15.84 per barrel in 1998. By 1999, however, oil prices were on the rise again. In 2004, oil averaged $43.17 per barrel, and by 2008 it averaged $91.77—almost as high as the peak year of 1980. A flood of new investment in the tar sands was unleashed, and as one might expect, rapid population growth resumed (Simmons and Bourne 2007). Population growth rates for the Calgary CMA were 3.2, 3.2, 2.8, and 2.9 per cent in 2005, 2006, 2007, and 2008, respectively (Statistics Canada 2009). During those four years alone, the city of Calgary added 139,028 new residents (Statistics Canada 2009). And while the city continued to spread outward, it did so at somewhat higher densities—typically between 6.0 and 7.2 du/acre—reflecting changing planning requirements, smaller household size, and rapidly escalating housing costs.

Although largely driven by the oil and gas industry, the Calgary economy does not fit the stereotypical image of an oil town. Calgary's role in the industry largely derives from the command and control functions of oil and gas industry head offices and the producer-service firms that support them. Indeed, Calgary has become a major centre for head offices.

Head office relocation to Calgary rapidly accelerated during the first decade of the new millennium, with a 60.3 per cent growth in major head offices and a 24.8 per cent increase in head office employment between 2002 and 2006 (Calgary Economic Development 2007). This growth far exceeded that of all other Canadian cities. Calgary now has the highest concentration of head offices per capita in Canada, with 9.8 per 100,000 population compared to 5.1 for Toronto, 3.6 for Vancouver, and 2.4 for Montreal. The bulk of Calgary's head office expansion has been in the 'energy' field—i.e., oil and gas. Indeed, the nine largest Calgary-based corporations, ranked by 2005 revenues, are all energy-related. Overall, 74 of the top 109 head offices are energy-related (Calgary Economic Development 2006). As a result of the expansion of well-remunerated employment, per capita income in Calgary rose to $47,178 in 2006, the highest in the country and substantially above that of Toronto, which weighed in at $35,774. Calgary's growth in per capita income has been rapid: a 25.2 per cent increase between 2002 and 2006 (Calgary Economic Development 2007).

While the oil and gas industry is clearly the engine of Calgary's growth, employment by industry figures for 2006 reveal a more diverse, if still oil and gas–related, economy (Table 11.1). The top industrial employment category is construction, a category that includes both oil and gas activity and residential, commercial, office, and industrial construction activity. Food services and drinking places constitute the second-largest employment category, while architects and scientific research and services—serving both the oil and gas and construction industries—comes in third. Jobs directly related to oil and gas are in fourth place, and oil and other mining services are seventh. Other employment categories with a strong connection to the oil and gas industry include

IT services (eighth) and management, scientific, and technical consulting services (fourteenth), while insurance, real estate agents and brokers (tenth) is strongly connected to real estate and construction. Location quotients for the largest employment categories show—no surprise—that jobs related to the oil and gas industry are the most overrepresented: oil and gas (LQ 1.61), architects and scientific research and services (LQ 1.28), IT services (LQ 1.25), and management, scientific and technical consulting services (LQ 1.24). While location quotients for real estate development and construction activities are typically slightly below 1, they nonetheless represent some of Calgary's most significant employment categories in terms of absolute numbers of jobs, including construction (first), architects and scientific research and services (third), insurance, real estate agents and brokers (tenth), and local, municipal, regional public administration and Aboriginal public administration (thirteenth). The real estate development and construction industry is not only a very significant, if secondary, sector of the Calgary economy. It is also one that is highly dependent on local, as opposed to global, economic conditions. Its very high degree of local dependence (Cox and Mair 1988) has led to its intense involvement in local and regional growth politics, far more so than the oil and gas industry. We will return to the role of the real estate development and construction industry in Calgary politics later in this paper.

While the boom of 2002 to 2007 was widely welcomed by business interests, it was not without its difficulties and contradictions—which are likely to resurface with the next sign of vigorous economic growth. By the middle of the decade, Calgary's rapid expansion came to be threatened by pervasive labour shortages and the pressure on living and business costs they produced. It became harder to attract labour to Calgary as a result, and while in-migration continued at a high rate, out-migration to cheaper parts of the country such as Saskatchewan and Newfoundland and Labrador has also emerged as a major concern for business.

Immigration has become a much higher-profile issue in Calgary in recent years because of the city's growing labour shortages. Calgary's rapid population growth has relied more on interprovincial migration than on international migration, with twice as large a net contribution between 1996 and 2001, whereas in Toronto net interprovincial migration contributed just over 8 per cent of the international migration increase in the same period (Hiller 2007; 2009). This calculation, however, neglects an important financial issue: secondary migration whereby immigrants land in one place but move to another later. The financial problem is that the federal grants for immigrant settlement are provided to the municipality where the immigrant lands, even if the period of residence is short. It is widely thought that because of its lower international profile compared to Toronto, Montreal, and Vancouver but its high demand for labour, secondary migration is a larger phenomenon in Calgary than in other Canadian cities (Smart 1994). One of our informants suggested that secondary migration might account for an additional 20 per cent above and beyond primary immigrant numbers. The growth of the Temporary Foreign Worker program in Alberta further exacerbates Calgary's immigrant settlement funding shortfall, since temporary foreign workers receive no settlement funding at all.

Immigration in general has become the driving process shaping the ethnic composition of Calgary. In 2006, Calgary received 4.6 per cent of Canada's immigrants

Table 11.1 Calgary Employment by Industry (and Location Quotient)

	1996	2001	2006
Total Employed Individuals (greater than 8000)	498,345	609,885	653,505
Architects and Scientific Research and Services	18,055 (1.32)	25,195 (1.25)	32,795 (1.28)
Construction	32,720 (0.98)	42,795 (0.96)	53,670 (0.93)
Food Services and Drinking Places	30,410 (0.96)	38,180 (0.96)	36,685 (0.99)
Hospitals	12,845 (0.87)	15,105 (0.81)	16,635 (0.81)
Insurance, Real Estate Agents and Brokers	7010 (1.05)	10,895 (1.08)	11,790 (1.09)
IT Services	8380 (1.23)	16,600 (1.16)	14,190 (1.25)
Local, Municipal, Regional Public Administration and Aboriginal Public Administration	7695 (0.90)	8715 (0.91)	9700 (0.88)
Management, Scientific, and Technical Consulting Services	12,765 (1.20)	9005 (1.30)	9425 (1.24)
Nursing, Residential Care Facilities, and Other Health Services	3040 (0.78)	10,155 (0.84)	13,970 (0.93)
Oil and Gas	20,515 (1.73)	20,700 (1.67)	27,455 (1.61)
Oil and Other Mining Services	5545 (0.95)	8425 (0.98)	14,340 (0.97)
Printing and Related Support Services	5915 (0.89)	5945 (1.04)	8030 (0.98)
Security, Investigations, and Other Business Services	n/a	12,860 (0.88)	14,420 (0.85)
Social Assistance	n/a	12,135 (0.96)	10,775 (0.94)
Truck Transportation and Support Activities for Road Transportation	7560 (0.83)	9395 (0.85)	9900 (0.83)

Source: Statistics Canada special compilations.

(11,635), up from 3.9 per cent in 2004. In descending order, the top sending countries in 2006 were India, China (dropping from first place for the first time in many years), the Philippines, Pakistan, and the United Kingdom. In 2001, 20.9 per cent of the Calgary CMA population was born outside of Canada. The foreign-born population increased by 28 per cent by 2006, reaching 23.6 per cent of the total CMA population, the fourth-highest percentage of foreign-born residents among Canadian metropolitan areas.

Economic immigrants accounted for 46.7 per cent of all immigration to Calgary (dropping from 52.3 per cent in 2005), compared to only 28 per cent for Canada as a whole (City of Calgary 2007). This difference reflects the employment opportunities of Calgary, and various groups such as the Calgary Chamber of Commerce and Calgary Economic Development have begun to encourage higher immigration levels to respond to Calgary's labour shortages. However, infrastructure constraints and increasingly expensive housing is limiting Calgary's ability to substantially and rapidly increase its labour force.

The problems caused by Calgary's rising cost of living and overburdened social and physical infrastructure are exacerbated by the perception that Calgary is not particularly oriented toward integrating immigrants and visible minorities. Respondents to the Calgary Foundation's 2007 Vital Signs survey gave Calgary a C– for 'valuing diversity', a mark that improved slightly to C by 2008 (Calgary Foundation 2007; 2008). A recent City of Calgary (2009a) report found that racialized persons comprised 40 per cent of the city's poor population in 2006, up from 30 per cent in 1996. According to the same report 'over half of visible minority people [surveyed] in Calgary were concerned with racism and discrimination. They also expressed more concern with unemployment, debt, food and housing than other Calgarians' (City of Calgary 2009a, 3). Moreover, racialized poverty has become increasingly concentrated in particular Calgary neighbourhoods; in 37 neighbourhoods, more than 60 per cent of the low-income population is racialized. Clearly, more effective social integration of immigrants and visible minorities, and better integration into labour markets that still generate wasteful downward mobility among well-qualified immigrants and visible minorities, are key challenges facing Calgary and Canada generally (Frideres 2008).

The narrow emphasis on immigrants as employees also neglects the possibilities offered by the transnational networks immigrants create. Many studies have demonstrated that such networks can lead to the development of new economic niches and further economic linkages with other locations (Luova 2009; Smart and Smart 1998). Calgary immigrants' wide range of languages and cultural knowledge could serve as key resources if efforts were made to diversify beyond traditional export markets in the United States. The prevalence of return migration among certain groups, such as Hong Kong immigrants after 1997 (Salaff, Greve, and Wong 2010), could also offer opportunities if efforts were made to foster and maintain linkages with former residents who are often involved in businesses in fast-growing markets, such as the Pearl River Delta in southern China (Smart and Hsu 2004). Transnational ties among migrants from the same hometowns have made important contributions to the vitality and diversity of Calgary—for example, through ethnic festivals and the multi-million-dollar donation from Hong Kong that financed the impressive Calgary Chinese Cultural Centre. The conditions of departure for refugees, however, can often make the maintenance of transnational linkages problematic (Tanasescu and Smart, forthcoming).

While immigrants and visible minorities have had difficulties coping with life in Calgary during a period of rapid economic growth, Calgary businesses have also expressed some very strong concerns. The Calgary Economic Development (2008) survey of 400 Calgary businesses showed very mixed views of Calgary's economic prospects. In 2007, the most common concern of businesses was the lack of skilled labour (42 per cent of businesses), and 70 per cent had difficulty hiring staff. High labour, operating, and office space costs were a concern to many businesses, with 54 per cent perceiving growth as 'challenging'. By 2008, the global recession had eased those concerns somewhat, with lack of skilled labour dropping to the fourth most frequent business concern (21 per cent of businesses). Transportation infrastructure (44 per cent), lack of affordable housing (31 per cent), and lack of social support infrastructure (23 per cent) became the top three business concerns. Tellingly, all three of the top business concerns in 2008 can be viewed as consequences of failure to manage and keep pace with the demands of rapid growth. A 2008 Calgary Chamber of Commerce membership survey yields very similar findings, with the labour short-age as the top business issue in both 2007 and 2008 and transportation infrastructure (2008a) and infrastructure deficit (2007) as the second most important business issues. Interestingly, when Chamber of Commerce members were asked to identify the top municipal (as opposed to business) issues, city expansion and growth management was the top issue of 2008, and transportation infrastructure was the top issue of 2007—again, issues of inadequate investment and inadequately managed rapid growth.

Housing, Poverty, and Homelessness

Calgary's rapid growth has been associated with increasing inequality. Statistics Canada figures for the Calgary CMA show inequality to be fairly stable through the 1990s and into the new millennium, with gini coefficients of 0.49, 0.51, and 0.49 in 1991, 1996, and 2001, respectively. By 2006, however, the gini coefficient for the Calgary CMA had jumped to 0.61. Increasing inequality, combined with rapidly escalating housing costs, has culminated in what is widely considered to be a housing afford-ability crisis, especially for Calgary's least well-off households. Calgary's shelter costs have vaulted past Toronto's to make it the second most expensive in the country for owner-occupied housing and *the* most expensive for two-bedroom apartment rentals. Homelessness is a major concern, with more than 4000 people on the street or in shel-ters on any given night in 2008 and numbers still rising (Calgary Homeless Foundation 2009). Tens of thousands of people are at risk of homelessness because of soaring rents and low vacancy rates. Indicative of this critical problem, recent City of Calgary studies (2008a and 2008b) found over 72,000 Calgary households—19 per cent of all Calgary households—spending more than they could afford on housing in 2006. Not surpris-ingly, the Calgary Foundation's 2007 Vital Signs report gave Calgary a D– overall for housing and an F for the number of homeless people. The city's marks improved only slightly in 2008, when Calgary earned a D for housing. Calgary showed some further improvement in 2009, earning a C as the economy cooled and the mismatch between housing supply and demand eased.

It is important to recognize that Calgary's housing has not always been expensive. Between 2000 and 2005, Calgary was among the most affordable of the four major Canadian cities (Calgary, Ottawa, Toronto, and Halifax) examined by the Calgary Chamber of Commerce (2008b). In 2005, Calgary's housing affordability—measured as median housing cost as a percentage of median total income—was similar to that of Ottawa and Halifax. But in 2006, the global price of oil rose rapidly, stimulat-ing increased investment in the oil patch, expanding employment opportunities and incomes, and increasing migration to Calgary; these changes drove up the cost of housing dramatically. (Table 11.2) Modest employment growth from 2002 to 2005—averaging just under 2 per cent annually—gave way to an 8 per cent employment growth spike in 2006, followed by employment growth just under 4 per cent in 2007 and 2008 (Canada Mortgage and Housing Corporation 2009). The Calgary housing market could not possibly keep up with such red-hot employment growth.

In 2006, the new housing price index skyrocketed 43.6 per cent, followed by another 16.2 per cent increase in 2007. From 2005 to 2007, Calgary experienced a 65 per cent per cent spike in the cost of new housing. While the rapid escalation of new housing prices is of great concern, even more disconcerting are the changes in the rental market. Calgary has long been an expensive rental market, but recent rent increases are unprecedented. The average rent for a two-bedroom apartment rose 42 per cent from 2005 to 2008. During the same period, the rental vacancy rate bottomed out in 2006 at 0.5 per cent, since then rising slightly to 2.1 per cent in 2008.[2] Perhaps most concerning is how few rental units are being built in Calgary. Only 368 rental units were built in 2008—the most since 2004. Only 229 rental units were built in 2005, 2006, and 2007 combined. Considering the loss of rental housing due to condo conversions—919 condo conversions in 2005 and 946 condo conversions in 2006 (Poverty Reduction Coalition 2008)—and additional losses due to gentrification and redevelopment in older neighbourhoods—Calgary has actually experienced a net loss in rental housing during the very period it has experienced explosive growth. Stroik (2007) documents a loss of 4764 rental units between 2001 and 2006, while the Poverty Reduction Coalition (2008) found that the stock of rental housing declined by 2.6 per cent in 2005 and 2006, leaving a total of 40,333 rental units.

There are a variety of reasons for the contraction of Calgary's rental market. The high cost of construction—driven by rapid economic growth—means that new rental construction is only profitable at high rents, yet it is difficult to see rents climbing much higher than they already are. Difficulty in obtaining financing for new rental construction, combined with a lack of federal and provincial credits for new rental housing construction, exacerbates the problem. The lack of restrictions on condo conversions and redevelopment of older affordable housing makes the problem even worse. Clearly, market mechanisms, as currently constituted, are inadequate to the task of providing adequate affordable rental housing in Calgary. In this context, the retrenchment of social housing programs by the federal and provincial governments, and the downloading of responsibility for affordable housing to municipalities, borders on disastrous. Fully, 37 per cent of all renters—38,610 households—spend more than they can afford on housing, putting them at risk of becoming homeless (City of Calgary 2009b).

The City of Calgary, through the Calgary Housing Company, does provide non-market affordable housing, but its stock is insufficient to meet the need. The Calgary Housing Company had more than 10,000 subsidized and affordable housing units available in 2010, constituting the majority of all non-market units. Another 5804 non-market units were available for seniors, and 1610 units were provided by other non-profit corporations as of 2005 (City of Calgary 2009b). Altogether, however, non-market housing represents only 4 per cent of the total housing stock in Calgary. Not surprisingly, the Calgary Housing Company has long waiting lists for its housing.

Secondary suites—small rental units within primary dwelling units—have been touted as a partial solution to Calgary's affordable housing crisis. Several affordable

Table 11.2 Calgary Housing Market Indicators

	Rental Vacancy Rate	Average Rent, Two-Bedroom Apartment	Rental Starts	Multi-Family Starts	Single-Family Starts	Multi-Family Starts as % of Single-Family Starts	Per Cent Change in New Housing Price Index	MLS Average Price
2008	2.1	1148	368	7051	4387	161	0.6	405,267
2007	1.5	1089	20	5728	7777	74	16.2	414,066
2006	0.5	960	188	6564	10,482	63	43.6	346,675
2005	1.6	808	21	4948	8719	57	7.0	250,832
2004	4.3	806	475	5775	8233	70	5.5	222,860
2003	4.4	804	243	5116	8526	60	5.2	211,155
2002	2.9	804	295	4926	9413	52	5.2	198,350
2001	1.2	783	463	3790	7559	50	2.5	182,090
2000	1.3	740	18	4344	6749	64	2.4	176,305
1999	2.8	739	237	3987	6613	60	4.7	166,110
1998	0.6	707	64	3276	9219	36	7.6	157,353
1997	0.5	635	128	2559	8656	30	6.7	143,305
1996	1.5	595	17	1249	5862	21	1.0	134,643
1995	3.6	584	26	1298	4387	30	0.8	132,114
1994	5.1	585	61	1698	5179	33	2.4	133,571
1993	5.9	584	247	1409	5220	27	3.1	133,998
1992	5.5	598	94	1052	5982	18	0.6	129,506
1991	3.7	599	147	600	4150	14	-2.7	128,255
1990	2.0	584	296	1440	5564	26	12.5	128,484

Source: Canada Mortgage and Housing Corporation 2009.

housing organizations as well as the Calgary Chamber of Commerce have promoted secondary suites on the grounds that they increase the supply of affordable rental housing, help homeowners pay their mortgages, increase urban density, and are cost-effective (Calgary Chamber of Commerce 2008b). Unlike many other Canadian cities, however, Calgary has adopted highly restrictive policies on secondary suites in established neighbourhoods, largely based on NIMBY fears of increased traffic congestion and parking problems.

There has also been discussion of addressing Calgary's housing affordability from the income side of the equation. Census figures classify a substantial portion of Calgary households as low-income: 13.4 per cent in 2006, 14.1 per cent in 2001, 19.8 per cent in 1996, and 17.1 per cent in 1991. Vibrant Communities Calgary (VCC) (2009) has determined that Alberta's minimum wage of $8.80 per hour falls far short of what is needed to meet basic needs, given the cost of living in Calgary. VCC has calculated that a person working full-time in Calgary would need to earn $12.25 per hour with benefits or $13.50 per hour without benefits (a 'living wage') to meet basic needs at a decent standard of living. In 2008, 65,000 employed Calgarians (10.8 per cent of the Calgary labour force) earned less than a living wage. In April 2009, three living wage policy options were presented to city council for consideration. City council declined to adopt any of these policy options.

But contrary to common stereotypes of ultra-conservative Calgary, a variety of public opinion surveys indicate substantial popular support for measures to address the affordable housing crisis. For example, a 2007 Canada West Foundation survey showed that 70.6 per cent of Calgarians considered addressing affordable housing a high or very high priority, while 66.9 per cent of Calgarians indicated reducing homelessness was a high or very high priority (2007a). The same survey showed that 39.9 per cent of Calgary respondents wanted support for homeless programs increased and 48.5 per cent wanted to increase affordable housing options (2007b).

The most encouraging development in Calgary's housing situation has been the province of Alberta's adoption, in 2008, of 'A Plan for Alberta: Ending Homelessness in 10 Years'. This plan provides direct financial and institutional support for the City of Calgary's own '10-Year Plan to End Homelessness'. Both plans adopt a housing-first approach, prioritizing the provision of permanent shelter to those who have none. The Alberta Ministry of Housing and Urban Affairs has committed to fund 11,000 new affordable housing units province-wide by 2012, and both the provincial and City of Calgary 10-year plans aim to end homelessness by 2018. These policy initiatives are undoubtedly rooted in compassion for the homeless but politically justified on fiscal grounds. As the Alberta Secretariat for Action on Homelessness (2008) states, the cost of 'managing' the homeless population for the next 10 years would be $6.65 billion, while the cost of ending homelessness would be $3.32 billion. In other words, ending homelessness would save taxpayers $3.33 billion.

While there has been a mixed public policy shift in response to Calgary's poverty and housing affordability issues, there has been a very clear and significant shift in the housing market. Calgary's housing market has long been dominated by single-family detached housing, and Calgary's development industry has argued vigorously to maintain this pattern of growth. Michael Flynn, executive director of Calgary's Urban

Development Institute, has asserted that what Calgarians want is 'a piece of grass and a garage' (Markusoff 2010). Data for most of the 1990s might be interpreted as supporting Flynn's assertion: less than a third as many multi-family as single-family dwellings were built during that decade. But beginning in 1999, as housing costs continued to rise and demographics began to shift, multi-family housing construction became far more common, with about two-thirds as many multi-family as single-family dwellings being constructed. Following the 65 per cent spike in new housing costs between 2005 and 2007, multi-family construction became the most common type of new construction in 2008. Part of the shift toward multi-family housing involved numerous speculatively built centre-city condo towers. This created a centre-city condo bubble that has since burst. But townhouse and low-rise multi-family construction remains very popular, signalling a significant shift in the Calgary housing market. The shift toward multi-family housing reflects many influences, including changing demographics, frustration with increasingly time-consuming commutes from the suburbs, and—most of all—the dramatically higher costs of single-family housing. The notion that Calgarians have an innate and immutable preference for single-family housing—a piece of grass and a garage—has proven highly questionable at best. On the contrary, evidence shows that demand for single-family housing is quite elastic. As costs climb, preferences shift to more affordable multi-family housing.

Geographies of Growth and the Politics of Planning

In popular discourse, Calgary's growth has been virtually synonymous with sprawl (Foran 2007). How accurately 'sprawl' describes Calgary's growth has been a matter of some debate. Calgary's growth has not been unplanned, nor has it been characterized by leapfrog development, two of the common criteria for sprawl. But it has been relatively low-density, based on segregated land uses, and automobile-dependent. By these latter criteria, most of Calgary's growth can be characterized as sprawl (Couroux et al. 2006). While the sprawl question continues to generate debate, there is no question the city has expanded greatly. Between 1951 and 2008, Calgary's built form has grown from 40 square kilometres to 463 square kilometres, and the area encompassed by Calgary's city boundaries has expanded from 104 square kilometres to 848 square kilometres, a more than seven-fold increase (City of Calgary 2009c) (Figure 11.2). The city's density is also a matter of some contention. Dividing 2008 population by area yields a density of 1256 persons per hectare, low for a large North American city. But if one subtracts major undevelopable lands within the city, such as the airport, the reservoir, and other atypical large urban land uses, density rises to 2308 persons per hectare—in the ballpark of many other large North American cities. Geographical variation in density across the city is substantial (Figure 11.2). Suburban developments built between the 1950s and 1980s are the lowest-density parts of the city, typically between four and six dwelling units per acre. Developments built since the 1990s are more dense—up to seven and in some cases as high as nine dwelling units per acre. Future greenfield development will be even denser. The older, pre-war, inner-city neighbourhoods are also comparatively dense. Considered together, these different historical development patterns create a counter-intuitive donut density pattern—a relatively dense core, an inner ring of older low-density suburbs, surrounded by comparatively higher-density new suburbs on the fringe of the city.

Employment patterns in the city are even more geographically distinct (Figure 11.3). The bulk of Calgary's employment is concentrated in four areas: the central business district, the Manchester industrial area immediately south of the central business district, industrial areas in the vicinity of the airport, and industrial areas in the southeast of the city. The overwhelming majority of Calgary's employment is concentrated on the eastern side of the city, while the majority of Calgary's residents live on the western side. This mismatch between employment and residential location has contributed to

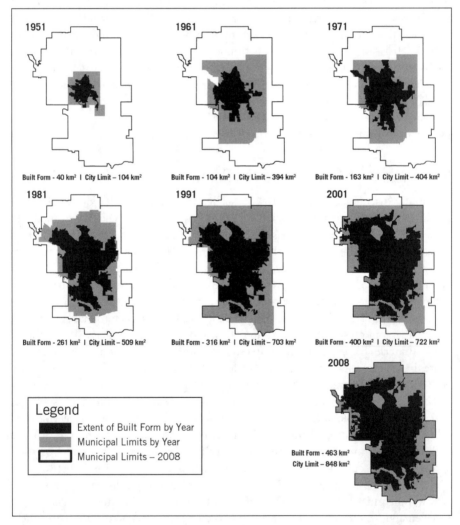

Figure 11.2 Calgary's geographical expansion, built form, and municipal limits, 1951–2008

Source: City of Calgary 2009c

the city's worsening traffic congestion as morning journeys to work are heavily west to east, with the reverse occurring in the evening. The lack of balanced two-way traffic flow was identified as a critical challenge in the 1995 Calgary Transportation Plan, a plan that attempted to rectify this problem, largely unsuccessfully. Over the past two decades, the city has attempted to establish major new suburban employment centres on the western side of the city, only to see their growth fall far short of expectations.

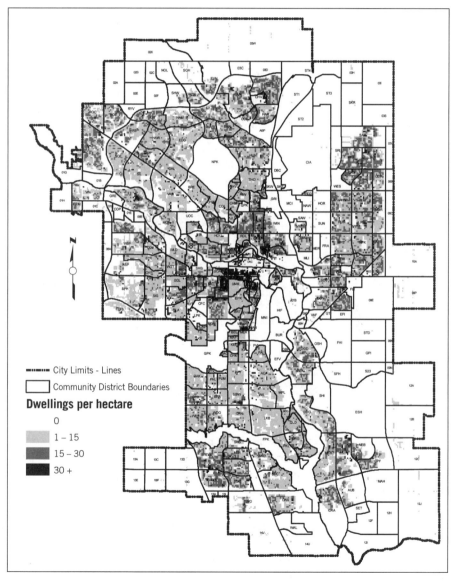

Figure 11.3 Geographical variations in Calgary's density, dwelling units per hectare

Source: City of Calgary 2009c.

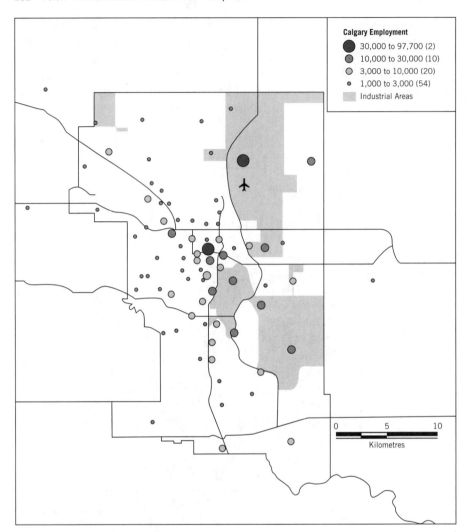

Figure 11.4 Calgary employment patterns.

At the heart of the problem is the city's reliance on the automobile as the primary mode of mobility. Suburban employment centres that are not located near the city's major north–south freeway and have no light rail transit access simply have been too inaccessible compared to the city's four major existing employment centres.

Increasing traffic congestion and commute times feed the widespread perception that Calgary's quality of life has deteriorated as the city has grown. Some studies lend credence to this perception. A study in the 1990s placed Calgary as having the highest quality of life among 13 Canadian cities (Giannias 1998), while a more recent international survey of 50 cities ranked Calgary twenty-fourth for quality of living, lowest among the five Canadian cities included (Mercer Human Resource Consulting 2007). The city's growth has generated huge pressures on every element of infrastructure so that in the past five years almost every poll asking Calgarians about the main issues facing the city has had infrastructure as the top answer (White 2007). Indeed, when Mayor Dave Bronconnier was first elected in 2001, he was widely associated with the slogan 'roads, roads, roads'. While expanding road capacity was Bronconnier's initial emphasis, his policy agenda evolved into something much broader. In 2004, the mayor and city council initiated a 'city-led, community-owned' participatory visioning process dubbed ImagineCalgary. The process, which took place over 2005 and 2006, involved almost 18,000 citizens and resulted in a highly detailed document calling for a much more sustainable, liveable, socially just, and citizen-driven city. The ImagineCalgary process was a clear expression of citizen dissatisfaction with the status quo and desire to move the city in a new direction. Unlike many other consultative processes that do little more than gather dust, ImagineCalgary became the touchstone document for a new municipal development and transportation plan. After three years of work by city land-use and transportation planners, Calgary City Council adopted the new 60-year land use and transportation plan, called Plan-It, in 2009. Plan-It reflects many, but certainly not all, of the citizen aspirations expressed through ImagineCalgary. The adopted version of Plan-It calls for half of all new growth to be within the existing developed footprint of the city. This stands in stark contrast to past development practices, which saw virtually all growth occurring in new suburbs. Plan-It calls for a more pedestrian-, transit-, and bicycle-friendly city, with complete mixed-use communities and close co-ordination between land-use and transit planning, promoting increased transit ridership and expanded transit service. Much of the city's redevelopment will occur along light rail lines in the form of transit-oriented development (Figure 11.5). Currently, 9 per cent of all trips in Calgary are taken by transit; Plan-It aims to increase that figure to as much as 20 per cent. It appears that Calgary has turned a major corner in its growth and development practices, but not without much political struggle involving a coalition of pro–Plan-It forces that might be characterized as 'strange bedfellows'.

Like many Canadian cities, Calgary suffers from a major infrastructure deficit. While older Canadian cities suffer infrastructure deficits largely because of aging infrastructure and maintenance issues, Calgary's $10.4-billion infrastructure deficit (City of Calgary 2009d) stems almost entirely from failing to keep up with the infrastructure investments required to meet the demands of rapid growth. The city's inability to keep up with the costs of rapid growth played a part in city council's ultimate support for Plan-It. It was clear that the city's approach to growth was leading to ever-higher demands on city coffers; new development did not pay for itself, and more compact transit-oriented forms of development would save the city substantial sums of money, both in infrastructure and operational costs. Indeed, a detailed cost analysis of different development scenarios performed for the city (City of Calgary 2009e) showed that the recommended scenario (the 50/50 redevelopment/greenfield development split that was eventually adopted) would cost the city 33 per cent less to build than the dispersed (status quo development practice) scenario. The difference came to $11.2 billion over the life of the plan. The pro–Plan-It 'business' argument, combined with strong citizen support for Plan-It, led to city council's unanimous vote to approve Plan-It, although only after a last-minute deal with development interests led to a reduction of minimum densities in new suburban development.

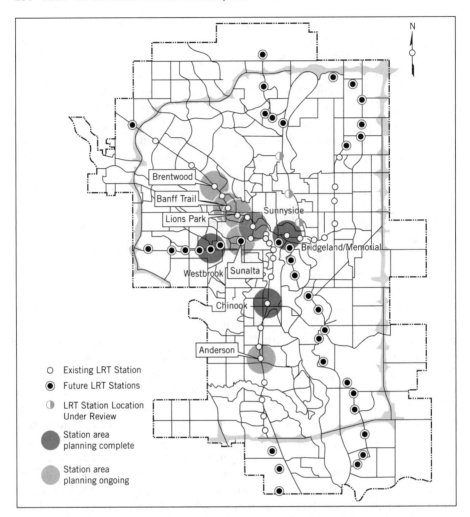

Figure 11.5 Light rail, transit-oriented development, and major employment centres as specified in Plan-It.

Source: Derived from City of Calgary 2009c.

Plan-It will put the City of Calgary on a more fiscally favourable footing, although not as favourable as it might have been at the even higher densities that were favoured by a number of citizen stakeholder groups. Ultimately, the city still wrestles with the seemingly intractable problem of financing the infrastructure and services a rapidly growing city requires.

Alberta's cities have few options for raising revenue other than property taxes, and the provincial government has been reluctant to increase revenue transfers since the 1980s. Indeed, the economic decline of the 1980s set the stage for the current infrastructure deficit through what LeSage and McMillan (2009) have called a 'sustained provincial disengagement from the municipal sector' in which cities received greater freedom of action but were 'also poorer and saddled with additional responsibilities'.

Calgary has long been a conservative city in a conservative province (Smart 2001; Miller 2007), but the current threats to the continuation of the economic boom are producing some significant shifts in political positions. The Calgary Chamber of Commerce, for example, has been advocating measures to address the growing problem of homelessness and affordable housing and supporting the expansion of public transit and mixed-use high-density neighbourhoods organized around transit hubs (Calgary Chamber of Commerce 2007; 2008c). Indeed, the Chamber of Commerce has been broadly supportive of Plan-It. Calgary's conservative politics, and in particular the strong influence of the corporate business sector, has meant that NGOs attempting to influence urban policy have had to adapt their tactics in order to be effective. We were told by our informants that to be effective advocates for social change in Calgary, NGOs have to make a case that action is in the interests of business or the city as a whole rather than 'merely' a matter of social justice. For example, corporations became quite supportive of the Calgary Homeless Foundation after research revealed that a very high percentage of the homeless population works full-time and that it would be less expensive to solve the homelessness problem than to continue the status quo. In general, for a social policy initiative to have a chance of success, a business case needs to be made demonstrating the returns on investment. NGOs also feel that without corporate support for their initiatives, it is difficult to get sufficient support from city council to move forward.

But business interests are almost never homogeneous. Different fractions of capital may have different material interests (Harvey 1985), and those in the development industry are often at odds with other fractions. Calgary's aldermanic and mayoral campaigns have been dominated by development industry financing. The industry's influence has been facilitated by the fact that Calgary has the least transparent system of campaign finance among Canada's major cities. Lack of campaign finance transparency attracted considerable attention during the October 2007 election, which was frequently described as 'wild west' (Guttormson 2007). Some have argued that Calgary's campaign finance system has encouraged the sprawling nature of urban growth, given the dominant role property developers play in electoral campaigns (Lorimer 1978; see also Austin and Young 2006). Foran (2007) has argued, however, that despite high levels of developer influence, limits are posed by the internal organization of city administration. The development industry's ability to guide development has also, in part, been a result of consumer demand for indirectly subsidized spacious single-family detached dwellings. But the resulting emphasis on suburban growth at the periphery has exacerbated transportation and fiscal problems, and this dynamic seems to have hit its limits. With broad business and citizen support for change in Calgary's growth and development practices, the old fiscal model is slowly being transformed, including the imposition of higher lot levies charged to property developers.

While the politics of growth and development have shifted, substantial tensions over the role of the province in financing urban growth have arisen in recent years. A high-profile feud between Mayor Bronconnier and Premier Stelmach dominated headlines for several months in the summer of 2007, with accusations that the premier had reneged on infrastructure commitments. A new 10-year commitment of $3.3 billion in provincial funds for Calgary resulted, but this caused some anger in Edmonton because the deal was seen as biased toward Calgary. Taking a hard stance against the province seemed to have resulted in considerable political dividends for the mayor in the October 2007 election (Braid 2007), and the mayor continued to push the cause of improved municipal finance in its aftermath. Bronconnier told a 2008 meeting of Rotarians that

> [m]unicipal finance has been studied time and again. And always the conclusion is the same. Municipalities like Calgary do not have the sustainable revenues required to build the infrastructure and provide the services we need now and will need in the future. Underfunding of municipalities is a reality and challenge for every mayor across Canada (Baker 2008).

This challenge became even greater in late 2009 when the province rolled back part of its 10-year $3.3-billion funding commitment, pledging instead to provide the funds over 11 years.

While Calgary now receives a small portion of federal gas tax revenue and some stimulus funds for affordable housing, the federal government plays a very small role in financing the city's infrastructure. Many Calgarians have long believed their concerns are not heard in Ottawa, and the long-standing feud between the Alberta government and the federal Liberals did not encourage close ties between the City of Calgary and the federal government. A senior planner told us that shortly after he started working for the city in the 1980s, he raised the idea of applying for federal funds but was told 'we don't do feds here.' The electoral success of the Conservative party in 2006 has modified this dynamic, but many feel that Stephen Harper has tended to take his base in Alberta for granted.

In short, rapid population and economic growth and its attendant pressures structure everything about contemporary Calgary. Rapid growth has resulted in considerable political efforts to obtain resources from the other scales of the state, as well as substantial shifts in municipal growth and development plans.

Calgary Looking Forward

Calgary's rise in the Canadian urban system can only be understood in terms of its extreme concentration in one particular industry, the oil and gas industry, and that industry's crucial role in the contemporary global economy. As the oil industry goes, so goes the Calgary economy. Indeed Calgary's economy, and that of Alberta generally, is so dependent on the fortunes of the oil and gas industry that Alberta could justifiably be characterized as an oil and gas rentier state. Provincial policy is designed first and foremost to promote the continued dominance and prosperity of the oil and gas industry. The drawbacks of policies that promote rapid growth in this industry seem to be poorly recognized—except when one examines how breakneck growth is affecting cities like Calgary. Tellingly, the most recent (2009) Vital Signs report, as well as the most recent City of Calgary (2009f) Citizen Satisfaction Survey, indicate a substantial increase in the perceived quality of life of Calgarians as the economy has cooled off.

Calgary's economic dynamics raise a couple of key questions. How long will the economy's petro-motor last? And is Calgary positioned for transition? The answer to the first question can only be speculative. Perhaps oil and gas will continue to drive the Calgary economy for a few decades, or perhaps for only one more. As for Calgary's ability to transition to a more diverse and sustainable economy, one can only observe that it is not happening now. The primary reason the Calgary economy is not becoming more diverse is that the high cost of living, high wages, and labour shortages make it very difficult for new industries to take root and grow. This does not appear likely to change anytime soon, but it should be noted that plans and policies enacted at the municipal level—in particular the adoption of Plan-It—represent attempts to control costs and put the city on a more sustainable footing. Indeed, Plan-It can be seen as an attempt to find a 'sustainability fix' (Jonas and Gibbs 2004) for the clearly unsustainable growth the city has experienced in recent years.

But much of the growth and governance of Calgary stems from provincial economic policy, beyond the purview of the local state. The work of cities, at least rapidly expanding ones like Calgary, is increasingly becoming one of influencing and channelling the consumption of the rentier classes—e.g., those in the oil and gas and real estate development industries—in ways that ensure the availability of sufficient bodies of labour, both skilled and unskilled. This is no easy task in an environment of high housing, food, and other costs. If a labour-short economy does become more common, attracting and retaining even unskilled workers could be a major challenge.

Calgary also serves to highlight the management challenges for a booming city in a world of labour shortages and global flows of real estate investment. High real estate prices tend to attract outside real estate investors. In London and New York, for instance, the cost of housing has relatively little connection with local wage rates, resulting in huge problems not just for the poor but also for the middle class. If options become available for good jobs in other cities, where lower salaries are compensated for by lower costs of living, labour shortages may become intensified. When, like Calgary, high-cost cities don't have major cultural attractions, the problem of retaining sought-after workers can become even greater. Higher wages may even make the situation worse: if economic migrants have a target they are working and saving for (a house back home, savings sufficient for them to retire early, their children's university tuition), paying them higher wages may mean that they can leave sooner.

Two trends, a shift toward more prevalent labour shortage problems and greater economic reliance on the consumption of rentier classes, are in many ways in conflict with each other. As a result, they create major challenges for urban managers. Making a city more attractive to the rentier classes, for example through high-end redevelopment initiatives, can exacerbate labour shortages because of the reduction in stocks of affordable housing and other increases in costs of living that may inhibit in-migration by less privileged workers, who are nevertheless in demand by the businesses that serve the newly rich. One possible path would seem to suggest a rather dystopian future of a tripartite division among 'gold collar' elites, those who service elite needs, and those who are not needed at all but simply represent social control problems to be managed. More progressive paths could also be imagined, perhaps based on the greater encouragement of local art, culture, and high-quality environmentally sustainable food to create the vibrant urban landscapes craved by supposedly mobile elites, coupled with a return to greater income redistribution and support for education and training.

But what about the commonly assumed desirability of growth? Leo and Brown (2000) and Leo and Anderson (2006) have compellingly challenged the assumption that urban growth necessarily leads to better cities. Leo and his colleagues argue that cities with populations that are not growing do themselves a disservice by adopting policies that attempt to foster growth at any cost. They make a strong case instead for a better quality of life in slow-growth cities and recommend policies that channel resources into improving the quality of life for the existing population. In a world increasingly concerned with environmental sustainability, we need to think more seriously about the advantages of slow-growth and no-growth cities and how they should be managed to best advantage. Such cities could be very positive examples for a labour-short future. Poor urban management, on the other hand, could result in decline that pushes young people to faster-growing cities rather than providing a higher quality of life affordable on lower incomes.

Calgary's infrastructure problems and the rapid inflation of construction costs during the boom years of 2005 to 2008 highlight the need to return to Keynesian economic policies. Planning for steady, or counter-cyclical, infrastructure construction is undeniably preferable to building during the height of a boom—which is expensive, inefficient, and only serves to further stimulate an already overheated economy. Unfortunately, this is what Alberta did during the height of the recent boom. Now that the economy has cooled, the province has enacted the inverse—and equally inexplicable—policy of cutting infrastructure spending just when the costs of infrastructure construction have dropped and the economy needs to be stimulated. Most maddening of all, we have seen this before. If during the 1980s bust Alberta's government had, instead of slashing expenditures, boosted employment by building the infrastructure it so badly needed, the long-term benefits would have been considerable. It did not, and now Alberta repeats the practice of hyper-stimulating booms and deepening busts. Only in an economy with a lucrative resource revenue stream could a government enact such policies and survive.

While more enlightened provincial urban policy does not appear to be on the immediate horizon, Calgarians clearly desire change. In the October 2010 mayoral election, Calgarians convincingly chose Naheed Nenshi, a 38-year-old professor of non-profit management and reformist candidate, over the presumed front-runner, Ric McIver, known for his conservative positions during his nine years on city council. Nenshi had been involved in a number of Calgary reformist initiatives, including ImagineCalgary, the Better Calgary Campaign, and CivicCamp Calgary. His platform (http://www.nenshi.ca) not only reaffirms support for Plan-It but spells out a number of clear reform positions: reform campaign finance; promote citizen participation; legalize secondary suites across the city; reduce poverty and promote social inclusion; make Calgary Transit the preferred mobility choice; make Calgary a city of sustainable, walkable, complete communities. Nenshi's clear victory over considerably better-funded opponents seems to confirm that Calgary has indeed turned a corner. What lies around that corner, however, is difficult to predict, given the role provincial, federal, and global forces will play in shaping the future of Calgary.

Acknowledgements

We want to express our gratitude to Mario Polèse and Cedric Brunelle for calculating the Calgary location quotients and to Scott Bennet for research and cartographic assistance.

References

Alberta Secretariat for Action on Homelessness. 2008. *A Plan for Alberta: Ending Homelessness in 10 Years.* Edmonton: Alberta Ministry of Housing and Urban Affairs.

Austin, S., and L. Young. 2006. *Political Finance in City Elections: Toronto and Calgary Compared.* Calgary: University of Calgary, Institute for Advanced Policy Research.

Baker, B. 2008. 'Calgary Mayor Dave Bronconnier guiding city through rapid growth'. http://www.citymayors.com/mayors/calgary-mayor-bronconnier.html.

Braid, D. 2007. 'Squeaky mayor gets provincial grease'. *Calgary Herald* 15 September: B1, B4.

Calgary Chamber of Commerce. 2007. *Renaissance Calgary: Blueprint for a 21st Century World-Leading Capital.* Calgary: Calgary Chamber of Commerce.

———. 2008a. *Membership Priorities Survey.* Calgary: Calgary Chamber of Commerce.

———. 2008b. *Municipal Land Development Policies and Regulations and the Impact on Calgary Housing Affordability.* Calgary: Calgary Chamber of Commerce.

———. 2008c. *Plan-It Calgary Consultation Response.* Calgary: Calgary Chamber of Commerce.

Calgary Economic Development. 2006. *Top Calgary-Based Corporations.* Calgary: Calgary Economic Development.

———. 2007. *Head Offices.* Calgary: Calgary Economic Development.

———. 2008. *2008 Business Survey.* Calgary: Calgary Economic Development.

Calgary Foundation. 2007; 2008; 2009. *Calgary's Vital Signs: Taking the Pulse of Calgary.* Calgary: Calgary Foundation.

Calgary Homeless Foundation. 2009. *The Homeless among Us.* Calgary: Calgary Homeless Foundation.

Canada Mortgage and Housing Corporation (CMHC). 2009. *Housing Market Outlook: Calgary CMA.* Ottawa: CMHC.

Canada West Foundation. 2007a. *Looking West 2007: Urban Policy Priorities and Assessing Governments.* Calgary: Canada West Foundation.

———. 2007b. *Caring Cities? Public Opinion and Urban Social Issues in Western Canadian Cities.* Calgary: Canada West Foundation.

City of Calgary. 2005. *Briefing Note: Residential Density.* Calgary: City of Calgary.

———. 2007. *Calgary and Region Social Outlook.* Calgary: City of Calgary.

———. 2008a. *Research Brief #03 Housing Affordability in Calgary.* Calgary: City of Calgary.

———. 2008b. *Fast Facts #09 Housing Need over Time among All Calgary Households.* Calgary: City of Calgary.

———. 2008c. *A Context for Change Management in the Calgary Regional Partnership Area.* Calgary: City of Calgary.

———. 2009a. *Inequality in Calgary: The Racialization of Poverty.* Calgary: City of Calgary, Community and Neighbourhood Services.

———. 2009b. *Fast Facts #04 Affordable Housing and Homelessness in Calgary.* Calgary: City of Calgary, Community and Neighbourhood Services.

———. 2009c. *Calgary Snapshots.* Calgary: City of Calgary, Land Use Planning and Policy.

———. 2009d. *Municipal Sustainability Initiative Investments in the City of Calgary.* Calgary: City of Calgary.

———. 2009e. *The Implications of Alternative Growth Patterns on Infrastructure Costs.* Calgary: IBI Group.

———. 2009f. *Citizen Satisfaction Survey.* Calgary: Ipsos Reid.

Climenhaga, D. 1997. 'The death and life of regional planning in the Calgary area'. (Master of Journalism thesis, School of Journalism and Communications, Carleton University, Ottawa).

Couroux, D., N. Keough, B. Miller, and J. Row. 2006. *Toward Smart Growth in Calgary: Overcoming Barriers to Sustainable Urban Development.* Calgary: Sustainable Calgary Society.

Cox, K., and A. Mair. 1988. 'Locality and community in the politics of local economic development'. *Annals of the Association of American Geographers* 78 (2): 307–25.

Foran, M. 2007. *Expansive Discourses: Urban Sprawl in Calgary 1945–1978.* Edmonton: Athabasca University Press.

Frideres, J. 2008. 'Creating an inclusive society: Promoting social integration in Canada'. In J. Frideres, J. Biles, and M. Burstein, eds, *Immigration*

Integration and Citizenship in 21st Century Canada, 77–101, Montreal: McGill-Queen's University Press.

Ghitter, G., and A. Smart. 2009. 'Mad cows, regional governance, and urban sprawl: Path dependence and unintended consequences in the Calgary region'. *Urban Affairs Review* 44 (5): 617–44.

Giannias, D.A. 1998. 'A quality of life based ranking of Canadian cities'. *Urban Studies* 35 (12): 2241–51.

Guttormson, K. 2007. 'Mayor vows review of election financing'. *Calgary Herald* 17 October: A1.

Harvey, D. 1985. *The Urbanization of Capital*. Baltimore: Johns Hopkins University Press.

Hiller, H.H. 2007. 'Gateway cities and arriviste cities: Alberta's recent urban growth in Canadian context'. *Prairie Forum* 1: 47–66.

———. 2009. *Second Promised Land: Migration to Alberta and the Transformation of Canadian Society*. Montreal: McGill-Queen's University Press.

Leo, C., and K. Anderson. 2006. 'Being realistic about urban growth'. *Journal of Urban Affairs* 28 (2): 169–89.

Leo, C., and W. Brown. 2000. 'Slow growth and urban development policy'. *Journal of Urban Affairs* 22 (2): 193–213.

LeSage, E.C., jr, and M.L. McMillan. 2009. 'Alberta'. In A. Sancton and R. Young, eds, *Municipal Government in Canada's Provinces*, 384–452. Toronto: University of Toronto Press.

Lorimer, J. 1978. *The Developers*. Toronto: James Lorimer.

Luova, O. 2009. 'Transnational linkages and development initiatives in ethnic Korean Yanbian, northeast China: 'Sweet and sour' capital transfers'. *Pacific Affairs* 82 (3): 427–46.

Mansell, R., and R. Schlenker. 2006. *Energy and the Alberta Economy: Past and Future Impacts and Implications*. Calgary: Institute for Sustainable Energy, Environment and Economy, University of Calgary.

Markusen, A. 1987. *Regions: The Economics and Politics of Territory*. New York: Rowman and Littlefield.

Markusoff, J. 2010. 'Calgary's next decade: Dust will still fly over skyline'. *Calgary Herald* 3 January.

Mercer Human Resource Consulting. 2007. *2007 Quality of Living Survey*. Mercer Human Resource Consulting.

Miller, B. 2007. 'Modes of governance, modes of resistance: Contesting neoliberalism in Calgary'. In H. Leitner, J. Peck, and E. Sheppard, eds, *Contesting Neoliberalism: Urban Frontiers*, 223–49. New York and London: Guilford Press.

Poverty Reduction Coalition. 2008. *Cementing Our Relationship: Private Sector Involvement in Affordable Housing*. Calgary: Poverty Reduction Coalition.

Reasons, C. 1984. *Stampede City: Power and Politics in the West*. Toronto: Between the Lines.

Salaff, J.W., A. Greve. and S.-L. Wong. 2010. *Hong Kong Movers and Stayers: Narratives of Family Migration*. Champaign, IL: University of Illinois Press.

Simmons, J., and L.S. Bourne. 2007. 'Living with population growth and decline'. *Plan Canada* summer: 13–21.

Smart, A. 1994. 'Business immigration to Canada: Deception and exploitation'. In R. Skeldon, ed., *Reluctant Exiles? Migration from Hong Kong and the New Overseas Chinese*, 98–119. Armonk, NY: M.E. Sharpe.

———. 2001. 'Restructuring in a North American city: Labour markets and political economy in Calgary'. In M. Rees and J. Smart, eds, *Plural Globalities in Multiple Localities: New World Borders*, 167–93. Lanham: University Press of America.

Smart, A., and J.-Y. Hsu. 2004. 'The Chinese diaspora, foreign investment and economic development in China'. *The Review of International Affairs* 3 (4): 544–66.

Smart, A., and J. Smart. 1998. 'Transnational social networks and negotiated identities in interactions between Hong Kong and China'. In M.P. Smith and L. Guarnizo, eds, *Transnationalism from Below*, 103–29. New Brunswick, NJ: Transaction Publishers.

Statistics Canada. 2009. *CANSIM II; Series V52008885*. E-STAT. Ottawa: Statistics Canada.

Stroik, S. 2007. *Homelessness: What Do We Know?* Presentation to the Community Summit on Calgary's 10-Year Plan to End Homelessness, Calgary, 23 April. Calgary: City of Calgary.

Tanasescu, A., and A. Smart. Forthcoming. 'The limits of social capital: An examination of immigrants' housing challenges in Calgary'. *Journal of Sociology and Social Welfare*.

Vibrant Communities Calgary (VCC). 2009. *Living Wage Fact Sheet*. Calgary: VCC.

While, A., A. Jonas, and D. Gibbs. 2004. 'The environment and the entrepreneurial city: Searching for the urban "sustainability fix" in Manchester and Leeds'. *International Journal of Urban and Regional Research* 28 (3): 549–69.

White, R. 2007. Shared vision needed for Calgary. *Calgary Herald* 13 October 2007: J2.

Notes

1. Historical Crude Oil Prices, http://inflationdata. com/inflation/Inflation_Rate/Historical_Oil_ Prices_Table.asp.

2. Late 2009 CMHC forecasts indicate some easing in the rental market, with the vacancy rate estimated at 4.0 per cent and rent for a two-bedroom apartment estimated at $1075 for 2009.

12 Vancouver: Restructuring Narratives in the Transnational Metropolis

Trevor Barnes, Tom Hutton, David Ley, and Markus Moos

Introduction: Signifiers of Change in the Vancouver City-Region

Vancouver is Canada's third city in terms of population and shares a sustained high-growth trajectory with certain Canadian city-regions, notably Toronto and Calgary. Vancouver's postwar development has followed important national trends, as observed in sequences of industrial restructuring, which have produced contractions in blue-collar employment, the concomitant rise of a 'new middle class' of elite service sector workers, and insistent social upgrading of residential neighbourhoods (Ley 1996). These signifiers of change can be identified across the national urban system, as demonstrated in the preceding case studies of Canada's other 'power cities': Montreal, Ottawa, Toronto, and Calgary.

But a more searching examination of Vancouver's growth and change since the 1970s discloses clear evidence of divergence and differentiation, including variation in business cycles relative to other Canadian cities.[1] Within Metro Vancouver we find the major infrastructural elements of a classic gateway city—functions arguably more important for Vancouver than for any other Canadian metropolis. But Vancouver's economy is increasingly animated by an SME (small- and medium-sized enterprise) economy of entrepreneurs, traders, technology and scientific personnel, and a 'new cultural economy' of creative workers.[2] At the same time, Metro Vancouver encompasses a large industrialized farming sector within its territory, again counter to the trends of other large Canadian cities, reinforcing its outlier status.

We argue here that this emergent economy of difference relative to other Canadian cities has been shaped by a confluence of distinctive multi-scalar processes. These processes operate at both the regional level and within international circuits, the latter process underscored by Vancouver's insistent integration within the markets, societies, cultures, and capital networks of the Asia-Pacific. More than perhaps most other Canadian urban regions, Vancouver exemplifies the city as a space of flows and recurrent restructuring rather than as a durable construct of stable industries, labour, social class, and communities.

Accordingly, this introduction will be followed by a succinct rehearsal of both pervasive and more regionally contingent factors driving growth and change in Vancouver. Next, we present an analysis of change in Vancouver's labour force, employment, and occupations, including implications for community reformation, followed by an account of the critical saliency of international immigration in the reconfiguration of Metro Vancouver's economy, employment structure, and society. A profile of the evolving space-economy of the Vancouver city-region follows, emphasizing exemplars of change within principal zones of the metropolis. The penultimate section presents a study of Vancouver's video game industry, which illustrates broader signifiers of innovation and change in the region as well as operating characteristics of an important cultural industry. Finally, the chapter offers a summary of key observations from the Vancouver case study.

Contours of Change in the Vancouver City-Region

Vancouver shares the development tendencies of other large- and medium-sized metropolitan cities within advanced societies, including industrial restructuring, which has produced a services-led economic growth trajectory and new divisions of labour. Momentum in the office-based producer services that led employment expansion in the 1970s and 1980s has appreciably slowed, especially in clerical and middle management positions, and has been followed by growth in scientific, technological, professional, and creative industry labour since the 1990s.[3]

Second, Vancouver's socio-economic trajectory is defined by professionalization but includes as well polarization and marginalization (see Walks, Chapter 6). Demand for professionals and other skilled labour has provided opportunities for highly remunerated work for many while serving to marginalize those deficient in skills (or those—notably a significant number of immigrant professionals—lacking credentials recognized by a Canadian accreditation agency). The stringent realities of the labour and housing markets, coupled with retrenchment in social programs and transfers, have generated a large population of working poor in the Vancouver region and a substantial and apparently growing underclass.

Third, the reconfiguration of the metropolitan space-economy includes rapid growth on the regional periphery; the redevelopment of inner suburbs, which encompass some of the principal centrepieces of the regional economy; and the comprehensive reconstruction of the urban core, a program that incorporates population growth and social change as well as the emergence of a new economy of specialized production, consumption, and spectacle.

Contingency in Vancouver's Development Trajectory

But this template of structural change takes us only so far into an appreciation of how Vancouver compares to other Canadian city-regions as well as to those situated in other national urban systems (see Chapters 2 and 3). The specifics of Vancouver's development, including labour formation and occupational change, as well as the mix of industries and enterprise that characterize the region's economy, are shaped by a distinctive mélange of forces and factors.

First, we acknowledge that Vancouver's development pathway has been shaped over the past quarter century not so much by a post-industrial as a *post-staples* economic trajectory. Vancouver developed outside the ambit of major Canadian industrial metropoles and instead specialized in industries associated with British Columbia's staples economy: resource processing and manufacturing, finance, administration, management, transportation, and distribution. The secular decline of Vancouver's staples vocation, including deep contractions in processing capacity and labour as well as an attenuation of head office functions, has served to decouple Vancouver from the resource economy in the BC hinterland (Hutton 1997).[4] Many of the engineering, surveying, and other specialized consultancies that catered to the staples economy developed core competencies, which have allowed them to serve clients in international markets (Davis and Hutton 1991; 1994), a parallel with the Montreal experience described in Chapter 8.

Second, and related to the changing fortunes of the city's staple functions, Vancouver presents a *post-corporate* stage of development within which much of the city's platform of head offices has been stripped away by successive rounds of globalization that have privileged first-order business centres for the concentration of 'command and control' functions widely seen as defining features of global cities. Vancouver remains the site of control for important public utilities and Crown corporations, as well as administrative offices of government, but now lacks the apex functions of corporate control boasted by other Canadian cities, notably Toronto (locus of control for Canada's five major banks as well as many other financial and industrial corporations) and Calgary (EnCana), as well as Vancouver's 'nearest neighbour' in the Pacific North-West, Seattle: the 'emerald city' boasts multinational corporations that project global cultural influence as well as high-order corporate stature.[5]

Vancouver thus lacks the corporate power projection of higher-order global cities, but for all that occupies significant niches within international circuits in terms of strategic gateway functions (including the Port of Vancouver, Canada's largest, and Vancouver International Airport) and an influential socio-cultural trajectory of *urban transnationalism*, driven by international immigration. In this, Vancouver's experience resembles that of other cities at the peak of the Canadian hierarchy, Montreal and Toronto. But while Toronto (like London and New York) has received significant flows of immigrants from each of the principal regions of origin, including Latin America, Asia, Africa, and Europe, Vancouver's immigrant profile presents a pronounced Asia-Pacific tendency. Vancouver has benefited from the relatively large inflows of immigrants within the business (investor and entrepreneurial) category compared to most other Canadian cities. Further, beyond the economic realm the Asia-Pacific orientation of Vancouver's immigration experience since the 1980s has generated a more comprehensive restructuring of housing markets, cultural signifiers, citizenship, and identity, as we will recount later in this chapter (Olds 1995).

A fourth feature of contingency underpinning Vancouver's distinctive development trajectory within the Canadian urban system comprises its broader regional setting and, more specifically, its bioregional characteristics (Oke, North, and Slaymaker 1992; Steyn et al. 1992). In historical terms, Vancouver's siting on the Pacific Coast, cut off from the prairies and the rest of Canada by the Rocky Mountains, moulded an identity for Vancouver and BC characterized in Jean Barman's well-known monograph as 'the West beyond the West' (1991). Vancouver's civic incorporation in 1886 was strategically enabled by the completion of the national rail system, linking the city with the agricultural economies of the prairies, with the dominant industrial metropoles of central

Canada, and with Halifax, Canada's principal Atlantic port, some 5000 kilometres to the east. The rail connection (and later the Trans-Canada Highway) undoubtedly stimulated Vancouver's growth, but the regional setting provides a potent geographical counterweight to the national influence, exemplifying the multi-scalar quality of contemporary urbanization. Vancouver has been positioned within 'Ecotopia', one of Joel Garreau's 'nine nations of North America', a bioregional and cultural territory bounded by Alaska's Yakutat Peninsula, the continental divide, and Cape Mendocino in northern California (Garreau 1981). A more modest conception of this geospatial ideal takes the form of 'Cascadia', a contemporary proxy for the Pacific North-West, which incorporates a more compact urbanized core (Figure 12.1) comprising the Vancouver and Seattle city-regions, with a combined population exceeding 5 million.[6]

The loftier aspirations of Cascadia as an integrated economic and trading zone have not been fully realized and perhaps cannot be fulfilled in at least the medium-term as the US continues its program of enforcing 'thicker' borders in the extended aftermath of 9/11. It is also the case that BC treaty claims with First Nations are yet to be resolved, including territory in the Vancouver region (Blomley 2004). Despite these developmental constraints, the distinctive environmental features of Cascadia represent both a principal signifier of cultural identity and a basis for recreational lifestyles in the region.[7]

Our final category of contingency influencing Vancouver's economic development consists of *policy factors*, including features of governance. A shift in the regional pattern of investments by British Columbia governments since the 1980s has accelerated Vancouver's growth momentum. From the 1950s the provincial government, led by premiers from Kelowna and backed by powerful ministers representing other interior constituencies, invested heavily in infrastructure designed to support the 'opening up' of a staples economy in the BC hinterland. This political economy of staples development was exposed by a particularly severe downturn in the early 1980s, and a succession of governments led by Vancouver-based premiers (including two former mayors of the City of Vancouver) have directed large quanta of capital into the Vancouver regional economy, notably in 'gateway' transportation infrastructure but also in tertiary education and housing. The provincial thrust has been complemented by a sequence of multilevel governance programs, the foremost being the 1986 International World Exposition (Expo 86) and the 2010 Vancouver Winter Olympic Games, as well as rapid transit systems and international marketing. Vancouver has been a major beneficiary of largesse from the federal and provincial governments under the rubric of these multilevel governance initiatives, contributing in no small way to the city's development (Hutton 2009).

There are also distinctive features of local government (metropolitan and municipal) that have imparted particularity to Vancouver's development trajectory. The regional governance structure takes the form of a confederation of its constituent 22 municipalities and electoral districts (Figure 12.2), with a selective repertoire of powers and responsibilities in such areas as parks, sewerage, and infrastructure, rather than an executive, directly elected regional government with comprehensive powers. Since the 1970s, the Greater Vancouver Regional District (GVRD, now Metro Vancouver) has promoted a regional structure shaped by (first) designated regional town centres (RTCs)

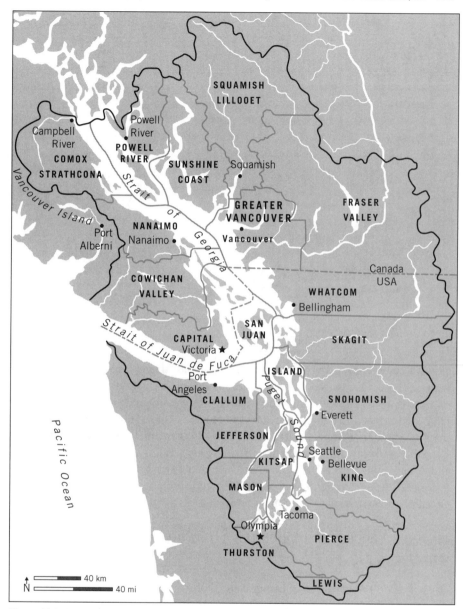

Figure 12.1 Vancouver in its regional setting: Cities, regional districts (British Columbia), and counties (Washington State).

within suburban municipalities and more recently by a program for 'compact and complete' communities in the region's core municipalities in the interests of deflecting development pressure in the city, reducing sprawl, and preserving critical ecological assets. In support of this program, more than $7 billion has been invested in fixed rail transit since the mid-1980s, including the initial Expo Line linking the downtown with key suburban RTCs, the inner suburban Millennium Line, and the new Canada Line

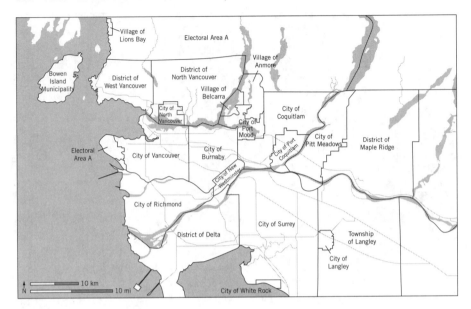

Figure 12.2 Metro Vancouver (formerly Greater Vancouver Regional District [GVRD]) and its constituent municipalities.

(August 2009) connecting Vancouver International Airport and Richmond Centre with downtown Vancouver. To an extent, this environmental mission has succeeded, reinforced by the Agricultural Land Reserve (ALR), administered by an Agricultural Land Commission (ALC) appointed by the provincial government. But this effort at shaping regional structure has been vitiated by the asymmetries in power over land-use policy and regulation between Metro Vancouver and the municipalities. Further, and despite numerous efforts to establish a regional economic development strategy over the past two decades, no metropolitan policy program with the power to shape Greater Vancouver's space-economy exists.[8]

Shifts in Vancouver's Industrial Structure, Employment, and Occupations

Does Anyone Work Here? Imageries of Vancouver's Economy

Visitors to Vancouver are often struck by the public life in this city: cafés, outdoor patios, public squares, and scenic beaches bustling with activity almost any time of day. Amid beautified streetscapes, markets, parks, and cultural amenities surrounded by the ocean and mountains, the city portrays images of relaxation, play, and leisure. In a remarkable development almost unimaginable 30 to 40 years ago when the central city was still the site of warehouse districts and rail yards, residential condominium towers now dominate the skyline (Figure 12.3). On foot one is more likely to encounter residents shopping and mingling in restaurants rather than the businessmen in suits that are emblematic features of downtowns in other large urban economies such as New York or Toronto. 'Does anyone work in this place?' a visiting relative once asked.

Figure 12.3 Residential towers in Vancouver's downtown. (Source: Markus Moos)

Descriptors of an emerging 'leisure' (Veblen 1899 [1953]) or 'creative' class (Florida 2002) come to mind when cities are characterized by their ability to provide consumption amenities that attract the workers of the new economy. The emergence of cities as sites of consumption among workers in new economy sectors helps us add meaning to Vancouver's changing urban economy, but the 'pooling of faceless talent captured by a café culture and breathtaking views' (Barnes and Hutton 2009) is hardly a sufficient explanation for the multi-faceted dimensions of the city's economy. At minimum we might also evoke notions of a 'dual city' by pointing to the growing homeless population and urban poor or the class of workers in personal service and retail industries catering to the new economy workers (Sassen 1990; Mollenkopf and Castells 1991). Yet these popular conceptualizations of economic change and their social consequences are broad brush, and it is instructive to consider Vancouver's changing enterprise and employment structure more specifically as it is constituted locally by a complex set of relationships related to the city's transformation to a post-staples economy integrated into the Pacific Rim network of capital and labour flows. In this section, we inquire about the material characteristics of Vancouver's employment and occupational structure, labour force, and enterprise profile.

The city experienced rapid growth in employment and population in the 1980s and 1990s (Table 12.1). The labour force grew by 35 per cent between the 1981 and 2001 censuses, exceeding growth in the rest of the country by 12 per cent. Vancouver's labour force grew by another 9 per cent between the latest two censuses as compared to 8 per cent for the rest of Canada. As is the case for all of Canada, Vancouver's population is aging, and the median age in 2006 for the census metropolitan area (CMA) was only slightly below (39.1) than for Canada as whole (39.5). Despite an aging population, it is the residential geography of young workers, particularly those in professional occupations, that is heavily concentrated in the inner city and creates demand for high-density construction.

Since the 1970s, Vancouver's economy has become increasingly decoupled from the provincial staples economy that once dominated economic activity in the city. Sawmills, fish packing plants, and the corporate offices of resource firms gave way

Table 12.1 Labour Force Characteristics, Vancouver CMA and Canada, 1981–2001

Variable	1981		2001		2006	
	Vancouver	Canada	Vancouver	Canada	Vancouver	Canada
Population	1,268,183	24,343,750	1,986,965	30,007,094	2,116,581	31,612,897
Population 15 and older	1,009,305	18,861,700	1,641,570	24,281,560	1,752,390	25,664,220
% Immigrants	29.7	16.1	37.7	18.3	39.3	19.6
% University-educated[a]	16.3	12.3	20.8	15.5	24.6	18.1
Total labour force	681,390	12,009,250	1,049,910	15,576,565	1,150,490	16,861,180
% Self-employed	8.5	9.1	13.7	12.0	13.7	12.1
% Working from home[b]	4.0	6.7	7.6	7.5	8.0	7.3
% Unemployed	5.0	7.3	7.2	7.4	5.6	6.6
Median household income ($)	58,884	52,236	56,148	52,563	55,231	53,634
Median personal income ($)	29,502	24,585	23,610	23,081	25,032	25,615
% below LICO (before tax)[c]	-	-	13.7	11.2	20.8	15.3

Source: Authors' calculations using Statistics Canada, Census of Canada, 1981, 2001, and 2006.

Notes:
a Population 15 years and older with at least one university degree.
b Working from home for pay excluding those in agricultural industries.
c Population 15 years and older earning less than Statistics Canada's Low Income Cut-off measure. All dollar figures adjusted for inflation using Bank of Canada rates ($ 2005).

to residential uses, leisure spaces, beautified waterfronts, and new economy clusters. While the city's ports continue to serve important trade functions for the resource economy, the dominant trend has been the tertiarization of the industrial structure and simultaneous transformations of the urban built environment and social structure. Between 1971 and 1991, the largest percentage growth in employment occurred in the community, business, and personal service industries (13 per cent), while the largest decline was in manufacturing (−7.7 per cent) (Hutton 1998). By 2006, Vancouver's industry profile reflected a division of the labour force into professional, scientific, and technical service industries on the one hand and retail trade, accommodation, and food services on the other (Table 12.2). While certainly serving local demand, the accommodation and food services sectors also depend on tourism. The cruise ship industry in particular has been a growing contributor to the local service economy. Construction, educational services, and health care/social assistance are also among the top industries as measured by percentage of the labour force (Table 12.3).

Table 12.2 Employment by Industry for Greater Vancouver, 1996, 2001, 2006

	Annual Averages (thousands of employees)		
	1996	**2001**	**2006**
Total employed, all industries	946.5	1039.1	1,187.1
Goods-producing sector	182.2	176.2	211.9
Agriculture	5.9	6.6	10.0
Forestry, fishing, mining, oil and gas	10.0	5.6	8.1
Utilities	5.3	5.5	3.7
Construction	59.4	53.5	85.3
Manufacturing	101.6	104.9	104.7
Services-producing sector	764.3	862.9	975.2
Trade	152.0	165.7	191.8
Transportation and warehousing	58.1	66.8	67.6
Finance, insurance, and real estate	78.2	77.8	88.0
Professional, scientific, and technical	74.1	95.8	112.0
Business, building, and support services	37.6	42.7	54.5
Educational services	55.8	72.5	92.4
Health care and social assistance	88.6	96.1	115.8
Information, culture, and recreation	50.4	66.3	70.3
Accommodation and food services	74.6	84.9	86.9
Other services	45.3	52.8	52.7
Public administration	49.5	41.4	43.3

Source: Statistics Canada.

Table 12.3 Industry Characteristics, Vancouver CMA, 2006

Industry	% of Labour Force	% Change 2001–6	% BA or Higher	Average FT Income ($ 2005)
11 Agriculture, forestry, fishing, and hunting	1.2	4.7	11.2	39,529
21 Mining and oil/gas exploration	0.4	95.1	46.2	119,659
22 Utilities	0.5	–8.1	36.0	71,719
23 Construction	6.4	36.4	10.9	54,693
31–33 Manufacturing	8.5	–1.3	18.8	53,726
41 Wholesale trade	5.4	10.1	23.2	57,353
44–45 Retail trade	10.9	7.2	17.3	41,078
48–49 Transport and warehousing	5.7	–0.2	15.5	53,723
51 Information and cultural	3.7	–5.0	34.4	65,147
52 Finance and insurance	4.8	2.8	37.6	70,219
53 Real estate/rental and leasing	2.6	19.6	27.4	62,014
54 Professional, scientific, and technical services	9.3	17.2	54.3	69,168
55 Management of companies	0.2	123.8	39.0	98,315
56 Administration/support and waste management	4.7	18.0	19.5	38,330
61 Educational services	7.2	11.7	66.0	53,292
62 Health care and social assistance	9.3	7.8	37.3	52,487
71 Arts, entertainment, and recreation	2.4	13.7	28.4	40,848
72 Accommodation and food services	8.0	12.3	12.5	29,928
81 Other services	5.1	14.4	21.0	38,960
91 Public administration	3.8	–1.1	36.6	61,063
All industries	**1,150,490**	**9.6**	**29.0**	**53,995**

Source: Statistics Canada, Census of Canada, 2006. Adapted from Spencer and Vinodrai 2008.

The Knowledge Economy and the New Middle Class

The changing industrial structure is correlated with the growth of knowledge workers who are less involved in producing things and instead generate, analyze, or disseminate information. The transformation can be understood as the result of economic and technological transitions from mass production in Fordism to flexible specialization in post-Fordism. Service-oriented cities such as Vancouver are characterized by a highly educated labour force (Table 12.1) working in managerial, professional, and technical occupations (Table 12.4). More than 30 per cent of Vancouver's labour force possessed some university education in 2006. There has been a strong correlation between the growth of the service sectors and urban expansion, which has resulted in cities being repositioned in the urban hierarchy (Coffey 1994). Vancouver, among other mid- and medium-sized metropolitan cities such as Ottawa and Calgary (see Chapters 9 and 11, respectively), has benefited most from this trend. Vancouver's occupational profile reflects this transition, with growth occurring in managerial, administrative, natural sciences, engineering, mathematics, social sciences, and related occupations as well as teaching and medical (Table 12.5). Occupations in processing, machining, product fabricating, and assembly as well as construction, transport and material handling, and related fields became of decreasing relative importance, although manufacturing employment over the past decade has remained relatively stable, in contrast to the major

Table 12.4 Location Quotients of Knowledge Workers by CMA

CMA	Location Quotient			Proportion of Workforce
	Managerial	Professional	Technical	
Halifax	1.552	1.200	1.456	0.013
Quebec City	0.979	1.317	1.129	0.024
Montreal	1.238	1.188	1.089	0.117
Sherbrooke/Trois-Rivières	0.811	1.063	0.895	0.010
Ottawa-Gatineau	1.743	1.682	1.354	0.039
Oshawa	0.760	0.721	0.596	0.010
Toronto	1.588	1.242	1.426	0.165
Hamilton	1.058	0.888	1.031	0.022
St Catharines/Niagara	0.621	0.694	0.637	0.012
Kitchener	0.947	0.934	0.961	0.015
London	0.782	1.068	0.891	0.015
Windsor	0.762	0.792	0.693	0.010
Sudbury/Thunder Bay	0.659	0.852	0.864	0.009
Winnipeg	0.844	1.029	1.115	0.024
Regina/Saskatoon	0.870	1.089	1.034	0.015
Calgary	1.288	1.230	1.393	0.036
Edmonton	0.926	0.976	1.077	0.034
Vancouver	1.229	1.164	1.504	0.066
Victoria	1.126	1.272	1.670	0.010
Rest of Canada	0.541	0.688	0.580	0.354

Source: Authors' calculations using Statistics Canada, Census of Canada, 2001.

Notes: Differences in distributions differ from zero at p<0.000. Knowledge workers include those with at least one university degree working in managerial, professional, or technical occupations.

Table 12.5 Employed Labour Force by Occupation, Vancouver CMA (%)

Occupation	1996	2001	2006
Management occupations	10.7	11.9	11.3
Business, finance, and administrative	21.1	19.8	19.1
Natural and applied sciences and related	5.6	7.2	7.4
Health occupations	4.8	5.2	5.4
Social science, education, government services	6.9	7.4	7.8
Arts, culture, recreation, and sport	3.5	4.1	4.1
Sales and service occupations	27.5	25.9	26.1
Trades, transport, and equipment operators	13.1	12.1	12.6
Primary industries	2.0	1.7	1.8
Processing, manufacturing, and utilities	4.8	4.7	4.3
All occupations	963,905	1,049,910	1,150,490

Source: Adapted from Statistics Canada, Census of Population, 1996, 2001, 2006, Catalogue no. 97-559-XCB2006012.

Notes: Includes employed population 15 years of age and older; 20% sample data.

contractions in Montreal and Toronto (see Chapters 8 and 10, respectively). According to the latest census, Vancouver's labour force continues to specialize in high-order service sector occupations, particularly in arts, culture, recreation, and sport (Figure 12.4).

The social correlate of the highly tertiarized nature of the economy is the emergence of an affluent 'new middle class' of professionals and managers working in the quaternary sectors of the economy (Ley 1996), the urban expression of the 'post-industrial society' projected in Daniel Bell's seminal social forecast (1973). The new middle class is implicated in the gentrification of the inner city (Hamnett and Cross 1998), and their disposable incomes and consumption-based lifestyles provide stimulus for employment in retail, cultural, and entertainment sectors. The occupational structure remains highly gendered along traditional divisions, with women overrepresented in sales, services, clerical, and administrative occupations. However, Vancouver's female labour force has a higher percentage of professionals, semi-professionals, and technicians (Table 12.6). The growth of single-earner, particularly female-headed, households in higher-earning occupations has contributed to the demand for inner-city residential housing development at higher densities. At the other end of the income spectrum is a growing number of workers in lower-level service occupations. The occupational distribution provides potential for income polarization commonly ascribed to global cities (Sassen 2001; Hamnett 2003), and certainly in recent years the percentage of those earning more than $80,000 and those earning less than $20,000 has increased, while the rest of the income distribution has remained remarkably stable. The proportion of workers below Statistics Canada's Low Income Cut-off increased from 13.7 per cent to 20.8 per cent between the 2001 and 2006 censuses, an increase of more than 6 per cent compared to 4 per cent for the rest of Canada (Table 12.1).

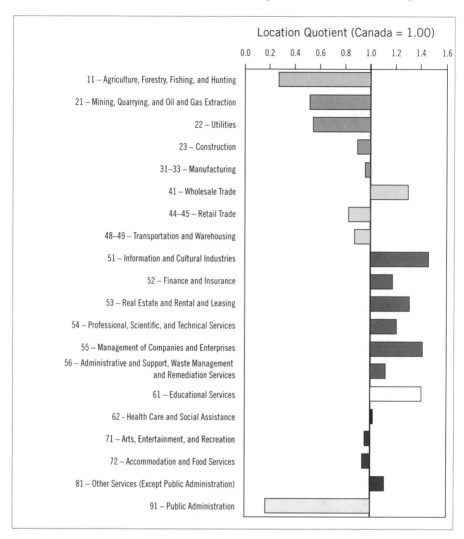

Figure 12.4 Industry specialization—establishment counts.

Source: From Spencer and Vinodrai 2008 using Statistics Canada Establishment Counts, 2008

Whether or not polarization can be ascribed to greater global integration is perhaps debatable, since Vancouver has been described as divided since its origin as a resource town. Nonetheless, the tensions created by an increasingly polarized income structure are vividly visible in the housing market. Dual-income earners and professionals can outbid lower-income households in the property market, reducing the supply of low-income housing, particularly rental stock. The City of Vancouver estimates that an income of $58,860 was required to service a mortgage of $200,000 (10 per cent down payment) in 2006 (Gray 2006). At 2006 market rates, that would be sufficient to purchase a one-bedroom condo. The result is that more than 50 per cent of Vancouver's labour force earns less than is required to afford a one-bedroom condo in the central

Table 12.6 Experienced Labour Force by Occupation and Gender, Vancouver CMA (%)

Occupation	Female	Male	All
Senior managers	0.7	2.2	1.5
Middle and other managers	8.0	12.7	10.4
Professionals	18.0	16.7	17.3
Semi-professionals and technicians	8.8	8.4	8.6
Supervisors	1.2	0.9	1.0
Supervisors: crafts and trades	0.5	2.1	1.3
Administrative and senior clerical personnel	10.1	2.1	6.0
Skilled sales and service personnel	3.8	5.2	4.5
Skilled crafts and trades workers	0.9	11.1	6.2
Clerical personnel	15.5	5.6	10.3
Intermediate sales and service personnel	17.3	8.8	12.9
Semi-skilled manual workers	3.3	12.0	7.8
Other sales and service personnel	10.4	8.2	9.2
Other manual workers	1.5	4.2	2.9
N-cases	14,876	16,199	31,075

Source: Authors' calculations using Statistics Canada, Census of Canada, 2001.

Notes: Includes employed labour force 15 years of age and older. Differences statistically different from zero at p<0.001.

city (Table 12.1), and that number increases further if we consider the suitability of housing for larger households. It is not surprising that most growth in the housing stock is occurring in the suburbs and that impoverishment is spreading beyond inner-city borders (Bunting, Walks, and Filion 2004). Average industry earnings demonstrate that affordability may be particularly an issue for those working in the accommodation and food services sector where average earnings are the lowest, especially since these jobs are often centrally located (Table 12.3).

While the emerging residential geographies are in part a reflection of the suburbanization of employment in manufacturing and transport-related industries, the new economy firms, many of them spun off by the process of vertical disintegration, are smaller in size and woven into the downtown urban fabric to produce new landscapes of economic production. Fifty-nine per cent of the metropolitan area's businesses had fewer than five workers, and only 1 per cent more than 100 (Table 12.7). Canada's enterprise profile is generally characterized by a high proportion of small- and medium-sized businesses, but the trend is more pronounced in Vancouver, chiefly in the central city. Industry profiles show a high concentration in information and cultural industries, educational services, real estate–related activities, and management of companies and enterprises (Figure 12.4). Small- and medium-sized businesses

Table 12.7 Percentage of Establishments by Number of Employees

Number of employees	Vancouver	Canada
1–4	59.0	55.0
5–9	18.5	20.2
10–19	10.9	12.1
20–49	7.2	7.8
50–99	2.5	2.7
100–199	1.1	1.2
200–299	0.6	0.6
500+	0.2	0.3

Source: Statistics Canada Establishment Counts, 2008. Adapted from Spencer and Vinodrai 2008.

are particularly prevalent in professional, scientific, technical, and other services. The city's enterprise and industrial profile is reflected in its built form. While the city has a sizeable central business district, few new office towers have been constructed since the 1990s, and indeed several commercial buildings in the CBD have been converted to residential use.

Economic restructuring associated with the tertiarization of employment has also led to increases in self-employment, part-time work, home-based work (telework), and other forms of flexible labour because of outsourcing and downsizing. The percentage of self-employed in Vancouver increased from 8.5 per cent in 1981 to 13.7 per cent in 2006. The increase in self-employment is related to the growth of small- and medium-sized businesses. Self-employment is particularly prevalent in non-manual occupations, constituting up to 40 per cent of that workforce (Moos and Skaburskis 2010). The percentage of those working from home for pay, excluding those in agricultural industries, doubled from 4 to 8 per cent. About half of the home-based workers are self-employed. Vancouver has the highest percentage of home-based workers among Canadian CMAs, and in some neighbourhoods the percentage of residents working from home full-time is as high as 30 per cent. The geography of home-based work and self-employment is associated with high-status neighbourhoods and the presence of natural amenities. The growth in home-based work arrangements is related to the emergence of a new economy labour force that is increasingly able to decide on their residential location based on quality-of-life factors, but importantly the trend also shows occupational polarization (Moos and Skaburskis 2007). The highest proportion of home workers is in the arts, culture, and recreation occupations, followed by management, financial, secretarial, and administrative occupations. Industrial home-based work has also seen a return, with evidence of low-income workers in Vancouver performing sewing and small-scale manufacturing from their own homes.

Concepts of home and work become blurred for many of the new economy workers, even if home-based work is not full-time. Production in the new economy is not as rigidly defined spatially and more fluid temporally. For some, this may mean greater opportunity to pursue recreational opportunities; for others, it implies growing insecurity in the labour market and an inability to find full-time work. In Vancouver, a rising homeless population and growing percentage of those below the Low Income Cut-off is indicative of an urban economy with employment insecurity, although part-time work arrangements in Vancouver have increased by only 2 per cent from 17.1 per cent between 1987 and 2008 (Metro Vancouver 2009). Part-time work is more common in the rest of Canada, and this could be related to the higher cost of living in Vancouver. Part-time work is most common in sales, service, clerical, and administrative occupations (Table 12.8), but it is interesting to note that part-time work is more common in managerial and professional occupations in Vancouver than it is in other major Canadian cities and the rest of Canada—a sign of a more leisure-oriented labour force perhaps, at least among those in higher-order occupations.

With an abundance of amenities catering to the residential new middle class as well as visiting tourists, Vancouver—particularly the inner city—has become a space of consumption that hides the material realities of production. Imageries of the new

Table 12.8 Percentage of Workers in Occupational Category Working Less Than 35 Hours per Week, Three Largest CMAs and Canada

Occupation	Vancouver	Toronto	Montreal	Rest of Canada
Senior managers	19.3	17.3	16.0	18.5
Middle and other managers	22.3	16.0	17.5	18.7
Professionals	33.1	27.5	30.8	30.7
Semi-professionals and technicians	39.9	34.7	35.0	37.5
Supervisors	22.1	24.1	20.7	26.7
Supervisors: crafts and trades	23.9	16.4	18.1	23.7
Administrative and senior clerical personnel	38.9	34.4	34.4	39.6
Skilled sales and service personnel	34.5	30.3	34.2	39.1
Skilled crafts and trades workers	28.5	18.7	21.4	25.0
Clerical personnel	39.8	35.7	36.9	38.7
Intermediate sales and service personnel	53.9	51.4	52.4	55.2
Semi-skilled manual workers	36.1	25.0	28.4	33.7
Other sales and service personnel	62.6	61.3	61.0	66.9
Other manual workers	47.4	34.5	37.4	44.8

Source: Authors' calculation using Statistics Canada, Census of Canada, 2001.

Notes: Includes employed labour force 15 years of age and older. Differences statistically different from zero at $p<0.001$.

economy are more tucked away, less visible than the large office towers or factories that continue to symbolize economic production in other cities.

Vancouver does not often make the list of global cities, remaining peripheral to the world's financial and business centres. Yet the city's economy has been transformed by global capital and labour flows. Once known as 'a village at the edge of the rainforest', Vancouver has become a major gateway to the Pacific Rim network (McGee 2001). Vancouver's Pacific Rim linkages and the city's reputation as an attractive place for global investment, buttressed by government campaigns, spurred initial growth in the inner-city housing stock. The sale of the Expo lands by the provincial government to the Hong Kong–based Li family following the 1986 World's Fair earmarked the beginning of modern-day Vancouver's identity as a place for global investment, particularly in real estate (Olds 2001). This occurred during a time of transition in urban policy in the 1980s, from social liberalism toward neo-liberalism, with all levels of government engaged in selling Vancouver as an attractive place for global investment dollars (Mitchell 2004). The influx of investment, and also thousands of skilled migrants and business entrepreneurs, prior to the Sino-British agreement on the reversion of the Crown colony of Hong Kong to China, provided Vancouver with the means of transforming its capital stock and urban infrastructure to attract new economy industries and workers. In the following section, we develop a more detailed account of immigration as a leading factor of growth and change in Vancouver's economy.

Immigration and the Vancouver Economy

Trade, Immigration, FDI, and the Metropolitan Vancouver Economy

Any discussion of Vancouver's metropolitan economy must acknowledge at the outset that it is an economy that has always been open to international opportunities and vicissitudes. As a resource-rich province, British Columbia has been in the business of exporting staples through Vancouver, its principal port and largest urban centre. An international gold rush was a trigger that brought British Columbia to formal colonial status in 1858. The gold rush diversified an earlier frontier economy based on sending furs to Europe and was followed by the world-famous canneries based on the vast Pacific salmon run. Wood products, coal, and other minerals, plus the Prairie wheat trade, all passed through the Port of Vancouver through the twentieth century. Thinking internationally, and more recently globally, has been part and parcel of doing business in BC.

Vancouver as a gateway city historically looked east to Europe, notably Britain, more recently south to the United States, and increasingly now west to Asia. If the free trade agreements with the United States in 1989, adding Mexico in 1994, have consolidated north–south flows, the linkages with Asia-Pacific have been major growth points and the source of much government coaxing, not least through the flamboyant Team Canada trade missions led by Prime Minister Chrétien in the 1990s. But globalization is not just a process of looking outwards. The mutuality of flows has constantly brought both investment and labour into Vancouver's economy. The most dramatic of these flows at present are from Asia.

The Asian connection, though always Vancouver's destiny, only recently became its desire. For most of British Columbia's history, the Pacific option was shunned, its labour closely controlled or excluded altogether in the case of Chinese during the quarter-century of the Chinese Immigration Act (1923–47) that sought to perfect what the invidious earlier head tax upon Chinese immigrants had achieved only imperfectly (Anderson 1991). It was only in the 1960s with the liberalization of immigration and the discovery of the putative Pacific Rim that economic energies turned to the west, Europe's 'far east'. The fast-deteriorating Canadian and British Columbia economy in the 1980s—the provincial economy shrank 8 per cent in 1982, unemployment reached 15 per cent in 1984 and remained above 10 per cent for most of the rest of the decade—made the attractions of the Asia-Pacific growth region an irresistible magnet. Trade missions to Asia-Pacific from all three levels of government repeatedly made the case that Vancouver and BC were open for business. The spectacle of the World's Fair, Expo '86, drew in the punters to Vancouver for a closer look. Liberalized immigration, including the unapologetically economistic business immigration program, opened the door wide for newcomers, especially wealthy newcomers, and the selection of Li Ka-Shing, patron of Hong Kong's wealthiest family, as land developer of the vast Expo site in 1988 accelerated an East Asian landfall of investment and immigration that has dramatically reshaped the city.

To understand Vancouver's economic geography today, it is essential to underscore its gateway function to Asia. This has been the source of extraordinary investment, abundant immigration, a substantial upturn of trade, and the abrupt remaking of the city's identity in less than a generation. These relations are dominated by ties to Greater China and South Asia, the world's two most populous and fastest-growing regional economies. But while ties to India and Pakistan have chiefly featured immigration of family members and some skilled workers, relations with Greater China have been altogether more propulsive.

The 2006 Census identified 40 per cent of the CMA population as foreign-born; 440,000, more than one-fifth of the entire population, had arrived during the previous 15 years. Immigrants now fill more than half of new jobs created in the metropolitan labour market (Hiebert 2009). There were some 380,000 people of Chinese ethnicity in Greater Vancouver in 2006, three-quarters of them foreign-born, amounting to nearly 20 per cent of the regional population. The number identifying with South Asian sources, while smaller, remained substantial at 207,000 (64 per cent of them foreign-born) or 10 per cent of the overall population. These sources and others in Asia have outnumbered earlier European and American flows: from 1991 to 2006, 80 per cent of immigrants to Greater Vancouver originated in Asia and only 10 per cent in Europe. Indeed, with the tendency for metropolitan out-migration of the native-born in recent years, the pan-Asian complexion of the Vancouver population has grown by leaps and bounds.

In the late 1980s, Japan was the Asian leader in foreign direct investment (FDI) in British Columbia. Coal exports to fuel Japanese steel mills were accompanied by investment in the tourist sector during the 1980s in the peak years of Japan's bubble economy. The Vancouver–Whistler region was the most favoured, with purchase of hotels, golf courses, and tourist resorts, comprising a vertically integrated tourist sector (Edgington 1996). With the bursting of the bubble, capital was withdrawn to prop up core business in Japan, but already the post-Expo inflow of capital was underway from Hong Kong and Taiwan. There were two major differences between this activity and Japanese FDI. First, Hong Kong and Taiwanese investment moved overwhelmingly into real estate, primarily in Vancouver; second, accompanying investment was large-scale immigration so that for a decade Hong Kong was the leading source of immigrants to Vancouver and to Canada as a whole. In the decade up to 1997, acute geopolitical concerns led to portfolio dispersal by wealthy Hong Kong families, while sabre-rattling by the Mainland across the Taiwan Strait led to anxiety among well-off families in Taiwan. East Asian business families were also drawn by Canadian education opportunities for their children (Waters 2008) and the reputed quality of life in relatively unpolluted Vancouver. What facilitated their migration was Canada's business immigration program, which caught the attention of wealthy and footloose émigrés from East Asia: almost 70 per cent of the more than 90,000 business immigrants who entered Vancouver from 1980 to 2001 were from Hong Kong or Taiwan (Ley 2003).

Foreign Direct Investment and Vancouver's Social and Economic Geographies

There has been immense speculation concerning the scale of capital transfers across the Pacific during this period, flows that attracted global hyperbole in the most intense years of activity from about 1988 to the mid-1990s (Gutstein 1990). While the full magnitude of money transfers is not knowable, they certainly amounted to several billion dollars annually from Hong Kong alone during the peak years (Mitchell 2004); the Canadian Imperial Bank of Commerce estimated the inward flow from Hong Kong to Canada at between $2 billion and $4 billion a year in the early 1990s (Symonds, Yang, and Zuckerman 1991). Very large sums were also transmitted from Taiwan after the relaxation of overseas exchange controls in 1987, releasing a huge pent-up flow of personal savings that were running at the remarkable level of 40 per cent of Gross National Product (Bradbury 1989). Vancouver was the primary beneficiary of these funds. A senior banker in Vancouver told one of us in the early 1990s, 'The banks have so much Asian money coming in, they don't know what to do with it' (Ley 2010). In mid-decade, a second senior Vancouver banker confided that US$100 million had entered his bank via one Taiwanese branch in a single month following provocative Chinese naval exercises off the coast of Taiwan.

Capital arrived both as investment from East Asia and in the suitcases of immigrants. Some 150,000 people of Chinese origin landed in Greater Vancouver from 1986 to 1996, and many of them were fabulously wealthy. From an assessment of total funds guaranteed before immigration officials offshore, it seems likely that the liquid assets of business immigrants alone landing in Vancouver between 1988 and 1997 amounted to $35 billion to $40 billion (Ley 2010). The arrival of this extraordinary wealth over a short period had inevitable consequences for Vancouver's property market. Prices soared in an almost perfect relationship with net immigration, and by 1991 Vancouver had vaulted ahead of Toronto to record the most expensive housing in Canada, a status that has been held to the present. Globally exposed, the market underwent significant oscillations in the 1988–2001 period as off-shore capital serially entered the regional market but then withdrew, either pursuing better returns elsewhere or, at the end of

1997, in response to the devastating downturn in the Asia-Pacific economy that caused the repatriation of funds to shore up home defences (Ley and Tutchener 2001).

With this immigration came a unique feature of Vancouver's economic geography, the emergence of the real estate market as a key element of regional development. Immigrants from Hong Kong and Taiwan had in many cases amassed their own wealth from real estate, and in imitating the investment pattern of the much-admired Li Ka-Shing, they *embodied* the territory's 'land (re)development regime' as a model for capital accumulation, a model that they carried with them to Vancouver (Tang 2008; Ley 2010). It was not only property sales that boomed; there were also significant purchases of high-end consumer goods and services, including luxury cars, home furnishings, electronics, and private education. So substantial were such expenditures that the arrival of wealthy migrants was widely credited with allowing the Vancouver region to escape the worst privations of the deep national recession in the early 1990s. It is difficult to overestimate the unique impact of off-shore capital, both impersonal and embodied, in fashioning the present landscape of the Vancouver region.

Yet as these millionaire migrants entered the regional economy in passive investment and active business ownership, their performance was remarkably weak. Confronting the uncertainty of an economic culture different from that of East Asia, Chinese entrepreneurs often entered the large ethnic enclave economies of Vancouver and Richmond. But facing hyper-competition in a saturated market, many entrepreneurs ran anemic and underperforming businesses and often resorted to transnational migration to replenish their fortunes from active business interests in East Asia while their Vancouver economic activities languished (Ley 2006). For many families, transnational migration became a first step to relocation, and after 1996 significant return migration, to Hong Kong in particular, has occurred.

This weak economic performance has continued with the more recent arrival in the past decade of young skilled workers from China. That nation took over from Hong Kong as the largest single immigration source at the end of the 1990s and maintained its primacy for the next decade. Despite high levels of human capital, workers are significantly underemployed, a repetitive and serious flaw in Canada's immigrant integration record (Reitz 2001; Bauder 2003): scientists and engineers commonly have factory or menial service jobs; doctors have to re-qualify to become nurses or care aides. The barriers appear to be, in no special order, the familiar trio: problems with the recognition of foreign credentials, less than satisfactory English, and the preference of employers for workers with Canadian experience (Teo 2007). Teo's interviews show real resolution among many Mainland skilled workers in Vancouver to make their Canadian residency work but at the same time growing exhaustion as they struggle with de-skilling. 'We compare ourselves,' says one, 'skilled immigrants who have come to Canada—to agricultural workers who come to Guangdong [province] to do those kinds of work.' Another skilled worker commented that working in Canada was similar to the Cultural Revolution when educated urban dwellers were sent to the country for hard labour and reform. A third observed, 'If you have a very good job, [Canada] is heaven. If you have no job then it is hell. It is not better than going back to China.' Among this sample of 80 skilled workers, Teo (2007) identified 'quiet courage amidst despair'. The model of transnationalism among this young Chinese group sees children sent to China to be looked after by grandparents or babysitters so that both parents can work in one or more jobs or go back to school. But in light of their circumstances, fewer than one-third of those interviewed have decided for sure on long-term settlement in Canada. They will stay to complete residency requirements for citizenship despite circumstances some call 'immigration prison', but then they will re-assess the situation.

If immigrants from East Asia endure the worst earnings losses, surveys and government databases confirm the difficult landfall of most immigrants from countries outside western Europe and the United States. Downward mobility and de-skilling are common, and earnings data confirm only a slow process of economic integration toward the mean level of Canadian earnings (Pendakur and Pendakur 1998; Picot 2004). Despite the rising skill level of immigrants, the return on human capital is falling, and poverty levels seem entrenched in a number of urban enclaves of concentrated immigrant settlement (Smith and Ley 2008). Immigrants as a whole in Greater Vancouver received average incomes in 2005 that were 75 per cent of the non-immigrant mean income, but this figure fell consecutively by period of arrival and for the most recent immigrants (2001–4) was less than half of the native-born average (Hiebert 2009). While immigrants are a mainstay of the labour and housing markets of Greater Vancouver, those who have arrived since 1991 experience on the whole precarious returns in the local economy.

The Reconfiguration of Vancouver's Space-Economy

So far we have identified a range of forces—market, social, cultural, and policy factors—that drive processes of transformation within metropolitan Vancouver as a whole. We now turn to a discussion of how these forces intersect to produce a distinctive economic geography in the Vancouver city-region.

The morphology of the Vancouver economy presents a complex palette of industries, institutions, and labour, configured by the ongoing internal industrial specialization within the metropolis as elucidated by Allen Scott (1988), as well as the spatial 'splintering' of the city driven by technological innovation as described by Stephen Graham and Simon Marvin (2001), and imparting (as Ed Soja observes) 'greater complexity and instability to the restructured social mosaic' (Soja 2000, 282).

There are at the same time important integrative elements that draw on the resources of the region as a whole and are manifested throughout the Vancouver metropolitan area. These elements include, notably, the regional labour and housing markets and commuter shed, which shape in many ways the morphology of the region. Second, *distribution* represents a major system within Greater Vancouver, strongly associated with the region's strategic transportation functions and wholesaling activity. Third, there are patterns of connectivity shaped by retail activity, although the proliferation of major shopping malls widely throughout the region has a tendency to generate sub-regional consumption behaviours. Fourth, the operation of the Agricultural Land Reserve has been fundamental to the preservation of increasingly high-yield farming throughout the Fraser delta, Canada's most fertile agricultural zone, which comprises much of the metropolitan territory. A fifth linking feature is represented by the *experiential* domain of the economy, comprised of the region's unique mix of ecological assets (parks, beaches, rivers, mountains) available for public recreation and a major draw for tourists as well as for local consumption. Finally, the very large illicit drug economy in Vancouver is widely dispersed throughout the region, organized by major gangs and syndicates, and comprises large-scale production (notably high-grade marijuana), distribution (Vancouver as entrepôt for hard-drugs like heroin and ecstasy), and local sales.

The Zonal Structure of Metropolitan Vancouver's Space-Economy

The diversity of the Greater Vancouver economy, and more particularly the SME profile of enterprise and employment, is manifested in an increasingly complex spatiality. The largest and most strategic elements of the space-economy include the Port of Vancouver, by far Canada's largest, with principal installations along the Central Waterfront (containers, bulk cargo, and cruise ships), along the Fraser River, and the Delta Superport specializing in coal exports; Vancouver International Airport, in Richmond; and the Central Business District (Figure 12.5). The regional plan (the Livable Region Strategic Plan, approved in 1996 and now in review) recognises the CBD as a unique cluster of specialized activity within Metro Vancouver and also incorporates eight designated regional town centres (RTCs) within suburban municipalities (see Figure 12.5). While there are some signal successes in this long-running program, a combination of weak local (municipal) policy commitments and market trends have produced a pattern of development characterized more by dispersion than multi-nucleation (MacMillan 2004). What follows is a selective discussion of processes and trends that reconfigure the space-economy of Metro Vancouver.

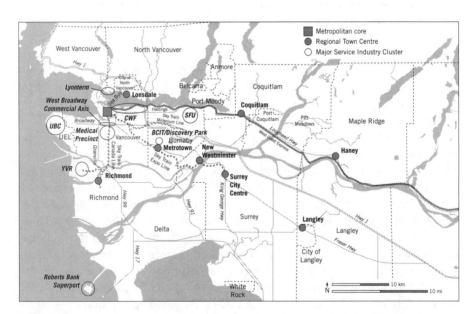

Figure 12.5 Regional town centres (RTCs) and principal service industry clusters in Metro Vancouver.

The Metropolitan Core: The 'New Economy' amid a Residentialized Central City

The metropolitan core (a zone taking in the old central area, in planning terms, and now incorporating an extended territory to the east of the downtown, to Clark Drive), although representing a steadily diminishing share of the regional economy as a whole, still includes about 220,000 workers and encompasses by far the largest array of specialized production industries and labour within Metropolitan Vancouver, as well as upscale consumption industries, higher education, and elements of the cultural economy. Although Vancouver has lost ground as a corporate control centre, there are still about 100,000 office workers in the core, principally in the CBD, as well as along the West Broadway office corridor, and including a host of producer-services firms and workers. Prominent among the intermediate service industries that congregate in Vancouver's core are the legal, accounting, and consultancy firms generic to the CBDs of most advanced cities, as well as financial service companies, many of which cater to the wealthy Asian immigrants described in the previous section. But since the 1990s, the monocultural office economy of the core has been supplanted in part by a more complex geography of production and consumption industries, arrayed within the CBD fringe and inner city (Figure 12.6).

The mix of industries within Vancouver's metropolitan core follows the relayering of capital and enterprise produced by successive phases of innovation and restructuring common to most cities, as well as more contingent local and regional features, including the major inner-city rezonings of the 1970s and 1980s, which positioned the city as an agent of post-industrialism (Hutton 2004).[9] The sectoral blend incorporates creative industries, institutions, and labour of the cultural economy, as well as a residual base of older, quasi-industrial enterprises—auto-related companies, supply and service firms, garment and food production, and even a few heavy-industry operations

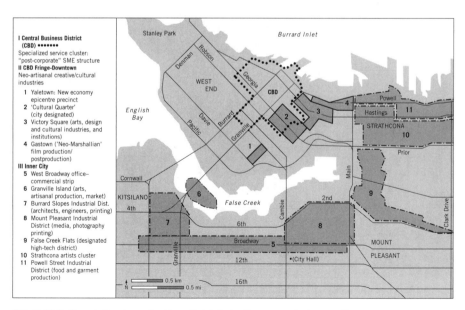

Figure 12.6 Specialized production districts in Vancouver's inner city.

Source: Hutton 2008

(for example, a large cement plant on Granville Island). The economic landscapes of the core may thus present as a promiscuous assortment of industries associated with successive development phases in the city. But we can also identify logics of location, which include the filtering effects of space, the adaptability of the built environment, and the dynamics of Vancouver's property market, as well as policy influences, which have produced a measure of localized industrial specialization within the core as well as affiliated social characteristics (Hutton 2008).

Thus Yaletown, a high-integrity heritage district ensconced between the Downtown South and the Concord Pacific mega-project across Pacific Boulevard, attracts leading-edge creative firms and labour (Figure 12.7) as well as upscale housing and commensurate consumption amenities, reflected in the very high rents prevalent within the district. Representative industries within the district's distinctive landscape of brick warehouses (Figure 12.8) include software designers, computer graphics and imaging firms, Internet development firms, architects, and interior designers. Yaletown projects the most glitzy and consumerist imagery of any district within Vancouver's inner city, combining as it does the epicentre of the city's new economy as well as the priciest loft conversions and most indulgent lifestyles. Indeed, such is the cachet of this micro-district for developers, new economy firms, lifestyle-seeking residents, and planners that a succession of planning area redesignations has produced a re-territorialization of 'Yaletown' to encompass about a quarter of the downtown (Figure 12.7) (Barnes and Hutton 2009).

In contrast, Victory Square and Gastown are situated in an altogether grittier precinct of Vancouver's inner city, occupying the western margins of the Downtown Eastside. Victory Square was the city's most prominent banking, commercial, and retail district in the early twentieth century, then experienced a long decline as the centre of gravity

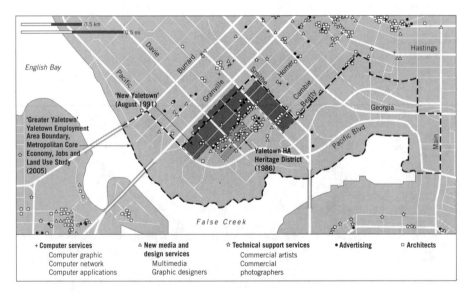

Figure 12.7 Yaletown as Vancouver's 'new economy epicentre': Industrial clusters and territorial expansion, 1986–2005.

Figure 12.8 Landscapes of the new economy: Yaletown, Vancouver.

for these functions shifted inexorably westward in the downtown; while Gastown, Vancouver's original townsite, suffered a similar downturn as its warehousing and wholesaling functions became obsolescent following the Second World War. But these conjoined districts have emerged as key sites of Vancouver's new cultural economy. Gastown, in particular, has been a bastion of Vancouver's film production and post-production industry for at least 20 years, domiciled within its heritage brick warehouses, complemented by the well-known Vancouver Film School situated on Hastings Street in Victory Square. The latter also incorporates a dense cluster of artists and design firms, architects, environmental consultancies, and NGOs, reflecting the exceptional diversity and complexity of the economy of the 'new inner city'. And while the area's redevelopment has proceeded at a more leisurely pace than that of Yaletown, which occupies a uniquely privileged quarter of the inner city, the opening of the landmark Woodward's project at Hastings and Abbott Streets within the interstices of Victory Square and Gastown promises to accelerate both industrial innovation and social upgrading within Vancouver's inner city. The city's credo of promoting 'revitalization without gentrification' in the Downtown Eastside is no doubt a commendable one, but the former aim may be easier to achieve than the latter (Barnes and Hutton 2009).[10]

These examples demonstrate the continuing vitality of Vancouver's metropolitan core as a zone of territorial innovation (after Morgan 2004), creativity, and specialized production. At the same time, to the normal pressures of market competition and the destabilizing effects of recurrent restructuring pressures we must add the dynamics of the property market to the mix of factors squeezing the (mostly) small firms operating in these inner-city sites. A combination of the city's comprehensive rezoning of 1991, which privileged new housing communities over commercial development, patterns of social demand, and contemporary building economics have produced an increasingly residential downtown, encapsulated in the city's 'Living First' program. The dominance of the office–commercial sector over new housing that characterized the 1980s has given way to an almost complete reversal of fortunes (Figure 12.9). While there is much to celebrate in what John Punter has called the 'Vancouver achievement' in high-density residential development urban design (2003), city planners and some elements of the business community have expressed concern regarding an imbalance of land use that is likely to constrain the generation of new employment in the core. In response, the city has undertaken a comprehensive review of land use in the central area (Vancouver Core Area Jobs and Economy Land Use Plan project), which has recently recommended holding the line on any further conversions of commercial/industrial land for housing, as well as proposing increased densities in designated commercial areas, implying a new episode in the developmental narrative of Vancouver's core.[11]

The Inner Suburbs: Redevelopment and Retrofitting

Metro Vancouver's inner suburbs include not only large residential populations but also strategic elements of the regional economy in such sectors as higher education, advanced technology research, and production, transportation, and wholesale trade. The City of Burnaby, immediately to the east of the City of Vancouver, with a residential population of 216,000, has transitioned from its initial vocation as industrial suburb of Vancouver to one strongly oriented toward higher education, research and development, and retail trade. The obsolescence of much of the City of Burnaby's industrial lands has stimulated a continuing program of zoning and land-use review, balancing issues of preserving a mix of industry types in the municipality with the need to attract advanced technology

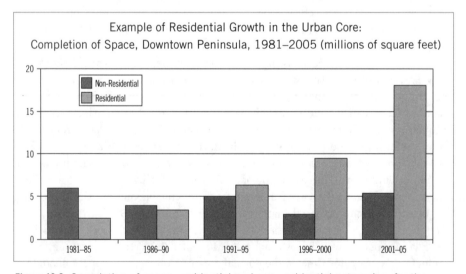

Figure 12.9 Completion of space, residential and non-residential categories, for the downtown peninsula, City of Vancouver, 1981–2005.

Source: City of Vancouver Planning Department

industries. With regard to higher education, Simon Fraser University's (BC's second-largest university) main campus is located on Burnaby Mountain in the northernmost quadrant of the municipality. Adjacent to the SFU campus is a major new project that combines a substantial new residential community with affiliated consumption and amenities ('UniverCity'), inserting an expression of contemporary urbanism into the landscapes of Burnaby Mountain. Burnaby also contains the British Columbia Institute of Technology and nearby Discovery Park, important institutions that shape the techno-logical inflection of the municipality's economy. It also encompasses 'Metrotown', the most successful regional town centre in Greater Vancouver, a product of a 30-year commitment to focusing growth in zones designated as priority development areas in the successive iterations of the regional plan.[12]

The City of Richmond (2006 population 174,000), south of the City of Vancouver across the north arm of the Fraser River, also projects a distinctive inner-suburban profile of contemporary redevelopment, including increased densities and recur-rent conflicts over land use. The centrepiece of Richmond's economy is Vancouver International Airport (YVR), Canada's most important air gateway to the Asia-Pacific and regarded widely as one of the world's most advanced airports in terms of design features. YVR is a propulsive feature of the Richmond (and Metro Vancouver) economy in its own right and also generates significant spill-over growth and benefits, as in ship-ping and air freight businesses and proximate business hotels catering to travellers. The City of Richmond is also home to a number of niche-level advanced technology firms in the transportation, aerospace, and sub-sea sectors. Richmond RTC projects a largely retail and local services ambience, but the recent opening (August 2009) of the Canada Line linking the centre to downtown Vancouver (23 minutes travel time) seems likely to offer new business development opportunities. Richmond's socio-ethnic profile has been comprehensively reconfigured by waves of immigration from Asia, most notably from Hong Kong and other Chinese societies, as described in the previous section, an experience that has served to insert an insistent transnational trajectory into Richmond's neighbourhoods, landscapes, and social morphology.

Growth and Change on the Metropolitan Periphery

Development pressure in the broader Vancouver region extends beyond the GVRD/Metro boundary to encompass rural (but rapidly urbanizing) communities such as Abbotsford, Matsqui, and Chilliwack, taking in the distinctive green spaces (wilderness, farming, and recreation) of the Fraser Valley, encompassing 150 kilometres from Richmond in the west to Hope at the entrance of the Fraser Canyon. It is within these extensive tracts of suburban and exurban space that the battle for sustainable development will likely be won or lost, since these are the principal zones of low-density development and single-family subdivisions, sprawl, and encroachment on the region's high-value green spaces.

Within Vancouver's outer suburbs, the City of Surrey is the largest municipality by area and carries a reputation as a somewhat rough-around-the-edges municipality with high incidence of poverty, crime, and sprawl. Surrey encompasses the low-density strip malls and business and industrial parks that are difficult to service by means of public transit. At the same time, Surrey boasts the highest growth rates of any of the region's municipalities, is already Metro's 'second city' after Vancouver in terms of population (433,000 as of the 2006 Census), and is widely touted by civic leaders as Vancouver's urban frontier. Surrey has transitioned from its origin as a site of principally European settlement to a multi-ethnic and multicultural community, observed notably in the large concentrations of South Asian immigrants that have transformed the municipality's community morphology, economy, and electoral politics.

The most consequential storyline in Surrey can be found in Surrey City Centre. The backdrop to this contemporary narrative of redevelopment and rebranding includes a 30-year effort by the regional planning authorities to promote a major regional town centre in Surrey, first as a counterweight to what was perceived as an over-concentration of growth in the City of Vancouver's central area, then as a means of reducing pressures of sprawl within Surrey itself, and more recently as a program for developing a second major urban centre in the region to complement Vancouver's urban core. This effort has long been frustrated by a mix of factors, including weak commitments (or even outright opposition) on the part of Surrey's councils to this keystone regional planning objective, failure to appreciate the power of agglomeration in the location of higher-order service functions, and a long-standing market preference for developing low-density industry and residential subdivisions in the suburbs generally.

Some of these factors persist. But there are signs of change, including the construction of a high-design tower in Surrey City Centre, which houses higher education and other services in a building that has won important international awards for design excellence. The mayor of Surrey, Dianne Watts, has proposed a vision of Surrey as a multicultural community incorporating progressive urbanism values, with a corollary more central role for Surrey in the evolving metropolitan setting. The policy approach includes neighbourhood intensification in the interests of sustainable development, an economic program that seeks to attract high-value businesses and talent, incentives for using public transit and other alternatives to private autos, and more stringent urban design guidelines.

Sector Case Study: Vancouver's Video Game Industry

We have elected to showcase Vancouver's video game industry because it demonstrates several cross-cutting features of the region's development. It is a 'new economy industry' but has strong links to longer-standing sectors in Vancouver such as film production. It illustrates in a particularly vivid way the evolution of new media in the cultural economy of the city, including the formation of 'neo-artisanal' labour (Norcliffe and Eberts 1999), and distinctive (and in many ways problematic) work practices and career trajectories (Peck 2005). Finally, although the industry exhibits strong concentrations within the inner city and thereby exemplifies our earlier point about the emergence of new specialized production spaces in the urban core, it also includes important sites within the inner suburbs, notably Burnaby (Figure 12.10), and therefore offers insight into the processes of internal specialization and inter-industry network formation in the contemporary metropolis.

Vancouver is now one of the global hubs for the video game industry, which in 2007 generated worldwide $41.9 billion in revenue (games, consoles, and hand-held devices). At the end of 2008, there were within Vancouver approximately 145 firms directly and indirectly involved in the video game industry, the vast majority concerned with game design and production. At that point, the industry employed roughly 3500 people,

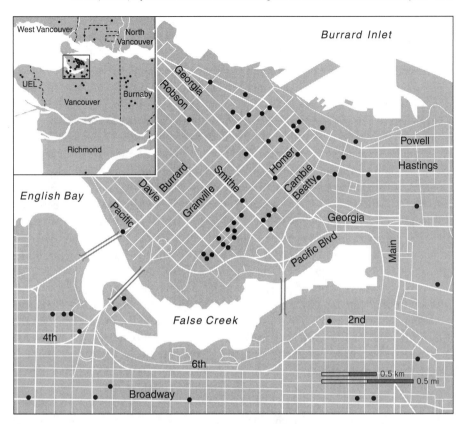

Figure 12.10 Distribution of video game firms in Vancouver (2010).

which represented a gain of approximately 1500 employees in just five years. In the past, the industry was tagged as 'recession-proof'. The per-hour cost of playing video games makes it one of the lowest-cost forms of paid entertainment available. At the beginning of the 2008 recession, the industry was even bullish about its prospects, predicting that it would lure yet more customers into low-cost gaming. That prediction has already proved incorrect, however, and the video game industry, at least in Vancouver, has suffered. At the end of 2009, rumour on the street had it that 800 video game workers were unemployed and that the largest video game firm in the city, Electronic Arts (EA), was poised to lay off a yet unspecified number of employees at its Vancouver site.

That said, the industry is clearly here to stay, having reached the threshold of sustainability and self-reproduction and with a hierarchy of firms already entrenched. It varies from one person making occasional educational games at the University of Victoria library while holding a Canadian government student loan to a branch of the largest video game company in the world, Electronic Arts, based at a suburban 'campus' in Burnaby. That suburban campus accommodates more than 1000 employees in a five-storey futuristic building, 'the magic factory', which opened in 1999.

The very first glimmerings of the video game industry appeared in the early 1950s, the result of whimsical play on the cathode ray tube on the part of some bored Cold

War Pentagon researchers. The first commercial sale of a game took place in the early 1970s. Since then, video game production has become one of the fastest growing of advanced capitalism's cultural industries, now exceeding the value of Hollywood films as well as the total value of US book sales, both hardcopy and electronic.

The industry's production chain is divided into hardware and software. The hardware comes in various forms, but the most important is the game console, the machine that allows a user to play the software. Three main consoles are currently sold, each by a different firm: PlayStation 3 (PS-3), made by Sony; the Xbox, made by Microsoft; and the Wii, made by Nintendo. The actual manufacturing of the hardware typically takes place in cheap-labour locations offshore. Further, the hardware is often sold at below-cost as a loss-leader—because the real goal is to hook consumers on a particular company-made machine so that they will buy the real money-maker: the software—the game itself.

It is on this end of the business, the development of the software, that Vancouver firms have focused. Again, there is an important back story. Video games are sold by 'publishers' who then own the intellectual property (IP) rights. Publishers are typically large multinational corporations. In the recent past, there was only one publisher in the Vancouver video game industry, EA. But that has changed over the past five years, with a number of large multinational-corporation publishers entering the city, often by taking over existing locally owned, medium-sized firms. The most recent was the acquisition by French corporate giant Ubisoft of Action Pants, a firm with about 100 employees, in February 2009. Earlier, in 2005, Activision bought Vancouver's Radical Entertainment, but Activision in turn was bought by the giant French conglomerate Vivendi SA in 2007. In addition, Microsoft and Disney, both of which are publishers, have moved into the city and are producing video games.

More generally within the Vancouver video game industry cluster, one can recognize three broad kinds of firms operating. At the top are the large multinational corporations like EA, Ubisoft, Microsoft, Disney, and Vivendi, which produce games but also publish them. The length of a game cycle varies, but it can be as short as three months or as long as five years and can cost anything from less than $1 million to more than $20 million. Then there are the SME independent developers such as Action Pants, the firm recently taken over by Ubisoft. Large publishers often contract these firms to produce games for them. Action Pants, for example, had been under contract to Ubisoft as an independent developer before Ubisoft decided it wanted a Vancouver studio of its own to complement its Montreal facilities (later, in July 2009, Ubisoft announced that it was also moving into Toronto, attracted by a $263-million subsidy from the Ontario Liberal government). At the bottom of the firm hierarchy are the small start-ups, such as the undergraduate working in the University of Victoria library for four months and using a government-provided student loan as financing. Another example of a start-up is a firm based in Burnaby operating out of a two-bedroom apartment. This firm has five partners, all located on Canada's east coast, who used savings from previous jobs to develop a game that they thought would 'change the world'. Like that of the undergraduate, their big dream was that a large corporation would become interested in their 'great idea' and take over their business. They all would become millionaires.

But why has Vancouver become the place to dream that dream? Why has the city become one of the global hubs of the video game industry along with Austin, Los

Angeles, Paris, Seoul, and Tokyo? According to the human resource managers of many of the firms located in Vancouver, it is because of the natural and social amenities of the city. This is a version of Richard Florida's (2002, 223) 'creative cities' argument that the geography of the new economy is set 'by the location choices of creative people—the holders of creative capital'. In this argument, talented people are drawn to 'cool' urban places like Vancouver (although that can be difficult to imagine on a rain-sodden Monday morning in late November).

The counter, and more plausible, argument is that Vancouver's success as a hub for the video game industry has little to do with the city's seawalls, mountain views, and club scene but rather with the spatially agglomerative, cluster effects initially generated in 1991 when EA and another firm, Radical Entertainment, were born in the same location within the city. Ann Markusen (1996), in a well-known typology, examined different forms of industrial clustering using the idea of 'sticky places': that is, the notion that some geographical sites seem better able to hold industry than others. Vancouver's inner-city location appears to be precisely such a 'sticky place'. Markusen offers a four-fold typology, one of which, the 'hub and spoke' district, comes closest to defining the video game industry cluster in Vancouver. In her original model, a large corporation is at the centre of the cluster around which satellite firms circulate. In Vancouver's case, EA is at the centre along with Radical, and then through a process of 'fission', successive generations of spin-off firms have arisen, beginning as start-ups and, if they are successful, later developing into SMEs.

The beginning of that fission process actually took place in 1983 with the founding of Distinctive Software Inc. (DSI) by two Vancouver schoolboys, Don Mattrick and Jeff Sember, who designed their first game, Evolution, in 1982. They were initially based in the basement of Sember's parents' house in suburban Burnaby, but after Mattrick bought out Sember in 1988, Mattrick moved to the up-and-coming but still edgy inner-city neighbourhood of Yaletown, then a warehouse district where rents were still relatively cheap and internal building spaces could easily be refurbished into studio space. DSI found success, especially through the late 1980s, and attracted EA, which bought it in 1991 for $11 million. But there were internal disputes over the takeover, leading to the formation of a second company, Radical Entertainment, by disgruntled DSI personnel unwilling to go to EA. Radical Entertainment remained the second-largest video game maker in Vancouver until it was taken over by Activision/Vivendi in 2005. The rest is history. Through the fission process, the Vancouver video game industry has now reached its fifth generation of spin-off companies within 20 years, and Yaletown, where DSI and later Radical located, has become the key industrial urban quarter for the industry, with a second subsidiary precinct in Burnaby around EA's 'magic factory'.

Conclusion: Observations from the Vancouver Case Study

We have been concerned in this chapter with presenting a profile of Vancouver that embodies transformative experiences since the 1970s. In both social and economic terms, and also in the remaking of the region's land use and built environment, the master narrative takes the form of a steep decline in the city's vocation as centre of a classic core–periphery staples economy and Vancouver's coincidental ascendancy as an outpost of the Asia-Pacific, fuelled by inflows of immigrants, capital, and culture from the 1980s onward. These events constitute the principal lineaments of Vancouver's experience of contemporary urbanism, mediated and shaped by distinctive policy and planning exercises that have inserted the public interest into the redevelopment process, notably within the metropolitan core. Enhancement of the public realm, policy commitments to residential development, investments in tourism infrastructure and marketing, and the proliferation of restaurants and cafés all serve to underscore Vancouver's exemplary status as a city of consumption (after Glaeser, Kolko, and Saiz 2001), setting it at least a little apart from other Canadian city-regions, which are more dependent on industrial production and exports.

That said, Vancouver is by no means immune to the forces of globalization and restructuring that impose themselves on both advanced and transitional cities alike, and we can readily identify inscriptions of these broader tendencies within the region. Thus, the sequence of abrupt industrial innovations that have characterized urban economic development since the 1980s, including the technology-driven 'new economy' of the 1990s and the emergence of a 'new cultural economy' of the city as described by Allen Scott (1997; 2006) and others, finds expression within the region. These experiences have transformed districts and sites within the region's economic landscape and have naturally influenced the trajectories of creativity and innovation widely seen as signifiers of advanced urban economies (Barnes et al. 2009) while generating evidence of dislocation and volatility, as our case study of the video game industry and vignettes of inner-city districts demonstrate.

More broadly within the region, there are pressures on many sectors, including the gateway functions and allied industries that form such a large part of the metropolitan economy (Gilbert and Perl 2008). Metropolitan Vancouver's economy has exhibited a degree of buoyancy since the deep recession of the early 1980s, aided by the resiliency of an entrepreneurial SME enterprise base and by inflows of FDI and senior government capital investments. But Vancouver is by no means a global economic powerhouse (Peck 2009) and can hardly count on another *deus ex machina* on the lines of the infusions of talent and capital of the 1980s. Foremost among the more problematic socio-economic issues is the persistent gap between very high housing prices and decidedly average earnings (by national standards), which places great stress on households and individuals and exacerbates the inequality and polarization tendencies described by Alan Walks in Chapter 6.

Vancouver continues to attract admirers and mimics within international circles, buoyed by the city's consistent high rating in global rankings of liveability and promoted by globe-trotting public officials, journalists, and urbanists at large. 'Vancouverism', projected as a synthesis of planning, urban design, and liveability, has assumed a prominent niche within the larger lexicon and discourses of contemporary urbanism. And for many, Vancouver offers remunerative employment as well as a rich array of environmental and consumption amenities. But there are other narratives that must be disclosed in the interests of achieving a semblance of balance. At one level, we can acknowledge a crude binary of the glittering inner-city landscapes of high-rise condos, loft conversions, upscale amenities, and new economy industries juxtaposed against the deprivation and disorder of the Downtown Eastside, perhaps the best known features of contemporary Vancouver and a focus of international scrutiny during the 2010 Vancouver Winter Olympics.

But this fixation on what is after all a restricted terrain within the metropolis should not blind us to the more problematic features that abound in the region as a whole, including the capacity of the economy to generate employment opportunities for an expanding labour force, as well as the complex social and environmental problems acknowledged throughout this chapter. At least one observer of the Vancouver regional planning experience has suggested that the 'easy gains' of the past 30 years are likely to be followed by stiffer challenges ahead (Tomalty 2002). Metro Vancouver (and its predecessor, the GVRD) has achieved worthwhile goals in land-use planning, within the limits of its jurisdiction and resources, necessarily proceeding on key issues by means of inter-municipal consensus. But more exigent issues—enforcing municipal compliance with regional growth management goals, addressing climate change, promoting affordable housing, and more effectively mobilizing the collective resources of the region for economic development within globalizing and increasingly competitive markets—will require stronger leadership, greater regulatory and policy powers, and a constructive relationship with the provincial government.

References

Anderson, K. 1991. *Vancouver's Chinatown: Racial Discourse in Canada, 1875–1980.* Montreal: McGill-Queen's University Press.

Barman, J. 1991 *The West beyond the West: A History of British Columbia.* Toronto: University of Toronto Press.

Barnes, T.J., A. Holbrook, T.A. Hutton, and R. Smith. 2009. 'Creativity and innovation in the Vancouver city-region'. Integration paper prepared for meeting of the Innovation Systems Research Network, University of Toronto, 5–6 November.

Barnes, T.J., and T.A. Hutton. 2009. 'Situating the new economy: Contingencies of regeneration and dislocation in Vancouver's inner city'. *Urban Studies* 46 (5 and 6): 1247–69.

Bauder, H. 2003 '"Brain abuse," or the devaluation of immigrant labour in Canada'. *Antipode* 35 (4): 699–717.

Bell, D. 1973. *The Coming of Post-industrial Society: A Venture in Social Forecasting.* New York: Basic Books.

Blomley, N. 2004. *Unsettling the City: Urban Land and the Politics of Property.* New York and London: Routledge.

Bradbury, N. 1989. 'Cash-rich Taiwan eyeing Canada as a place to invest its money'. *Chinatown News 3 September*: 30 (reprinted from *The Financial Times of Canada*).

Bunting, T., A. Walks, and P. Filion. 2004. 'The uneven geography of housing affordability stress in Canadian metropolitan areas'. *Housing Studies* 19 (3): 361–93.

Coffey, W.J. 1994. *The Evolution of Canada's Metropolitan Economies.* Montreal: Institute for Research on Public Policy.

Davis, H.C., and T.A. Hutton. 1991. 'An empirical analysis of producer service exports from the Vancouver Metropolitan Region'. *Canadian Journal of Regional Science* 14: 375–94.

———. 1994. 'Marketing Vancouver's services to the Asia Pacific'. *The Canadian Geographer* 38: 18–28.

Edgington, D. 1996. 'Japanese real estate involvement in Canadian cities and regions, 1985–1993'. *The Canadian Geographer* 40 (4): 292–305.

Florida, R. 2002. *The Rise of the Creative Class—And How It's Transforming Work, Leisure, and Everyday Life.* New York: Basic Books.

Garreau, J. 1981. *The Nine Nations of North America.* Boston: Houghton Mifflin.

Gilbert, R., and A. Perl. 2008. *Transport Revolutions: Moving People and Freight without Oil.* London: Earthscan.

Glaeser, E.L., J. Kolko, and A. Saiz. 2001. 'Consumer city'. *Journal of Economic Geography* 1: 27–50.

Graham, S., and S. Marvin. 2001. *Splintering Urbanism, Networked Infrastructures, and the Urban Condition.* London and New York: Routledge.

Gray, C. 2006. 'Affordable and modest market housing: Sub-area 2A of southeast False Creek (the Olympic Village)'. City of Vancouver Administrative Report A11. 28. September. Vancouver: Director of the Housing Centre. http://www.city.vancouver.bc.ca/ctyclerk/cclerk/20061017/documents/a11.pdf.

Gutstein, D. 1990. *The New Landlords: Asian Investment in Canadian Real Estate.* Victoria: Porcepic Books.

Hamnett, C. 2003. *Unequal City: London in the Global Arena.* London: Routledge.

Hamnett, C., and D. Cross. 1998. 'Social polarisation and inequality in London: The earnings evidence 1979–1995'. *Environment and Planning C* 16: 659–80.

Hiebert, D. 2009. *The Economic Integration of Immigrants in the Metropolitan Vancouver Area.* Montreal: Institute for Research on Public Policy.

Hutton, T.A. 1997. 'The Innisian core–periphery revisited: Vancouver's changing relationships with British Columbia's staple economy'. *BC Studies* 113: 69–100.

———. 1998. *The Transformation of Canada's Pacific Metropolis: A Study of Vancouver.* Montreal: Institute for Research on Public Policy.

———. 2004. 'Post-industrialism, Post-modernism and the reproduction of Vancouver's central area: Retheorising the 21st-century city'. *Urban Studies* 41 (10): 1953–82.

———. 2008. *The New Economy of the Inner City: Restructuring, Regeneration and Dislocation in the Twenty-First-Century Metropolis.* London and New York: Routledge.

———. 2009. *Multilevel Governance and Urban Development: A Vancouver Case Study.* Vancouver: Centre for Human Settlements, University of British Columbia.

Ley, D. 1980. 'Liberal ideology and the post-industrial city'. *Annals of the Association of American Geographers* 70: 238–58.

———. 1996. *The New Middle Class and the Remaking of the Central City.* Oxford: Oxford University Press.

———. 2003. 'Seeking *homo economicus*: The Canadian state and the strange story of the business immigration program'. *Annals of the Association of American Geographers* 93 (2): 426–41.

———. 2006. 'Explaining variations in business performance among immigrant entrepreneurs in Canada'. *Journal of Ethnic and Migration Studies* 32 (5): 743–64.

———. 2010. *Millionaire Migrants: Trans-Pacific Life Lines.* Oxford: Blackwell-Wiley.

Ley, D., and J. Tutchener. 2001. 'Immigration, globalisation and house prices in Canada's gateway cities'. *Housing Studies* 16 (2): 199–223.

McGee, T. 2001. 'From village at the edge of the rainforest to Cascadia: Issues in the emergence of a livable subglobal world city'. In F.C. Lo and P.J. Marcatullio, eds, *Globalization and the Sustainability of Cities in the Asia Pacific Region*, 428–454. Tokyo: The United Nations University.

McMillan, S. 2004. 'Toward a livable region? An evaluation of business parks in Greater Vancouver'.

(Master's thesis, School of Community and Regional Planning, University of British Columbia).

Markusen, A. 1996. 'Sticky places in slippery spaces: A typology of industrial districts'. *Economic Geography* 72: 293–313.

Metro Vancouver. 2009. 'Labour force and employment activity, annual average 1987–2008'. http://www.metrovancouver.org/about/publications/Publications/lfs-annual-average.pdf.

Mitchell, K. 2004. *Crossing the Neoliberal Line: Pacific Rim Migration and the Metropolis.* Philadelphia: Temple University Press.

Mollenkopf, J.H., and M. Castells, eds. 1991. *Dual City: Restructuring New York.* New York: Russell Sage Foundation.

Moos, M., and A. Skaburskis. 2007. 'The characteristics and location of home workers in Montreal, Toronto and Vancouver'. *Urban Studies* 44 (9): 1781–1808.

———. 2010. 'Workplace restructuring and urban form: The changing national settlement patterns of the Canadian workforce'. *Journal of Urban Affairs.*

Morgan, K. 2004. 'The exaggerated death of geography: Learning, proximity, and territorial innovation systems'. *Journal of Economic Geography* 4: 3–21.

Norcliffe, G., and D. Eberts. 1999. 'The new artisan and metropolitan space: The computer animation industry in Toronto'. In: J.-M. Fontan, J.-L. Klein, and D.G. Tremblay, eds, *Entre la métropolisation et la village global : les scènes territoriales de la reconversion*, 215–32. Quebec City: Presses de l'Université du Québec.

OECD (Organisation for Economic Co-operation and Development). 2000. *Science, Technology and Innovation in the New Economy.* OECD Policy Brief. Paris: OECD.

Oke, T.R., M. North, and O. Slaymaker. 1992. 'Primordial to prim order: A century of environmental change'. In G. Wynn and T. Oke, *Vancouver and its Region*, 147–70. Vancouver: University of British Columbia Press.

Olds, K. 1995. 'Globalization and the production of new urban spaces: Pacific Rim megaprojects in the late 20th century'. *Environment and Planning A* 27 (11): 1713–43.

———. 2001. *Globalization and Urban Change: Capital, Culture and Pacific Rim Megaprojects.* Oxford: Oxford University Press.

Peck, J. 2005. 'Struggling with the creative class'. *International Journal of Urban and Regional Research* 29: 740–70.

———. 'Remaking the Vancouver model'. Hampton Fund Proposal. Vancouver: Department of Geography, University of British Columbia.

Pendakur, K., and R. Pendakur. 1998. 'The colour of money'. *Canadian Journal of Economics* 31 (3): 518–48.

Picot, G. 2004. 'The deteriorating economic welfare of Canadian immigrants'. *Canadian Journal of Urban Research* 13 (1): 25–45.

Punter, J. 2003. *The Vancouver Achievement: Urban Planning and Design*. Vancouver: University of British Columbia Press.

Reitz, J. 2001. 'Immigrant skill utilization in the Canadian labour market: Implications of human capital research'. *Journal of International Migration and Integration* 2: 347–78.

Sassen, S. 1990. 'Economic restructuring and the American city'. *Annual Review of Sociology* 16: 465–90.

———. 2001. *The Global City: New York, London, Tokyo*. 2nd edn. Princeton, NJ: Princeton University Press.

Scott, A.J. 1988. *Metropolis: From Division of Labor to Urban Form*. Berkeley and Los Angeles: University of California Press.

———. 1997. 'The cultural economy of cities'. *International Journal of Urban and Regional Research* 21: 323–39.

———. 2006. 'Creative cities: Conceptual issues and policy questions'. *Journal of Urban Affairs* 28: 1–17.

Smith, H., and D. Ley. 2008. 'Even in Canada? The multiscalar construction and experience of concentrated immigrant poverty in gateway cities'. *Annals of the Association of American Geographers* 98 (3): 686–713.

Soja, E. 2000. *Postmetropolis: Critical Studies of Cities and Regions*. Oxford: Blackwell.

Spencer, G., and T. Vinodrai. 2008. 'Where have all the cowboys gone? Assessing talent flows between Canadian cities'. Presentation to the annual meeting of the Innovation Systems Research Network, Montreal, 30 April–2 May.

Statistics Canada. 1995. *Labour Force Annual Averages 1989–1994*. Ottawa: Minister of Industry, Science and Technology.

Steyn, D.G., M. Bovis, M. North, and O. Slaymaker. 1992. 'The biophysical environment today'. In G. Wynn and T. Oke, eds, *Vancouver and Its Region*, 267–89. Vancouver: University of British Columbia Press.

Symonds, W., D. Yang, and L. Zuckerman. 1991. 'Chinese immigrants bring capital and entrepreneurial spirit to Canada'. *Chinatown News* 18 October: 3 (reprinted from *Business Week*).

Tang, W.-S. 2008. 'Hong Kong under Chinese sovereignty: Social development and a land (re)development regime'. *Eurasian Geography and Economics* 49 (3): 341–61.

Teo, S.Y., 2007. 'Vancouver's *newest* Chinese diaspora: Settlers or "immigrant prisoners?"' *GeoJournal* 68: 211–22.

Tomalty, R. 2002. 'Growth management in the Vancouver region'. *Local Environment* 7 (4): 431–45.

Veblen, T. 1899 [1953]. *The Theory of the Leisure Class: An Economic Study of Institutions*. New York: Modern Library reprint/Transactions publishers, UK.

Waters, J. 2008. *Education, Migration, and Cultural Capital in the Chinese Diaspora*. Amherst, NY: Cambria Press.

Notes

1. As a demonstration of this point, the recession of the early 1980s generated 14 per cent unemployment in Vancouver, associated with particularly deep commodity price shocks, but produced less acute socio-economic impacts on Toronto and Montreal. In contrast, the downturn of a decade later generated significant labour shedding in Toronto (contraction of 5.6 per cent overall from 1989 to 1994, with the loss of 106,000 manufacturing jobs in the labour force) and Montreal (2.2 per cent overall employment decline in the same period, including the loss of 67,000 manufacturing jobs). Over the same period, Vancouver's labour force expanded by 11.7 per cent overall, led by an increase of 39,000 construction jobs and including a modest gain of 8800 jobs in manufacturing (Statistics Canada 1995).

2. The OECD defines the 'new economy' as follows: 'The new economy describes aspects or sectors of an economy that are producing or intensely using innovative or new technologies. This relatively new concept applies particularly to industries where people depend more and more on computers, telecommunications and the Internet to produce, sell and distribute goods and services' (OECD 2000). Ars Electronica elaborates upon this theme by emphasizing the role of the Internet and ICT in the emergence of a new cultural economy based on 'commons-based peer production', as opposed to earlier forms within which the firm represented 'the primary vehicle for innovation and development', and a focus on 'the notion of property as the anchor and pillar for the creation of value' ('A new cultural economy'. http://www.mediafuturist.com/2008/08/ars-electronica.html).

3. Capital substitution for labour has cut deeply into the ranks of clerical workers, formerly one of the largest aggregations of labour in the urban

workforce, while successive rounds of corporate restructuring and downsizing have significantly reduced the numbers of middle managers, another major occupational cohort of the late twentieth century services economy.

4. The fortunes of three major enterprises underpinning BC's staples economy, each located in Vancouver—MacMillan Bloedel, the Vancouver Stock Exchange, and the Bank of British Columbia—are instructive in this regard. MacMillan Bloedel was British Columbia's most important corporation, with its provenance in the formation of the HR MacMillan Export Company (1919) and more fully realized as an important international integrated forestry company via mergers with the Powell River Company (1951) and the Bloedel Welch Company (1959). Its pre-eminence was underscored by the construction of an iconic building at Georgia and Thurlow in Vancouver's downtown, designed by Arthur Erickson, BC's leading architect. Success in European and American markets generated a capital value of some $4 billion. But a succession of commodity price shocks, a failure to manage stocks of fibre, and increasing competition all contributed to MacMillan Bloedel's decline over the 1990s, leading to its takeover by American forestry giant Weyerhaeuser in 1999. This calamity was mirrored by analogous trends and events in the realm of finance. The Vancouver Stock Exchange (VSE), founded to generated venture capital for BC's resource economy, eventually came to be dominated by penny-stocks and was compromised by a series of fraudulent dealings that attracted international notoriety (for example, an article in *The Economist*, 'An Augean Stable: The Vancouver Stock Exchange', 16 January 1993) that led to its takeover by the Alberta Stock Exchange domiciled in Calgary and finally by the Toronto Stock Exchange. The Bank of British Columbia, established in Vancouver in 1966 by the Social Credit government of W.A.C. Bennett, was designed in part at least to generate capital for the expansion of resource industries in BC but failed to achieve viable levels of liquidity and was ultimately acquired in 1986 in a highly symbolic takeover by the Hong Kong Shanghai Banking Corporation (now HSBC Bank Canada).

5. The roster of major Seattle-based multinationals includes Weyerhaeuser, the world's largest integrated forestry company; Boeing, a globally leading aerospace corporation with senior management, design, fabrication, and assembly in the Seattle–Puget Sound region (as well as

many parts and service input providers overseas and a small head office of about 400 employees located in Chicago); Microsoft, the world's largest computer software corporation; Starbucks, the largest retail coffee company in the world; and Amazon, the global leader in online marketing and sales.

6. A more recent integrative bioregional signifier spanning the Cascadia finds expression in the designation of three bodies of water—the Georgia Strait (wholly in BC), the Strait of Juan de Fuca (shared between BC and Washington State), and Puget Sound (wholly within Washington) as the 'Salish Sea', acknowledging the pre-colonial interdependency of First Nations on the maritime resources of this inner coastal zone. Under this proposal, each of the three constituent bodies of water would retain its name for official purposes.

7. We can also identify economic uses of the environmental resources and signifiers of the region, including both design and material inputs for furniture, garment production, graphic design, and interior design and decor.

8. The singular fragmentation of Metro Vancouver's political jurisdictions, business community, and social forces no doubt plays into this long-term failure to produce a workable regional economic strategy, as does the long period of general economic buoyancy since the deep recession of the early 1980s. The most recent effort, led by VanCity and a number of senior local politicians (including then-mayor of Vancouver City, Larry Campbell, now a Liberal member of the Senate), presented what many saw as a coherent and progressive strategy, but this document has been effectively shelved along with the residues of numerous predecessors over the past 30 years.

9. The Vancouver experience of land-use change in the urban core especially underscores post-industrialism as a political ideology and policy preference as well as a socio-economic process. In a well-known article in the *Annals of the Association of American Geographers* in 1980, David Ley observed that the reformist TEAM (The Electors Action Movement) city council of the early 1970s rezoned False Creek South from heavy industry to medium-density, mixed-income housing, signifying 'the most dramatic metaphor of liberal ideology, of the land use implications of the transition from industrial to post-industrial society, from an ethic of growth and the production of goods to an ethic of amenity and the consumption of services' (Ley 1980,

252). A more recent article in *Urban Studies* acknowledges the recurrent role of planning innovation and public policy in the reconfiguration of Vancouver's metropolitan core from the classic features of the post-industrial city to a reordered terrain of new residential communities, creative industries, and high-amenity public realm (Hutton 2004).

10. The Downtown Eastside's social upgrading trajectory has been accelerated by developments at either end of the district. The Woodward's project, on the western margins of the DTES, takes the form of a comprehensive redevelopment of the former department store that anchored the important retail strip of West Hastings until the area's decline and includes 536 market housing units and 225 social housing units, as well as the Simon Fraser Centre for the Contemporary Arts and a host of complementary retail and social services. A kilometre or so to the east of the Woodward's site lies Strathcona, a historically low-income community of immigrants and working-class residents now experiencing gentrification, with at least one house selling for more than $1 million in recent years (Adele Weder, 'Block buster', *Globe and Mail*, 22 January 2010: S4).

11. The City of Vancouver Planning Department maintains a website comprising detailed area analyses of employment and economic trends, planning proposals, and city council decisions: www.vancouver.ca/corejobs.

12. Burnaby's singular commitment to the tenets of the regional plan and to its own aspirations for the Metrotown Regional Town Centre (RTC) was demonstrated in city council's rejection in 1984 of the Ghermezian brothers proposal for a West Edmonton Mall–scale mega-project well outside the perimeter of Metrotown, which, had it been built, would have seriously compromised the development prospects of the RTC.

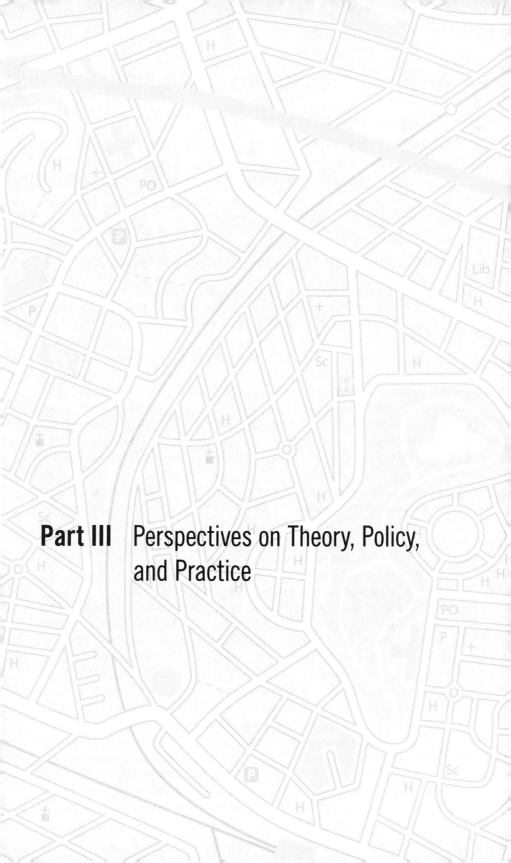

Part III Perspectives on Theory, Policy, and Practice

Perspectives on Theory, Policy, and the Future Urban Economy

Tom Hutton, Larry S. Bourne, Richard Shearmur, and Jim Simmons

Introduction: Economic Change and Urban Transformation

Our final task in this volume is to assess the theoretical and policy implications of the analysis of growth and change in urban Canada at three spatial scales: the national urban system, the city-region, and the internal structure and spaces of the metropolis. This multi-scalar framework will be complemented by a temporal span that takes in the 'long-wave' developmental sequences of the staples economy, industrialization, and the emergence of the urban service sector, as well as the more abbreviated episodes of innovation and restructuring that have characterized the past two decades in Canada and other advanced societies and cities. For this exercise, we review the findings of Chapters 2 to 6 for evidence of generalized tendencies, trajectories, dynamics, and growth characteristics at the level of the national urban system as a whole. We revisit the five case studies, both to confirm the imprints of general processes of change and to identity more contingent features of development in each city region. We also draw on the experiences of analogous research projects involving our contributors and on contemporary discourses and debates concerning urban growth and change.

The chapter opens with a discussion of the residual effects of staples development for Canadian cities. We provide evidence of continuing interdependencies, as well as the legacies of the resource economy for contemporary new industry formation, incorporating recent debates about the merits of the classic 'lock-in' version of path determinism versus evolutionary models of economic development.

Second, we rescale the territorial setting of contemporary economies of Canadian cities, from urban–regional industrial districts and clusters characterized by proximate production networks and circumscribed labour markets to the multi-scalar worlds of advanced economies shaped by global processes and the search for competitive advantage.

Third, we position Canadian cities and their constituent economies within evolving discourses of globalization and transnationalism. We acknowledge both the forceful insertion of city-regions within international circuits of finance, production, and culture and the strategic international gateway roles played by the largest cities, as well as the peculiarly bilateral structure of trade that in large part characterizes the export orientation of Canadian cities.

Next, we explore the robustness of principal late-twentieth-century models of urban economic transformation as applied to the development trajectories of Canadian city-regions, notably post-industrialism and post-Fordism. This discussion is informed by the empirical analysis and case studies presented in this volume and includes commentary on conceptual propositions associated with the emergence of the 'new cultural economy' since the 1990s. We also discuss the influence of new industry formation and allied social factors on the space-economy and morphology of the twenty-first-century metropolis.

In the final section, we offer conjecture on the prospective contours of the future Canadian urban economic landscape, necessarily limited by uncertainty but incorporating an inventory of observations from this volume that offer a platform for scenario-building and for identifying possibilities for innovation in governance, institutions, and policy that may enhance the resiliency of cities, their economies, and civil societies.

Retheorizing the Canadian City across Development Periods: Legacies, Lock-in, and the Limits of Path Dependency

Saskia Sassen (2006) has exhorted scholars to consider the deep economic histories of cities as a means of appreciating important continuities as well as identifying the many aspects of disjuncture that abound in this age of widespread industrial restructuring and recurrent economic shocks. For our investigation of change in the Canadian urban system, an appreciation of the 'deep economic histories' of cities necessarily means acknowledging the patterns of settlement and growth associated with the country's colonial status and specifically the extended epoch of resource extraction and trade, formulated brilliantly in Harold Innis's seminal theory of staples development (1933). Innis depicted the asymmetric spatial features of the classic core–periphery staples economy, comprising conditions of chronic underdevelopment in the vast Canadian resource hinterland that emanated from its dependence on a relatively compact urban core region situated in southern Ontario and along the upper St Lawrence, encompassing the great industrial metropoles of Montreal and Toronto together with their satellite manufacturing towns. According to the Innisian thesis, the developmental advantages enjoyed by the industrial cities of southern Ontario and the Laurentian region over the hinterland, augmented by the tariff barriers to imported manufactures established by the country's nineteenth-century National Policy, included higher value-added production, more stable labour markets, advanced transportation and communication networks, and, over time, the emergence of specialized services (banking, finance, management)—industries generally deficient within the peripheral regions.

Over the past quarter century, many of the interdependencies linking Canadian cities and their resource hinterlands have weakened appreciably with respect to resource processing and manufacturing, transportation and distribution, and, for some cities (Montreal, Vancouver, Winnipeg), head office functions and finance. The social correlates of the staples processing economy, represented by substantial communities in cities such as Ottawa, Montreal, and Vancouver, have all but disappeared. But we can readily identify important carryovers of staples development for contemporary Canadian cities, both in terms of established economic structures and functions and as legacies that have helped to shape the new economy of cities. As we observed in Chapters 2 and 3, Toronto continues to perform head office functions for resource sector corporations, while both Montreal and Vancouver accommodate the management of major resource-based utilities. Vancouver in particular still derives considerable employment and revenue from the seaborne export of natural resources. And the rise of Calgary as the ascendant metropolis of the 'New West' (Chapter 11) is inextricably bound up in its role as management, services, and financial centre for Alberta's energy sector.

Although the connection between resource hinterlands and Canadian cities has weakened, other types of connection are strengthening: in particular, increased personal mobility, the possibilities opened up by the Internet, and demographic changes such as an aging (and thus a retiring) population. These tendencies have increased the consumption, leisure, recreational, and residential role of the immediate hinterlands surrounding each metropolitan area, as shown in Chapter 5. Further, some basic connections remain unchanged: each city's hinterland continues to provide agricultural (climate permitting) and low-value resource (e.g., sand, gravel) products, as well as sheltering certain manufacturing activities that have been priced out of cities. Indeed, the very idea of what constitutes a city may be changing as physical criteria such as density and continuous built environment give way to less tangible criteria based on frequency of interaction, on the increasing separation between where people live and where they earn their income, and on the nature and geography of interdependency between the metropolitan core and its surrounding areas.

Within the manufacturing sector, and notwithstanding the move of much production activity to areas outside of (but accessible to) Canada's major cities, the head start enjoyed by (first) Montreal and then Toronto in the latter part of the nineteenth and the twentieth centuries, resulting from their lead roles as advanced manufacturing centres within the national staples economy, has never been overcome by other cities. This is readily seen in the durability of industrial production in the Toronto city-region, which ranks at the top of the larger metropolitan areas both in Canada and the US in the percentage of the workforce engaged in manufacturing, relative to the truncated manufacturing bases of other large city-regions such as Vancouver, Ottawa-Gatineau, and Calgary. In a number of urban centres, notably Vancouver, the staples economy constrained progress to more advanced forms of production through the shadow effect of high resource-industry wages on labour costs for manufacturing industries.

These legacies are well known, but there are others. The development of certain cities as regional cores for staple-based hinterlands gave rise to the formation of specialized intermediate service industries and employment, including finance (banking, stock exchange, and venture capital functions), engineering, surveying, geological services, management consulting, and other business services. Much of this 'core competency' in specialized services has survived the relative decline of the urban staples economy and has been redeployed with some success in international markets, as acknowledged in the Montreal (Chapter 8) and Vancouver (Chapter 12) case studies.

Second, the staples era for many Canadian cities included the formation of a distinctive built environment within central areas, as seen in stations and terminals, processing mills, dense congeries of warehousing, wholesaling, and distribution facilities, and allied housing, as well as office towers constructed wholly or in part to domicile resource corporations. These older structures were built to robust construction standards and are conducive to adaptive reuse, so many have been recently converted to space for a new economy of cultural industries, institutions, and labour. The heritage built environment in the inner cities of Toronto (for example, Liberty Village and the Distillery), Montreal (Mile End and the Plateau), and Vancouver (Yaletown, Victory Square, and Gastown) now form part of the essential fabric of the cultural economy, as our volume has shown. (The same can be said, albeit on a smaller scale, for heritage districts in Halifax, Ottawa, and Victoria.) Here, we can reference Jane Jacobs's (1961, 188) prescient observation on the utility of the urban fabric in industrial innovation: 'Old ideas can sometimes use new buildings. New ideas must use old buildings'—an apposite commentary on the present-day synergies between 'old' (heritage) buildings and the 'new economy'. These changes, often heralded as signs of urban renewal, are also accompanied by the displacement of people who do not benefit from them. The general context within which the latest phase of urban renewal is occurring, one of cutbacks in social services and inadequate provision of affordable housing (see Chapter 6 by Alan Walks), means that the advent of the cultural economy is also conjuring up a somewhat less discussed low-wage, low-opportunity economy of contract and contingent workers (Moulaert, Rodriguez, and Swyngedouw 2002; Peck 2005).

A third legacy of the staples era for Canadian cities consists of the psychic connections between urban populations and the natural environment. The acceleration of staples extraction in the postwar period produced conflicts over shares of resources at the regional level and concerns regarding resource depletion and environmental degradation, stimulating the formation of environmental NGOs and other social movements. The rise of the urban sustainability movement, which increasingly finds expression in city policies and programs for 'green urbanism' and for a 'green economy', represents an important residual of the traditional interdependency between Canadian cities and their resource hinterlands. However, when contrasted with many similarly developed European cities, Canadian cities remain highly sprawled and energy-intensive: their deep dependency on an historically abundant (but rapidly depleting) natural resource base has also bred a profligacy in resource utilization that has yet to be reined in (World Bank 2010).

Interrogating the 'Lock-in' Model of Path Dependency

While our analysis and case studies disclose the continuing influence of resource extraction on the economies of Canadian cities, the relationship between staples extraction and contemporary urban economic development presents a more complex relationship than one of simple path dependency. Ron Martin (2010) recently presented a critique of the classic interpretations of path dependency, which insist on developmental stasis and conditions of equilibrium once a region (or technology, industry, or industrial district) has 'locked-in' to a specific developmental pathway. In the canonical versions, elucidated notably by Paul David (2005; 2007) and Brian Arthur (1989; 1994), path dependency takes the form of a deterministic condition within which (for the object under study—region, industrial district, or industrial complex) 'history comes to an end, and stasis rules, until such time that an exogenous disturbance moves the system onto another structural or technological path' (Martin 2010, 8). Further, the 'lock-in' mechanism of path dependency produces over extended time 'more of the same or the simple reproduction of what is already there' (2010, 10).

But as Martin points out, such conditions are rarely observed in real economic land-scapes within which institutions typically comprise complex composites of actors and entities, where innovation at the micro-scale is nearly constant, and where competi-tion and other exogenous pressures produce adaptation and evolution of production systems and trajectories. Here Martin quotes Schneiberg, who observes that 'change can emerge within existing pathways from a number of endogenous institutional processes, from bricolage, recombination or assembly of fragments of alternative industrial orders, to the borrowing, transportation and elaboration of more or less coherent secondary paths' (Schneiberg 2007, 70, quoted in Martin 2010, 16). Further, new industries produced by episodes of innovation and entrepreneurship can develop alongside established ones, configuring a relayering of the regional–industrial econ-omy, including the reformation of long-established industrial districts (see Belussi and Sedita 2009). As an exemplar of this process of recombination, relayering, and evolution, Martin describes the development pathway of the high-technology cluster in Cambridge (UK), observing that at least six discernible industrial–technological pathways have emerged since the establishment of Cambridge Consultants in the early 1960s. He concludes that it would be difficult to argue that the Cambridge cluster has locked into any stable development path.

This evolutionary model of path dependency surely offers a better fit with the eco-nomic development experiences of Canadian cities than the canonical version that insists on stability and lock-in. The empirical analysis presented in Part I disclosed sig-nificant divergence within the Canadian urban system, while the five case studies in Part II revealed very different trajectories of growth and change at the upper echelons of the urban hierarchy, derived from the local–regional contingencies observed in each city. The dominance of a few major sectors has been accompanied (or in some cases—nota-bly Vancouver—supplanted) by the emergence of smaller and more diverse industries, enterprise types, and labour, following the episodes of innovation and restructuring that have characterized the past two decades and vividly evoking the processes of innov-ation and recombination elucidated by Martin, Harald Bathelt (notably in his work on the media sector in Leipzig; see Bathelt and Boggs 2003), and others. At another level, Mario Polèse and Richard Shearmur's (2004) account of the reversal of fortunes of Canada's two largest cities, Montreal and Toronto, during the 1970s demonstrates the possibilities of major change in the trajectories of urban development, while the 1980s saw the comprehensive shift of Vancouver's vocation from urban core of a regional resource periphery to incipient Asia-Pacific city and society (Hutton 1998).

The Role of Governance and Policy in Reshaping Development Trajectories

Governance, institutions, and policy are factors promoting economic change in many Canadian cities. To a degree, this follows the strategic shift from 'managerial' to 'entre-preneurial' local governance identified by David Harvey two decades ago (1989), as well as pressures associated with more recent rounds of innovation, globalization, and restructuring. Governance reform is in part of course a result of the search for more efficient local service delivery, but in some cases it also reflects a preference for institutions that can support economic development: the multi-layered institutional structure of development agencies in Montreal provides a vivid example of this. Most cities in Canada, at least those near the top of the hierarchy, are engaged in the pursuit of competitive advantage as they seek to attract larger shares of capital, enterprises, and talent. Some Canadian cities, like many internationally, are 'recombining' attributes of place, imageries, design, and human capital in an effort to achieve a rebranding of civic identity, including, in some cases, 'hallmark events' such as international cultural and sporting festivals. Even within the prosaic domains of land-use policy and zoning, Canadian cities are endeavouring to introduce innovative policy clusters to pave the way for new industries such as biotechnology, film and video production and other new media, and green technologies, among others—a departure from the older (and in many cases obsolescent) activities still present in some form. These policy factors, even in the aggregate, do not of course produce totalizing change, but they do exert a constant tendency favouring evolution—as does the operation of the market with its insistent competitive pressures and near-constant restructuring effects.

Recombination is in part derived from endogenous attributes. But transformative change within the economies and labour markets of Canadian cities is also attributable to the interplay of forces operating within inter- and intra-metropolitan space, as we shall now demonstrate.

Situating Canadian Cities within Discourses of Globalization and Transnationalism

Our examination of Canadian cities in a changing international arena amply demonstrates the power of exogenous influences for national urbanization processes, in both economic and social terms. With regard to the former, a near-doubling in the proportion of Gross Domestic Product accounted for by exports since 1971 reflects Canada's status as a major trading nation (Chapter 2), while in social terms international immigration is clearly a key driver—and in certain cases the most important propulsive force—of urban population and labour force growth. Further, the flows that shape the contours of urban growth and change in Canada increasingly include the two-way exchange of key growth factors typical of all advanced economies: capital, culture, technology, and information.

That said, the detailed empirical analysis presented in Chapters 2 and 3 also reveals just how selective this experience of globalization and transnationalism is for Canada's urban system and its constituent city-regions. More particularly, the overwhelming preponderance of Canada–United States commodity flows configures a bilateral trading relationship rather than a comprehensive globalization of external exchanges. As further elaboration, our analysis underscores the increasing weight of sub-continental trade flows taking the form of compartmentalized trade, business, and personnel flows between pairs of transborder cities: Toronto and New York, Montreal and Boston, Vancouver and Seattle. Thus, what on an initial examination presents as an international trading posture reveals itself more as a high-volume cross-border phenomenon: a continuation of the national economy's branch plant status in many ways. This trend is hugely important for Canada and for Canadian cities, one implication being high levels of regional dependency as well as a spatial fragmentation of national trade linkages.

Yet there is no doubt about the enormous—and increasing—importance of international immigration in determining rates of urban growth and change and in encouraging a concentration of growth. Simply stated, Canada is among the two or three global leaders in accepting international immigrants, along with the US and Australia, and is the leader in per capita immigrant flows. As for our earlier discussion of trade relations, the inward flow of human capital in the form of immigrants is highly focused, in this case on Canada's largest cities and more especially on our 'power metropolises': Toronto, Vancouver, Montreal, Calgary, and Ottawa-Gatineau. If we add Edmonton to make a 'big six' city cohort, then this group, constituting about 55 per cent of the national population, attracts about 82.5 per cent of Canada's immigrants. Transnationalism in Canada, then, both as a theoretical construct (Smith 2001) and an empirical reality (Ley 2010), within which ethno-cultural communities straddle national boundaries, is a signifier of metropolitan—not merely 'urban'—status in Canada.

To place this metropolitan bias in further perspective, we can compare the percentage of foreign-born populations in Canadian cities with that of those in the US, another of the world's leading receptor societies. On this measure, Toronto (with 45.7 per cent of its population foreign-born) and Vancouver (39.6 per cent) rank first and second, respectively, with American cities (in order, Miami [37 per cent foreign-born], Los Angeles [34.2 per cent], San Francisco [29.6 per cent], and New York [28.2 per cent]) occupying the next four places and with the top 10 rounded out by Calgary (23.6 per cent), Montreal (21.2 per cent), Edmonton (18.5 per cent), and Ottawa-Gatineau (18.1 per cent) (all data for metropolitan areas). As the forecasts presented in Chapter 7 indicate, these proportions are almost certain to increase in the future. Of course, these figures do not do justice to another important window to the world that is opened by Quebec's (and particularly Montreal's) role within the global Francophonie: indeed, Montreal is the world's second-largest francophone city after Paris: more than 70 per cent of its population declares French as its principal language.

Our analysis reveals another dimension of this transnationalism that shapes growth among the largest Canadian cities: an analysis of immigration by ethnicity, complemented by an examination of intercity air travel patterns, discloses particularly intimate relations between Montreal and other centres of la Francophonie, notably Paris and Brussels, and between Vancouver and key Asia-Pacific centres such as Hong Kong, Tokyo, Beijing, and (in south Asia) Delhi. The Toronto region, which often presents itself as Canada's candidate global city, reflects both defining internal dimensions (foreign-born population; coincident professionalization and polarization: see Walks and Bourne 2006) and the external measure of its diverse international air connections—perhaps a scaled-down version of New York or London, cities that by consensus occupy the peak of the global urban hierarchy.

Urban Economic Development in Multi-scalar Space

Just as we can demonstrate the pervading influence (and limitations) of economic regimes spanning temporal periods, we can also acknowledge the intersections of developmental tendencies across space, as follows:

1. the waxing power of key processes (economic, cultural, political) operating at the global scale: cross-spatial trends that have reshaped urban hierarchies and systems as well as privileging higher-order metropolitan cities, as elucidated in the world city model of Peter Hall and the global city theses of John Friedmann and Saskia Sassen;

2. national urban system dynamics, which feature complex economic, socio-cultural, and political relationships between cities, including those that link the 5000-kilometre-long archipelago of Canadian cities examined in Chapters 2 and 3 of this volume;

3. extended regional space, including multi-centred urban conurbations such as the southern-central Ontario city-region or the 'Megalopolis' of the US northeast seaboard, the Ruhr, and the Randstad; these regions incorporate complex sprawling production networks and related labour, service, and social organizations;

4. the metropolis or city-region, comprising an urban core, inner and outer suburbs, and peri-urban space (see Figure 5.1), which, although an imperfect container of urban populations and employment, encompasses substantial regional labour and housing markets (see, for example, Scott 1988) as well as spatial (and political) expressions of increasing diversity (Andrew 2004) and socio-economic polarization and disparity (Walks 2005; Bunting, Walks, and Filion 2004); and, finally,

5. the complex internal space-economy of the city-region, comprising clusters of specialized industries and labour as well as more diffusive activity and exhibiting new forms of interaction between actors in economic and social space within the metropolis (Indergaard 2004; Lloyd 2006; Catungal, Leslie, and Hii 2009).

It is the intersection of forces operating within and between these overlapping scales that act to stimulate urban development and define growth trajectories. As an example, we can reference foreign direct investment (FDI): although emanating from varied international sources, it typically flows through national or regional institutions and is then applied to a project or site within a Canadian region. This in turn generates more diffuse patterns of growth through secondary rounds of expenditure (although we acknowledge that a significant proportion of this investment is directed toward the purchase of existing firms, with profits then flowing out of the country). Then there is the example of Canadian immigration policy, which we earlier described as an implicit national urban development policy. This is a policy enacted by federal (and in the case of Quebec, provincial) agencies, mediated through provincial programs (e.g., programs to attract business immigrants or certain skilled trades) and influenced again by distinctive programs at the local level (e.g., those of Calgary and Edmonton): these programs help to shape the highly uneven flows of immigrants to particular cities and towns in Canada.

The operation of advanced production systems again underscores the complexity of spatial flows in the modern urban economy. At a sectoral level, the intricate intertwining of inputs—services as well as goods—in Ontario's strategically important auto manufacturing and parts industries provides a well-known example, with Canada's most important industrial province deeply implicated in the complex continental system of parts manufacturing, assembly, and end-product distribution (see

Chapter 3 and Chapter 10). At the level of the firm, the video game company Radical Entertainment offers a telling example: the enterprise is ensconced within Vancouver's inner city, is owned by Vivendi of France, creates games for Los Angeles publishers, uses the Internet for recruiting worldwide, and out-sources some of its illustration and artwork to companies in China. At the level of the individual, residential choices that reflect lifecycle, frequency of access to metropolitan markets, and amenity preferences are increasingly leading some people to live at a considerable distance from metro-politan areas—facilitated by ease of transport and communication—while remaining intimately connected to them for services, income, and contact with global networks (see Chapter 5). These examples demonstrate the complex spatial networks within which the economies of Canadian cities operate, a feature that may confer benefits in terms of market reach and input sourcing but may also portend increasingly com-petitive environments and greater economic instability and volatility, as well as more complex spatial, social, and technical divisions of labour.

Restructuring in the Canadian Metropolis: From Post-industrialism and Post-Fordism to the 'Cognitive-Cultural Economy'

The general form of late-twentieth- and early-twenty-first-century urban restructuring comprises, as is well known, (first) contractions in manufacturing capacity and labour and (second) the rapid (though uneven) rise of an urban service economy and society, combined with a privileging of scientific knowledge and specialized information. The emergence of a new cohort of technology and information economy capital-holders and executives, and an elite social class of information workers and design profession-als, has partly supplanted the capital-holders of manufacturing corporations who made up the traditional social upper tier (and political elite) of the industrial city, although traditional corporate elites, in particular in the form of resource and energy capital, continue to carry considerable influence. Beyond these material considerations, the force of the restructuring experience also served to undermine the integrity of urban communities that had formed the constituent elements of the Chicago School of social ecology, as well as to sweep away much of the 'factory world' of manufacturing, allied industries, and labour that had been the focus of industrial urbanism as a field of social science scholarship.

In spatial terms, industrial restructuring comprehensively transformed the eco-nomic structure, systems, and functions of the entire city region but especially that of the central city (Hutton 2004). Restructuring was driven by office investment in the CBD and produced landscapes of disinvestment, and structural unemployment and compromised working-class communities, within the inner city. This bifurcation of the core was linked to changes in the larger city-region, since in many cases the distress of industrial decline in the inner city was accompanied by new industries and labour formation in suburban municipalities. In some instances, city governments and other agencies fought a rearguard action against the tide of urban industrial decline, while certain national governments, notably Margaret Thatcher's Conservative adminis-tration in Britain and Ronald Reagan's presidency in the US, (in)famously appeared to write off traditional industries and constituent labour and social groups, both for ideological reasons and to clear the way for a brave new world of market-based services (Massey and Meegan 1980; Bluestone and Harrison 1982; Christopherson 2003). Post-industrialism in this context therefore included a contested political agenda.

Post-Fordism was introduced as an alternative descriptor to post-industrialism, connoting the rapid decline of mass-production industries in the West and the concomitant decline in the social security apparatus and relative job security that accompanied them. The decline of this system is attributable to a complex mélange of forces and factors that included a shift of production capacity and employment to (especially) East and Southeast Asia, thus creating a 'New International Division of Labour' (NIDL) (Fröbel, Heinrichs, and Kreye 1980). Marxists and critical studies scholars deliberately avoided the term post-industrialism, both because of its ideological connotations and because of its implied privileging of services industries and labour over manufacturing and the blue-collar workforce, preferring instead post-Fordism as a term to describe the emergent industrial and social regime and flexible specialization as the production modality. But the conditions of restructuring in many metropolitan areas included a general decline of manufacturing enterprises and employment over a quarter-century, not just the shrinkage of Fordist industries.

In the Canadian case, we can readily identify the baseline effects of industrial restructuring in the 1970s and 1980s, expressed in terms of a decline in manufacturing firms and labour, accompanied by a rise of services and more particularly the intermediate (producer) services that led employment growth in many Canadian cities for much of this period. The restructuring processes of the 1970s and 1980s also produced important secondary effects in Canadian cities. These effects included the emergence of a 'new middle class' of managers and professionals—the urban expression of Bell's post-industrial society (1973)—the growth of the corporate office complex within city centres (CBDs), and the modernist imageries of high-rise office towers that transformed the landscapes of the post-industrial city in Canada, underscoring the comprehensive nature of post-industrialism. The late twentieth century also saw the growth of a high-wage manufacturing industry in southern Ontario cities such as Windsor and Oshawa under the initial stimulus of the 1965 Canada–US auto pact. In fact, employment in manufacturing in Canada rebounded substantially throughout the 1990s and early 2000s. However, this sector is now once again (post-2004) undergoing a painful restructuring, part and parcel of a larger crisis in automobile production throughout North America (see Chapter 3). Despite this variability, overall the template of a coincident decline of (some) blue-collar industries and the rise of a new service economy offered at least a rough fit for the experience of many metropolitan Canadian cities (Coffey 1994).

While the 'new middle class' and high-wage manufacturing are the sunny side of the new economic regime, the stagnation of real incomes, an increase in low-wage service work, job instability, and reduced pension coverage are its downside (Walks 2001; Morissette and Johnson 2005). This new working poverty is sometimes concentrated within cities (Chapter 6) but may also develop largely unobserved in suburban and peri-metropolitan areas (Chapter 5), where low-wage new-economy jobs combine with the remains of the old economy that continues to supply basic commodities and manufactured goods such as cars, clothing, food, and low-value resources. The Marxist discomfort with the 'post-industrial' turn of phrase is thus to some extent associated with the fact that industry has not yet disappeared from Canada or from its cities and with the fact that questions regarding the distribution of the benefits and costs of economic change remain unresolved.

So to what extent do post-industrialism and post-Fordism, the critical framing concepts of urban and regional restructuring in the late twentieth century, maintain a level of robustness and relevance in the contemporary setting? Certainly, some of the defining conditions have changed significantly, calling into question the continued value of these terms. First, during the 1990s when the industrial production sectors of older metropolitan cities, in the UK and the US especially, had to a large extent run down, and when the world-leading manufacturing economy of Japan was experiencing a deep 'hollowing out' of labour, manufacturing in Canada was experiencing growth, a trajectory continuing well into the present century (Figure 3.4). Indeed, among large North American cities, Toronto ranked first (in 2006) in the proportion of metropolitan employment accounted for by manufacturing, at 14.1 per cent, with Montreal (12.8 per cent) and Winnipeg (12.4 per cent) occupying the fifth and sixth positions, respectively (see Table 2.5). Second, the rapid growth of speculative office construction in the 1970s and 1980s, built to accommodate the large management and clerical staffs that represented a major new division of labour in advanced economies and an emblematic feature of the post-industrial era, has appreciably slowed in the 1990s and 2000s, reflecting, among other factors, the substitution of capital for labour and corporate downsizing. Third, as Peter Hall recently observed (personal communication), the descriptor of post-industrialism has lost much of its force, as a 'new economy' synthesizing services, culture, qualities of place, and high-value goods production appears to have risen out of the ashes of the old manufacturing economy. Hall calls for new terminology to fit what he sees as a far richer, more complex, and 'more interesting' urban economic condition, one marked by the ascendancy of recombinant industries and labour, than its late-twentieth-century predecessor.

At the same time, economic geographers have embarked on a process of retheorization with a view to developing models of more contemporary application and explanatory power than that of post-Fordism. Allen Scott, for one, asserts that post-Fordism and flexible specialization were models specific to a particular economic era that has now passed and has proposed instead the idea of a 'cognitive cultural economy', a concept that embodies both the cultural inflection of production and labour practices and the centrality of knowledge and information, recalling Bell's original idea of the primacy of scientific knowledge and the rise of an affiliated social class from the early 1970s. Scott's proposition, which asserts the saliency of regional production networks, labour markets, and traditions of art and design in the formation of the new cultural economy (Scott 2008; Storper and Scott 2006), can be interpreted as an oppositional response to Richard Florida's (2002) well-publicized musings on the rise of a 'creative class': a descriptor that has caught the attention of the media and the broader public, as well as legions of public officials keen to promote catchy new visions and 'cool' industries in order to reverse pathways to decline. Florida has undoubtedly contributed to an important new urban discourse, including that originating in Canada, and has stimulated programs designed to capture the growth potential of creative industries, institutions, and labour as instruments of urban regeneration.

For many, Florida's prescription for culture-based regeneration seems too redolent of universalism and hyperbole to offer a credible new policy paradigm. The implicit discounting of the importance of local/regional production systems and labour, and the corollary over-valuing of amenity attributes and other elements of his 'Bohemian index', undermines the theoretical appeal of the creative class proposition. There are also problematic features in Florida's treatment of industries and labour aggregates that raise questions about the theoretical purchase of the creative class. He includes many of the long-established producer (intermediate) service industries and allied professional employment categories as a large subset within his expansive concept of a creative economy, labour force, and class in an effort to project the latter as the largest element of the contemporary economy.

But a priori there seems no logical reason why creative industries and labour should not instead be considered an element of the intermediate services industries and labour force. Indeed, our analysis of employment growth in Canada over the period 1989 to 2008, a span that coincides roughly with the emergence of the 'new cultural economy', shows that the professional, scientific, and technical services category not only includes substantially more employees than the information, culture, and recreation group but has grown significantly faster (see Figure 3.1). Further, although Canada's major cities scored high location quotients for the telecom, media, and arts industries, they were equally prominent in the professional and scientific services (Table 3.7). It seems more than a stretch to make a claim that the cultural economy and creative workforce, as important as they clearly are to advanced urban economies, have overtaken or supplanted the intermediate services and professional/managerial labour force as elements of advanced economies. The Toronto chapter (Chapter 10) makes this same point in its discussion of new emerging industrial clusters.

Whatever the current form of the economy's cutting edge, this vanguard is only a small part of a wider whole. As the chapters in this book show, economic activity in Canadian cities consists of a layering of all that has come before: heavy industry, light industry, resources, Fordist, post-Fordist, and post-industrial activities, together with the consumer, producer, and government services that necessarily accompany and frame a functioning economy. Creativity and the cognitive cultural economy—to the extent that they indeed capture some new trends—do not have much immediate impact on the wages, job stability, and benefits of most Canadians, nor on the type of activity that they pursue on a daily basis. However, they may slowly insinuate themselves across and within all these dimensions: change only happens when the cutting edge becomes the norm, and this can only be assessed in retrospect. Thus, without underplaying changes in the economy—transitional processes and outcomes that are of course important and which we have emphasized throughout—these changes must be understood in the context of the humdrum yet crucial activities and tensions that shape the current economy of city-regions.

Economic Change and the Internal Structure of the Metropolis

The modeling of internal structures and spaces of the metropolis represents a defining tradition of urban scholarship over the past half century (Bourne 1982). This critical domain of urban studies includes theories from land economics and social morphology (as in the Chicago School of social ecology) and entails study of the space-economy of the city-region, including the zonal structure of activity, land values, and labour. Each of these models was compromised by the sweeping industrial restructuring experiences of the late twentieth century, leading to new propositions designed to account for (in general terms) the decline of traditional manufacturing and the ascendancy of a specialized services economy, labour force, and social class(es).

By the early 1990s, this particular phase of industrial restructuring had essentially run its course. It has in turn given way to a period of 'short-wave' innovation and restructuring, encompassing a sequence that includes the knowledge-based economy, the technology-driven new economy, and the 'new cultural economy'—each of which has left its imprint on the structure and form of the metropolis (see Graham and Marvin 2001 and Hall 2006). However, economic restructuring has been accompanied by growing social and spatial inequality, as demonstrated in Chapter 6. The very volatility and instability that have characterized this short-wave restructuring tendency, including the more complex and apparently diffusive spatial patterns generated by abbreviated processes of change, produced a more problematic theoretical condition for many scholars. This view of an inchoate urban morphology was epitomized by Michael Dear and Stephen Flusty's somewhat whimsical view of chaotic and centre-less economic development they elected to term 'postmodern urbanism' (1998). A decade or so on from this paper, we can draw on a wealth of case studies to identify emergent logics of location underpinning the more complex economic (and social) geography of the contemporary metropolis and to suggest perhaps more compelling models of the emerging structure and morphology of this metropolis.

The starting points for our reflections here are two-fold: first, the contextual analysis of government, urban politics, and public policies that shape Canadian cities as outlined in Chapter 4 and second, the model of urban spatial structure described in Chapter 5, the latter based on an appreciation of commonalities and contrasts in the regional morphology and labour markets of Canadian cities. The corollary issue concerns the extent to which the detailed case studies presented in Chapters 8 to 12 offer opportunities for a further elaboration or specification of these abstract spatial models of concentricity and polarization.

Notwithstanding significant contrasts in the structures of the respective cities, the received idea of the monocentric city is supplanted by a more complex and spatially extended city-region, with a preponderance of growth weighted toward the suburbs and (in terms of growth rates if not absolute numbers) the peri-urban spaces and exurbs (Chapter 5). The suburban areas of the region now include not just low-density residential neighbourhoods, final-demand services, and space-extensive manufacturing and industry but, in most cases, some of the key centrepieces of the regional economy, including international airports, universities, science parks, and major secondary office complexes.

However, the CBDs of Canadian cities still retain key primary functions and project power across and beyond their regions: Bay Street (Canada's principal financial district) is in Toronto's CBD; the federal Parliament defines Ottawa's national political primacy; the headquarters of oil and gas companies, together with their high-order consultants, dominate the centre of Calgary; and downtown Montreal houses Quebec's principal resource, utilities, media, and global consulting headquarters. It is only in Vancouver that the CBD has perhaps ceased to play such a dominant economic role (even though Vancouver's core still includes about 200,000 jobs, about 20 per cent of the regional total; see Chapter 12). Thus, while the monocentric city, to the extent that it ever existed, is no more, this does not mean that Canadian cities no longer have one centre that dominates the others, projecting economic and symbolic power. Rather, Canadian cities, at least toward the higher levels of the urban system, are 'regional cities', defined in part by their uniquely specialized CBDs and in part by clustering, dense and complex production networks, innovation diffusion, long-distance commuting, and consumption and by overlapping labour markets that extend over significantly larger territories.

Our empirical studies of contemporary Canadian cities also call for some rethinking of the traditional mid-twentieth-century notion of the metropolis, a concept implying a dominant agglomeration of higher-order services in the CBD (and a defining imagery of skyscrapers concentrated exclusively in the CBD), with both upscale and low-income populations residing in separate neighbourhoods close to the core and a steep falling away of density gradients as one travelled outward from the core. The restructuring of the city-region set out in the case studies suggests the contours of more extended city-regions in which the CBD, even if dominant, is no longer the only centre capable of projecting power. There also remains an association between the idea of the 'metropolis' and that of cosmopolitan urbanism and lifestyles, connoting not just a richer socio-economic mix in the core and a much more variegated ethno-cultural quality—after all, these are now also part of the modern suburban reality in most larger Canadian cities—but also a distinctive quality of amenity, consumption, and culture that persist as signifiers of the new central city.

Synthesizing Industries, Labour Markets, and Society in the Urban Space-Economy

Herein perhaps lies one of the principal theoretical contributions of our study of urban change in the Canadian urban system, which has emphasized the importance of labour markets, employment growth, and occupational restructuring as principal markers of urban development, as well as offering entrées to wider analysis of changing urban economies and living conditions. Specifically, the five in-depth city-region case studies presented in this volume, coupled with the detailed empirical analysis of the first part of the book, suggest the possibility of synthesizing the discrete traditions of industrial urbanism and the social ecology school. Connecting the points between economic restructuring, employment, and occupational change and the social and ethno-cultural transformation of the contemporary city-region can be set out as follows:

First, at a macro-regional level, the morphology of the emerging metropolitan city is largely a product of the interdependencies between labour markets and housing markets, each framed by policy decisions on transportation infrastructure, planning, and the provision of services. The aggregated 'preferences and choices' of workers for employment opportunities are strongly mediated by 'options and constraints' in the residential market, and these in turn are filtered through conditions of differential access and mobility. The classic industrial city presented higher densities and a typical tight co-location of manufacturing employment and workers' housing. The post-industrial forces of the postwar period sundered the relationships and interdependencies that characterized this model of proximity, replacing it with a new morphology that included the rise of the CBD office complex, social upgrading in the core, and the suburbanization of population growth and new household formation (Ley 1996; Walks 2003; Hutton 2004). This in turn has generated corollary employment growth in the suburbs, as well as a long-distance (and problematic, from a planning perspective) commuting syndrome. But the scenarios of spatial change recounted in our case studies indicate a significant modification of this model, notably the coincident reindustrialization, gentrification, and social reconstruction of the inner city and the maturing of the economies of the suburbs, together with more complex ethno-cultural residential patterns that increasingly shape urban labour markets. These changes have also been accompanied by growing income inequality within metropolitan areas (Chapter 6) and increasingly within suburban and peri-metropolitan spaces (Chapter 5). All of these trends have been heavily conditioned by the structure of government and by a myriad of public policy decisions, as Chapter 4 describes.

A second—and related—interdependency between economic development and social change in the shaping of the city-region can be seen in the linkages between production and consumption in the contemporary cognitive–cultural economy. While perhaps the more exuberant claims of consumption as the driver of the urban economy are somewhat overstated, it does seem evident—as the conceptual model in Chapter 1 suggests—that consumption represents a major sector of the economy in cities such as Toronto, Montreal, and Vancouver (as well as the economies of Halifax, Victoria, and St John's). Consumption is strongly implicated as a factor in the emergence of creative industries, even if the Floridian 'amenity as destiny' argument for explaining urban growth seems exaggerated. In addition, consumption is reshaping some peri-metropolitan areas: retirees, as well as economically active individuals who derive their income from metropolitan areas, are increasingly residing in the peri-metropolitan zone and reshaping its economy by their patterns of consumption. Such 'residential economies'—i.e., localities whose economic base is residents with outside incomes—have been analyzed and theorized in France (Davezies 2009) but not yet in a Canadian context.

Third, the contemporary economy of the city, while still to be sure including 'spaces of enclosure' (notably factories) for certain kinds of work, is also recognizably a 'social economy' (Thrift and Olds 1996; see also Scott 2008). Such an economy is marked by the recurrent exchange of tacit knowledge (Gertler 2003) observed in dense patterns of social activity and exchange in specialized clusters and districts—the CBD, the inner city, and some areas of the suburbs, notably major town centres—and in proximity to universities and other knowledge-based clusters. As one moves beyond these suburbs and on to peri-metropolitan locations, 'spaces of enclosure' become more common, partly because the 'old economy' is more pervasive there but partly, also, because 'new economy' functions that require less social interaction choose to locate there (McCann 2007), a phenomenon observed by Allen Scott three decades ago (Scott 1982), demonstrating another important aspect of continuity in the economy of the metropolis. Our case studies show that the enhanced 'social density' of large cities represents a significant factor of comparative advantage for the development of knowledge- and contact-intensive industries and firms in Canadian cities, as it is in cities such as London, New York, Chicago, Paris, Barcelona, and Milan.

Fourth, the economic geography of employment in the contemporary Canadian city is increasingly physically intertwined with that of its changing social geography. A feature observed in most cities—Ottawa, Vancouver, and Montreal's CBDs and Toronto's inner rings—is the increase in population relative to jobs despite increasing job levels (Chapter 5). This important new spatial reality reflects the incursion of condominiums into the CBD and other formerly non-residential nodes, the growth of live/work elements in industrial and commercial areas, and the adaptive re-use of obsolescent commercial buildings for housing. A less visible but nonetheless consequential illustration of the physical intertwining of social and economic worlds in the city can be found in the growth of home-based work—now (as shown in the Vancouver case, Chapter 12) a significant element of the urban economy.

Finally, the development of new residential areas, including condominium districts, adaptive re-use and conversions, and medium- and high-density neighbourhoods, has become an instrument of economic development policy in its own right, as evidenced in Ontario's 'Places to Grow Plan' (see Chapter 10) and in Vancouver's 'Living First' program (Chapter 12). In this interpretation, and as already mentioned with respect to peri-metropolitan areas, residential development stimulates the economy, not only in the one-time effects of construction but also in the increasing demand for consumption goods and spaces as well as in its role in attracting 'talent' and high-end human and social capital for creative industries in older districts. In addition, residential intensification and social diversity are intended to encourage balanced communities with a mix of jobs and housing in more transit-supportive environments.

Economic Change and the Future of Canadian Cities: Synopsis and Conjecture

What broad generalizations can we extract from the above review? This entire volume has been about the insights that we can draw from a detailed analysis of the ongoing economic transformation of Canadian cities viewed at different spatial scales—from the national urban system, through the case studies of metropolitan regions, to the internal structure, growth, and changing character of individual city-regions. Changes—but also continuities—at each of these scales are taking place in an era of rapid economic restructuring, rising global uncertainty, and deepening fiscal challenges. The external drivers and the internal dynamics of urban economic growth in Canada have different configurations—that is, different sources and impacts—at each spatial scale, but all are closely interrelated. Our understanding of growth and change depends on careful examinations of all scales in concert.

In this section, we offer a number of generalizations drawn from the empirical analyses and theoretical debates of the earlier chapters that might serve as points of departure for subsequent discussions and debates. They are presented in succinct bullet form, since the empirical evidence, theoretical background, and necessary nuances that support and surround these key points can be found throughout this book. Our purpose here is not to go over these important elements again but to synthesize the key lessons that have emerged in a straightforward way. Given the complexity of city-regions, which are economic and social systems, contain economic and social systems, are part of wider economic and social systems, and are cross-cut by yet other such systems, even an abbreviated list of key points cannot be short and will contain apparent contradictions, since what is observed for one city or at one scale may not be observed at another.

So what matters if the economic change and future of Canadian cities is to be studied and understood?

1. Context matters: Canada's changing global context continues to be crucial. Although some globalization processes may be in temporary decline, the economy of Canada and its cities is still relatively open to external influence and in particular has become increasingly continentalized through integration with the much larger US economy.

2. Big cities matter: There is clear evidence throughout this book that Canada's economy and its major players have become increasingly concentrated in a small number of very large metropolitan areas, or what we have called city-regions (for an assessment of the developmental saliency of metropolitan cities both in the national and international contexts, see Polèse 2010).

3. Diversity matters: Despite the overriding importance of the national economy and the homogenizing influence of trade and capital flows, new technologies, and inter-regional integration, there is immense diversity in the economic structures and trajectories of individual Canadian cities and city regions.

4. Places matter: Urban economies in Canada are deeply rooted and functionally embedded in their own particular settings—in terms of the effects of local histories, politics, cultures, and inherited institutions of government and civil society, as well as with respect to the operation of local/regional networks of firms, labour markets, and systems of economic organization. Our five case studies clearly illustrate the importance of situating the local economy within these settings. Montreal is not Toronto, Ottawa is not Calgary, and Vancouver stands apart. They are individually distinct in terms of spatial structure, industrial mix, social morphology, and political affiliation and identity but are still part of the country's urban system.

5. Contingency matters: Urban economies are not only complex and diverse but constantly reinvent themselves in ways that both destroy and selectively build on inherited structures from the recent past. The new economy contains elements of the old and vice versa. Yet the directions of change are not everywhere the same, nor always evident in aggregate analyses of shifts in sectors and occupations, and are subject to the influence of agency (including top-down official directives or bottom-up initiatives), which has the potential to alter (or reinforce) a particular trajectory.

6. Policy matters: The spatial structure of the economies of Canada's larger cities, perhaps more than that of their American counterparts, reflects the imprint of governments and specifically of public policy decisions made over long periods of

time by different levels of government and for different reasons. Obvious examples include the effects of land-use planning and infrastructure investment decisions at the local, and especially the provincial, level; trade, taxation, innovation, and immigration policies emanating primarily from the federal level; and the unique role of the federal government in shaping the economy of Ottawa-Gatineau (as we saw in Chapter 9).

7. Geography matters: The distribution of employment in Canadian cities, although illustrating general tendencies common to most cities in developed countries—including the low-density dispersal of employment in new suburbs and the relative decline in jobs in the older suburbs, combined with pockets of new activities in the inner city—is still remarkably variable from city to city. This variability mirrors differences in historical investments in transportation and the built environment, physical constraints on development, and the mix of incentives and disincentives offered by local (and provincial) political structures and regulatory regimes.

8. The centre matters: Employment is growing faster in the suburbs, and key metropolitan-level facilities such as airports, employment centres, universities, shopping malls, and leisure complexes are increasingly located there and exert influence over surrounding areas. Despite this, the core districts of Canadian cities, their CBDs, remain distinct: they still retain the highest-order and most strategic command functions of the metropolitan economy, project economic, political, and symbolic power, and are key sites of economic and social innovation.

9. The old economy matters: The economies of Canadian cities still mirror their traditional dependence on the resource base to a significant extent—or on staple products—and in some regions on older manufacturing industries. There seems no escape from the clutch of commodities and the variability of prices set in external markets beyond the control, or even the influence, of governments, but at the same time, as argued above, there is no single path of development. Many peri-metropolitan spaces retain important traditional functions such as agriculture, low-value resources, and manufacturing. Even their leisure, residential, and retirement functions are not new but are being intensified and modified by improvements in transport and communications.

10. Timing matters: Considerable evidence has been presented above with respect to the importance of periodicity and cycles in urban development and specifically on the impacts of short and long waves of development on the urban landscape.

11. The nation-state matters: Despite globalization and ever-closer integration with the US and the economies of nearby US regions, the national government (and the provincial governments) do play a significant role in shaping urban economic development, although perhaps largely indirectly. Governments define boundary conditions, set (albeit within strict limits) monetary and fiscal policies, invest in new infrastructure, cultural facilities, and technologies, manage Crown corporations, negotiate international trade and labour exchange agreements, and set immigration targets. These actions, in combination, remain significant drivers of urban economic growth in Canada, and their impacts are markedly uneven across the country. Unfortunately, it is not clear that national or even provincial governments recognize the extent of the urban impacts of their policy decisions or

even the crucial role of cities, especially the larger cities, in defining the country's competitive position and future prosperity.

12. Finally, Canadian cities remain distinct—typically characterized as having stronger downtowns, higher average population densities, more transit usage, larger local governments, tighter planning controls, more viable inner-city areas, lower (but increasing) income inequalities, and so on, relative to many comparable US cities, but these differences are not as strong or as consistently evident as one might expect.

Although the focus of this book has been on economic change in Canada's metropolitan areas, it is not possible to isolate the economy and operation of markets from social and equity considerations. The space-economy of metropolitan areas evolves with and reflects wider social processes, and as it evolves new types of social tension and inequality are created as others are resolved. Thus, while all of the elements mentioned above are important dimensions that should be considered when thinking about economic change in metropolitan areas, one final point needs to be emphasized: the impacts of spatial–economic change on people matter.

These impacts have only been touched upon in this book and are not the volume's main theme. However, economic restructuring and industrial change is accompanied by processes of gentrification, gender imbalance, income polarization, changing accessibilities, and inequality: these are not static, and their nature, location, and consequences change as the economy and society evolve. As cities' economies change, as the nature and location of work evolves, as certain neighbourhoods and peri-metropolitan areas rise and others decline, people are forced to adapt. Those least able to adapt are rarely of prime concern to futurologists or to analysts of the cutting edge of economic change, but the question of who benefits (and who doesn't) from the spatial–economic changes we have been describing is a question that matters.

Perspectives on Our Urban Economic Future

Where are Canadian cities heading? At one level, we must confess that we do not know. Given the wide array of scales, processes, and complex interactions to be considered, it is not possible to reply with confidence, and the more we look at metropolitan areas, the less we seem to know about them (or we are struck by the more there is yet to know).

However, at another level, and taking a step back from the complexity that has been unveiled throughout the chapters of this book, it is not beyond the capacity of urban scholars to provide some informed opinion about the economic future of Canadian cities.

The Future Shape, Size, and Geography of Canadian Cities

The population projections recently released by Statistics Canada (2010) and summarized in Chapter 7 provide one scenario of the future growth, size, and cultural makeup of Canada, its cities and metropolitan areas. Given past trends, it is likely that we will see at least two different Canadas: a few growing metropolitan areas with advanced economies and highly diverse (visible minority) populations and the rest of the country, which to date has attracted few immigrants and is as a result facing a future of no growth (or population contraction) and an aging labour force. Both Canadas will face serious challenges, one resulting from the effects of relative decline and the other resulting from the consequences of rapid growth—for example, in terms of congestion, pollution, social service needs, and infrastructure costs (Slack, Bourne, and Priston 2006).

But it is easier to forecast population change than it is to predict the structure and composition of the urban economy, since the former is increasingly attributable to immigration—which is primarily the result of policy/political decisions. The economy is not as regulated or predictable. Jobs come and go with surprising frequency. While we might expect certain sectors to show employment growth in rough proportion to population growth—for example, in the local services and consumption sectors—and we might assume that certain components of the old economy will decline or disappear, it is much less clear what sectors and industries will drive the new economy over the next few decades.

The extent of overall employment growth will depend on the demographic factors just mentioned, but other factors will also come into play. After increases in employment and activity rates since the 1970s, brought about by immigration, working-age baby boomers, and the feminization of the workforce, employment growth is likely to be much slower over the coming decades. Activity rates will no longer be boosted by women entering the workforce, and the baby-boom generation is entering retirement. While immigration will ensure continued employment growth in metropolitan areas, this growth is likely to be slower than over the past decades.

Even without knowing which sectors will be at the economic cutting edge and which sectors will grow or decline, it is possible to envisage what the geography of Canadian cities will be in 10 or 20 years: by and large, it will resemble what we see today. The CBD will still be the CBD—perhaps slightly more or slightly less dominant than today but still a major focus of the metropolitan economy. Each city's major suburban employment centres will remain major suburban employment centres, and most will probably have grown in size. In particular, the employment centres focused on each city's airport will have consolidated. Some of the smaller suburban centres will have grown, but some will have become part of 'edgeless cities' (Lang 2003; Shearmur et al. 2007)—i.e., concentrations of jobs not in suburban centres but along major highways, either radial or between major suburban centres. Peri-metropolitan areas will retain their diverse functions, with increasing transfers from metropolitan areas (where command functions and high-value economic activity will continue to provide incomes) to residential and/or retirement communities (where income is spent).

In short, if evidence of the past 20 years is anything to go by, metropolitan areas will be recognizable in 20 years' time. Their spatial development is strongly constrained by existing infrastructure, rights of way, buildings, and property rights. This accumulation of physical capital and of rights is what prevents cities from evolving at the whim of short- or medium-term economic changes. Epochal changes such as the Industrial Revolution (with its concomitant rural to urban shift) and revolutions in transport technology (especially the advent of individualized methods of transport, currently the automobile) and building technology (the high-rise) have indeed brought about major reconfigurations of the city and new forms of growth, particularly in the suburbs where constraints are necessarily weaker. But the current revolution in information technology, while it is changing the way information is shared and is enabling some extension of the city into peri-metropolitan spaces, does not fundamentally alter the way that we relate to space. Without such a change (which may of course occur if individualized transport is rendered prohibitively expensive) and without a seismic economic shift (such as agriculture becoming a high-employment sector shifting power to rural areas), then our metropolitan areas will continue to evolve in a relatively stable manner. The need for agglomeration and face-to-face contact will remain, and they will continue to be tempered by costs, amenity preferences, and congestion. This does not mean that the economy of metropolitan areas will not alter, nor that neighbourhoods and buildings will not continually be re-used, redeveloped, and rethought: only that the geographic patterns now observed will continue to be observed with changes at the margins, in much the same way that medieval street patterns and rights of way still structure many European cities.

The Future Composition and Nature of the Economy of Canadian Cities

The devil, of course, is in the detail, and the detailed evolution of metropolitan economies cannot be predicted. So rather than speculate on specific sectoral trends beyond those identified in the chapters in Part I and in the case studies, an alternative approach is to pose a series of questions for students of the future city; the way we answer these questions will define our economic future. In Box 13.1 (Framing Questions for the Economic Future of Canada's Cities) below, we pose a series of questions concerning the evolution of Canada's city-regions, their constituent economies, labour markets, and societies, that flow out of the research presented in this book and might form the agenda for a new project.

These questions relate in part to uncertainties about the continuation of long-established trends, notably the half-century expansion of the urban services sector, which may now be entering a mature phase of more modest growth, coupled with (as Linda McDowell and Susan Christopherson [2009] have recently argued) growing disparities in terms of incomes, job satisfaction and advancement possibilities, and the security of work, especially among women. The text box also poses questions regarding immigration—clearly required to support the development of urban labour markets but increasingly associated with mixed outcomes in terms of the quality of jobs and levels of remuneration experienced by some of the most recent immigrants, that are addressed in Chapter 6 and in the case studies. There may also be legitimate questions about the capacity of Canadian cities to accommodate higher levels of immigration without placing more exigent stresses on local systems and services, including education, housing, and welfare. We also acknowledge uncertainties about the role of cities in Canada's engagement within the global economic arena—including the possibility of increasing exports in high-value-added products and services, a long-held aspiration to be sure, but one far from fully realized.

Finally, we pose a question concerning the sustainability of Canada's cities in an era of climate change and pervasive environmental degradation and resource depletion.

This meta-question appears beyond the scope of this volume. But we need only recall Canada's historical (and present-day) developmental dependence on resources and high per capita consumption of energy and other biophysical stocks and assets to appreciate the centrality of this question to the future welfare of Canada and the city-regions that constitute the largest share of its population and workforce. To date, cities have made incremental progress in this domain, notably in the fields of industrial ecology, green building technologies, and eco-density. But more far-reaching actions will be required, including significant reductions in per capita resource consumption in Canadian cities, which ranks among the highest in the world. The goals of urban sustainability (incorporating social as well as environmental values) and economic viability will assuredly stretch the policy capacity of Canada's cities in the present century.

Box 13.1 Framing Questions for the Economic Future of Canada's Cities

Is the recent decline of employment in traditional manufacturing in Canadian cities in response to a highly competitive global environment cyclical or structural? Is the decline likely to continue? Is there a role for new or niche manufacturing, perhaps set in localized clusters of integrated firms and supporting institutions?

Will the output from the resource sector continue to provide a substantial component of national and provincial GDP? Will it also provide significant employment opportunities, and at what skill and wage levels? Can that sector be used more effectively as a basis for a uniquely Canadian form of process innovation and the introduction of new technologies? What roles might First Nations play in resource development?

Has the tertiarization of the economy—the long-term growth of the services sector—effectively run its course? Is the proportion of employment in the services industries more or less at a maximum? If so, what sectors and industries might replace the high level of growth in services recorded in past decades?

Will the wider public sector (e.g., public administration, education, health) continue to offer significant employment opportunities, which it has to date in all Canadian cities? Or have we reached the limits of state provision?

Will components of the 'new' economy, however defined (e.g., as the knowledge-based or cultural economy), continue to expand? And if so, will they be of sufficient scale and scope to replace the loss of employment in other sectors?

Will immigration continue to drive the growth of Canada's labour markets and those of its major cities in particular? Will the flows of immigrants change skill levels, or more broadly, human capital levels, within the labour force, and will the distribution of those flows define the future geography of wage levels, labour supply, and worker mobility? Is there a limit to the capacity of Canada's major cities—in terms of their labour markets and social service needs—to absorb large numbers of new immigrants?

What are Canada's comparative advantages on the contemporary world stage? Can they be used to generate local economic initiatives and encourage inward investment? What roles will Canada's cities play in the global pursuit of competitive advantage?

What can individual cities, and the extended city-regions that have been the primary focus of this book, do to generate sustainable economic growth without deepening existing regional and local inequalities?

References

Andrew, C., ed. 2004. *Our Diverse Cities*. Ottawa: Metropolis.

Arthur, B. 1989. 'Competing technologies, increasing returns, and "lock-in" by historical events'. *Economic Journal* 99 16–131.

———. 1994. *Increasing Returns and Path Dependence in the Economy*. Ann Arbor: University of Michigan Press.

Bathelt, H., and J. Boggs. 2003. 'Toward a reconceptualization of regional development paths: Is Leipzig's media cluster a continuation or a rupture with the past?' *Economic Geography* 79: 265–93.

Bell, D. 1973. *The Coming of Postindustrial Society: A Venture in Social Forecasting*. New York: Basic Books.

Belussi, F., and S.R. Sedita. 2009. 'Life cycle versus multiple path dependency in industrial districts'. *European Planning Studies* 17: 505–28.

Bluestone, B., and B. Harrison. 1982. *The Deindustrialization of America: Plant Closing, Community Abandonment, and the Dismantling of Basic Industry*. New York: Basic Books.

Bourne, L.S., ed. 1982. *Internal Structure of the City: Readings on Urban Form, Growth and Policy*. 2nd edn. New York and Oxford: Oxford University Press.

Bunting, T., A. Walks, and P. Filion. 2004. 'The uneven geography of housing affordability stress in Canadian metropolitan areas'. *Housing Studies* 19 (3): 361–93.

Catungal, J.-P., D. Leslie, and Y. Hii. 2009. 'Geographies of displacement in the creative city: The case of Liberty Village, Toronto'. *Urban Studies* 46 (5/6): 1095–1114.

Christopherson, S. 2003. Review of J. Cowie and J. Heathcott, eds, *Beyond the Ruins: The Meanings of Deindustrialization* (Ithaca, NY: Cornell University Press, 2003). *Journal of the American Planning Association* 70: 487.

Coffey, W.J. 1994. *The Evolution of Canada's Metropolitan Economies*. Montreal: Institute for Research on Public Policy.

Davezies, L. 2009. 'L'économie locale 'résidentielle'. *Géographie économie et société* 11 (1) : 47–53.

David, P.A. 2005. 'Path dependence in economic processes: Implications for policy analysis in dynamical systems contexts'. In K. Dopfer, ed., *The Evolutionary Foundations of Economics*. Cambridge: Cambridge University Press.

———. 2007. 'Path dependence and historical social science: An introductory lecture'. Paper presented at the Symposium of Twenty Years of Path Dependence and QWERTY Effects, Russian University, Higher School of Economics, Moscow, 13 May 2005. SIEPR Policy Paper no. 04–022: 2007.

Dear, M., and S. Flusty. 1998. 'Postmodern urbanism'. *Annals of the Association of American Geographers* 88: 50–70.

Florida, R. 2002. *The Creative Class—And How It's Transforming Work, Leisure, Community and Everyday Life*. New York: Basic Books.

Fröbel, F., J. Heinrichs, and O. Kreye. 1980. *The New International Division of Labour*. Cambridge: Cambridge University Press.

Gertler, M. 2003. 'Tacit knowledge and the economic geography of context, or the indefinable tacitness of (being) there'. *Journal of Economic Geography* 3: 79–99.

Graham, S., and S. Marvin. 2001. *Splintering Urbanism: Networked Infrastructures, Technological Mobilities, and the Urban Condition*. New York and London: Routledge.

Hall, P. 2006. 'The polycentric city'. PowerPoint presentation, The Bartlett School, University College London.

Harvey, D. 1989. 'From managerialism to entrepreneurialism: Transformation in governance in late capitalism'. *Geografiska Annaler Series B-Human Geography* 88B: 145–58.

Hutton, T.A. 1998. *The Transformation of Canada's Pacific Metropolis: A Study of Vancouver*. Montreal: Institute for Research on Public Policy.

———. 2004. 'Post-industrialism, post-modernism, and the reproduction of Vancouver's metropolitan core: Retheorizing the twenty-first-century city'. *Urban Studies* 41 (10): 1953–82.

Indergaard, M. 2004. *Silicon Alley: The Rise and Fall of a New Media District*. London and New York: Routledge.

Innis, H. 1933. *Problems of Staple Production in Canada*. Toronto: Ryerson Press.

Jacobs, J. 1961. *The Death and Life of Great American Cities*. New York: Random House.

Lang, R. 2003. *Edgeless Cities*. Washington: The Brookings Institute.

Ley, D.F. 1996. *The New Middle Class and the Remaking of the Central City*. Oxford: Oxford University Press.

———. 2010. *Millionaire Immigrants: Transpacific Lifelines*. Oxford: Blackwell-Wiley.

Lloyd, R. 2006. *Neo-Bohemia: Art and Commerce in the Postindustrial City*. London and New York: Routledge.

McCann, P. 2007. 'Sketching out a model of innovation: Face-to-face interaction and economic geography'. *Spatial Economic Analysis* 2 (2): 117–34.

McDowell, L., and S. Christopherson. 2009. 'Transforming work: New forms of employment and their regulation'. *Cambridge Journal of Regions, Economy and Society* 2 (3): 332–45.

Martin, R. 2010. 'Roepke Lecture in Economic Geography—Rethinking regional path dependency: Beyond lock-in to evolution'. *Economic Geography* 86 (1): 1–28.

Massey, D., and R. Meegan. 1980. 'Industrial restructuring versus the cities'. In A. Evans and D. Eversley, eds, *The Inner City: Employment and Industry*. London: Heinemann.

Morissette, R., and A. Johnson. 2005. 'Are good jobs disappearing in Canada?' *Economic Policy Review* 11 (1): 23–56.

Moulaert, F., A. Rodriguez, and E. Swyngedouw, eds. 2002. *The Globalized City: Economic Restructuring and Social Polarization in European Cities*. New York: Oxford University Press.

Peck, J. 2005. 'Struggling with the creative class'. *International Journal of Urban and Regional Research* 29: 740–70.

Polèse, M. 2010. *The Wealth and Poverty of Regions: Why Cities Matter*. Chicago: University of Chicago Press.

Polèse, M., and R. Shearmur. 2004. 'Culture, language and the location of higher-order service functions: The case of Montreal and Toronto'. *Economic Geography* 80: 329–50.

Sassen, S. 2006. 'Chicago's deep economic history: Its specialized advantage in the global network'. In R.P. Greene, M.J. Bouman, and D. Grammenos, eds, *Chicago's Geographies: Metropolis for the 21st Century*. Washington, DC: Association of American Geographers.

Schneiberg, M. 2007. 'What's on the path? Path dependence, organizational diversity and the problem of institutional change in the US economy'. *Socio-economic Review* 5: 47–80.

Scott, A.J. 1982. 'Locational patterns and dynamics of industrial activity in the modern metropolis: A review essay'. *Urban Studies* 19: 111–42.

———. 1988. *Metropolis: From the Division of Labor to Urban Form*. Berkeley and Los Angeles: University of California Press.

———. 2008. *Social Economy of the Metropolis: Cognitive–Cultural Capitalism and the Global Resurgence of Cities*. Oxford: Oxford University Press.

Shearmur, R., W. Coffey, C. Dubé, and R. Barbonne. 2007. 'Intrametropolitan employment structure: Polycentricity, dispersal and chaos in Toronto, Montreal and Vancouver, 1996–2001'. *Urban Studies* 44: 1713–38.

Slack, E., L.S. Bourne, and H. Priston. 2006. 'Large cities under stress: Challenges and opportunities'. Background Report. Ottawa: National Advisory Committee on Cities and Communities.

Smith, M.P. 2001. *Urban Transnationalism: Locating Globalization*. Oxford: Blackwell.

Statistics Canada. 2010. *Projections of the Diversity of the Canadian Population 2006–2031*. Catalogue no. 91-551-X. Ottawa: Ministry of Industry.

Storper M., and A.J. Scott. 2009. 'Rethinking human capital, creativity and urban growth'. *Journal of Economic Geography* 9 (2): 147–67.

Thrift, N., and K. Olds. 1996. 'Refiguring the economic in economic geography'. *Progress in Human Geography* 20: 311–37.

Walks, A. 2001. 'The social ecology of the post-Fordist/global city? Economic restructuring and sociospatial polarisation in the Toronto urban region'. *Urban Studies* 38 (3): 407–47.

———. 2005. 'The city-suburban cleavage in Canadian federal politics'. *Canadian Journal of Political Science* 38 (2): 383–413.

Walks, A., and L.S. Bourne. 2006. 'Ghettos in Canadian cities? Racial segregation, ethnic enclaves, and poverty concentration in Canadian urban areas'. *The Canadian Geographer* 50 (3): 273–97.

World Bank. 2010. 'World development indicators: Energy use (kg of oil equivalent per capita)'. http://data.worldbank.org/indicator/EG.USE.PCAP.KG.OE.

Index